INTERNATIONAL

REVIEW OF CYTOLOGY

SUPPLEMENT 11B

Perspectives in Plant Cell and Tissue Culture

INTERNATIONAL

Review of Cytology

EDITED BY

G. H. BOURNE
St. George's University School of Medicine
St. George's, Grenada
West Indies

J. F. DANIELLI
Worcester Polytechnic Institute
Worcester, Massachusetts

ASSISTANT EDITOR

K. W. JEON
Department of Zoology
University of Tennessee
Knoxville, Tennessee

SUPPLEMENT 11B

Perspectives in Plant Cell and Tissue Culture

EDITED BY

INDRA K. VASIL
Department of Botany
University of Florida
Gainesville, Florida

ACADEMIC PRESS 1980
A Subsidiary of Harcourt Brace Jovanovich, Publishers
New York   London   Toronto   Sydney   San Francisco

ACADEMIC PRESS, INC.
111 Fifth Avenue, New York, New York 10003

United Kingdom Edition published by
ACADEMIC PRESS, INC. (LONDON) LTD.
24/28 Oval Road, London NW1 7DX

LIBRARY OF CONGRESS CATALOG CARD NUMBER: 74–17773

ISBN 0–12–364372–4

PRINTED IN THE UNITED STATES OF AMERICA

80 81 82 83   9 8 7 6 5 4 3 2 1

*To Vimla,
a valuable professional colleague
in plant cell and tissue culture research,
and my wife,
and to Kavita and Charu,
our daughters*

Contents

10. Isolation and Culture of Protoplasts

INDRA K. VASIL AND VIMLA VASIL

11. Protoplast Fusion and Somatic Hybridization

OTTO SCHIEDER AND INDRA K. VASIL

12. Genetic Modification of Plant Cells Through Uptake of Foreign DNA

C. I. KADO AND A. KLEINHOFS

13. Nitrogen Fixation and Plant Tissue Culture

KENNETH L. GILES AND INDRA K. VASIL

14. Preservation of Germplasm

LYNDSEY A. WITHERS

15. Intraovarian and *in Vitro* Pollination

M. ZENKTELER

16. Endosperm Culture

B. M. JOHRI, P. S. SRIVASTAVA, AND A. P. RASTE

17. The Formation of Secondary Metabolites in Plant Tissue and Cell Cultures

H. Böhm

18. Embryo Culture

V. Raghavan

19. The Future

Georg Melchers

List of Contributors

Numbers in parentheses indicate the pages on which the authors' contributions begin.

H. Böhm (183), *Institute of Plant Biochemistry of the Academy of Sciences of the GDR, DDR-4020 Halle/Saale, Weinberg 3, German Democratic Republic*

Kenneth L. Giles* (81), *Genetics Department, Iowa State University, Ames, Iowa 50010*

B. M. Johri (157), *Department of Botany, University of Delhi, Delhi 110007, India*

C. I. Kado (47), *Department of Plant Pathology, University of California, Davis, California 95616*

A. Kleinhofs (47), *Department of Agronomy and Soils, Washington State University, Pullman, Washington 99164*

Georg Melchers (241), *Max-Planck-Institut für Biologie, Tübingen, Federal Republic of Germany*

V. Raghavan (209), *Department of Botany, The Ohio State University, Columbus, Ohio 43210*

A. P. Raste (157), *Department of Botany, Deshbandhu College, University of Delhi, Kalkaji, New Delhi 110019, India*

Otto Schieder (21), *Max-Planck-Institut für Züchtungsforschung (Erwin-Baur-Institut), 5000 Cologne, Federal Republic of Germany*

P. S. Srivastava (157), *Department of Botany, S.G.T.B. Khalsa College, University of Delhi, Delhi 110007, India*

Indra K. Vasil (1, 21, 81), *Department of Botany, University of Florida, Gainesville, Florida 32611*

Vimla Vasil (1), *Department of Botany, University of Florida, Gainesville, Florida 32611*

Lyndsey A. Withers (101), *Friedrich Miescher-Institut, Postfach 273, 4002 Basel, Switzerland*

M. Zenkteler (137), *Institute of Biology, Adam Mickiewicz University, Poznan, Poland*

* Present address: Department of Life Sciences, Worcester Polytechnic Institute, Worcester, Massachusetts 01609.

INTERNATIONAL REVIEW OF CYTOLOGY

SUPPLEMENT 1

Z. Hruban and M. Recheigl, Jr., Microbodies and Related Particles: Morphology, Biochemistry, and Physiology, 1969

SUPPLEMENT 2

Peter Luykx, Cellular Mechanisms of Chromosome Distribution, 1970

SUPPLEMENT 3

Andrew S. Bajer and J. Molè-Bajer, Spindle Dynamics and Chromosome Movements, 1972

SUPPLEMENT 4

G. H. Bourne, J. F. Danielli, and K. W. Jeon, eds., Aspects of Nuclear Structure and Function, 1974

SUPPLEMENT 5

G. H. Bourne, J. F. Danielli, and K. W. Jeon, eds., Aspects of Cell Control Mechanisms, 1977

SUPPLEMENT 6

G. H. Bourne, J. F. Danielli, and K. W. Jeon, eds., Studies in Ultrastructure, 1977

SUPPLEMENT 7

G. H. Bourne, J. F. Danielli, and K. W. Jeon, eds., Neuronal Cells and Hormones, 1978

SUPPLEMENT 8

G. H. Bourne, J. F. Danielli, and K. W. Jeon, eds., Aspects of Genetic Action and Evolution, 1978

SUPPLEMENT 9

J. F. Danielli and M. A. DiBerardino, eds., Nuclear Transplantation, 1979

SUPPLEMENT 10

Warren W. Nichols and Donald G. Murphy, eds., Differentiated Cells in Aging Research, 1979

SUPPLEMENT 11A

Indra K. Vasil, ed., Perspectives in Plant Cell and Tissue Culture, 1980

SUPPLEMENT 11B

Indra K. Vasil, ed., Perspectives in Plant Cell and Tissue Culture, 1980

Preface

The understanding of plant growth, development, and differentiation remains one of the most basic and challenging problems in plant biology. Plant tissue culture techniques, among many other methodologies, provide a powerful and unique tool for the study of these problems. Serious efforts to grow plant tissues in culture began with the pioneering efforts of Gautheret, Nobecourt, and White in the late 1930s. Discovery of the cytokinins, itself a direct result of plant tissue culture studies, greatly stimulated research on the role of plant growth-substances in cell division, differentiation, and morphogenesis, and led to the successful regeneration of plants from isolated tissues in many species. The regeneration of entire plants from isolated cells, in 1965, affirmed the totipotency of plant cells. Then, during the short span of 7 years, 1966–1972, three major developments took place in quick succession: regeneration of haploid plants from highly specialized gametophytic cells (microspores), regeneration of plants from isolated protoplasts, and finally, regeneration of somatic hybrid plants following protoplast fusion. It is these discoveries and developments, which together, form the basis of the current widespread interest in the utilization of plant tissue culture techniques for the study, resolution, and attainment of a wide variety of research problems and objectives, including large scale clonal propagation of desirable genotypes, elimination of viruses from plants or the selection of disease-resistant varieties, biosynthesis of pharmaceutically useful metabolites from plant cells in culture, preservation of germ plasm, isolation and characterization of mutant cell lines and their use in plant improvement, overcoming problems of sexual incompatibility in the production of desirable hybrids, and most importantly, the modification of the genetic content of plant cells by parasexual methods resulting in the production of improved and/or novel genotypes.

Some of these objectives have already been achieved during the last decade, and are being successfully exploited commercially, as the elimination of viruses and the large-scale clonal propagation of selected horticultural plants. Progress made in the pursuit of the remaining objectives is encouraging. I believe that a majority of these too can be achieved by the end of this century, provided the current level of interest is not only sustained but augmented, there is a substantial increase in the level of support for such research, and a greater degree of emphasis is placed on the study and the resolution of the basic problems currently hindering full development and exploitation of these techniques. The present shortage of trained personnel will be remedied in the near future as a result of excellent courses and training workshops now being organized in many universities, and by many national and international organizations.

The current two companion volumes on *Perspectives in Plant Cell and Tissue Culture* provide a discussion of the basic techniques, problems, and potentials of plant tissue culture. There is ample evidence in all the contributions here that plant tissue culture technique is not an end in itself, as is sometimes alleged, but that it is a vigorous and living science, which has greatly improved, and will continue to improve, our understanding of plant biology. It has an important role to play in the future improvement and utilization of plants by man. During recent years there has been too much, and perhaps premature, emphasis on the applied and commercial aspects of plant tissue culture. For this reason, discussions in these two volumes are devoted to basic science, with relevant but only brief references to current and possible future applications.

Responsibility for the selection of various subjects that are discussed in these companion volumes, and their authors, is mine, and the choices clearly reflect my biases. All contributors have honored my request to work within the prescribed guidelines. However, treatment of the subject matter, and its interpretation, are the sole responsibility of the respective authors, and naturally reflect their own biases. Every attempt has been made to present a complete discussion of all of the issues and problems involved. It is never possible to maintain a uniform style of language and presentation in a contributed volume of this kind. I have, therefore, attempted to maintain the typical flavor of each contribution, which maintains the individual style and prose of each author. There is a small amount of unavoidable, and probably useful, overlap between some of the chapters, but that too serves to indicate the interdependence and interrelationships of the topics of discussion.

As one of those who appreciates the history of science, I believe it is as important for a student or research scientist to be aware of the most current information in his or her chosen field of research, as it is to be familiar with the evolution and development of research ideas. Too often these days only the current status of a field of research is discussed and valued, as though it were the whole and the final truth, and had no past. We fail to benefit from the efforts of countless scientists of earlier years who have played significant and often decisive roles in the development of modern ideas. A study of the history of science shows that scientific discoveries must always be treated as theories in development, for it is not too uncommon for today's dogmas to become the discredited or modified theories of tomorrow. The generation of scientific ideas and theories, and their constant reevaluation and modification based on new findings, is one of the greatest pleasures of our profession. This we must share with the scientists of tomorrow. It is for this reason that most of the chapters also provide a historical perspective.

The meticulous care taken by all the contributors in the preparation of their chapters, and their adherence to time schedules, have made my editorial responsibilities lighter and a very pleasant experience. For these, I am grateful and thankful to all of my distinguished colleagues, many of whom had to write in an unfamiliar language. For me, personally, it has been a uniquely useful learning experience, and I hope that it will be the same for the readers of these volumes.

INDRA K. VASIL

Contents of Supplement 11A

Chapter 10

Isolation and Culture of Protoplasts

INDRA K. VASIL AND VIMLA VASIL

Department of Botany, University of Florida, Gainesville, Florida

I. Introduction

The term "protoplast" has been defined as that "part of the plant cell which lies within the cell wall and can be plasmolysed, and which can be isolated by removing the cell wall by mechanical or enzymatic procedures. The protoplast is, therefore, only a naked cell—surrounded by the plasma membrane—which is potentially capable of cell wall regeneration, growth, and division" (Vasil, 1976). The absence of the cell wall makes the protoplast suitable for a variety of experimental manipulations that are not possible with intact cells, such as the uptake of cell organelles, microorganisms, or foreign genetic material to produce genetically modified cells and to form somatic hybrid cells and/or plants by fusing protoplasts of diverse genetic characteristics. Within the past decade, these remarkable and unique attributes of plant protoplasts have made them the most important experimental tool in the hands of cell biologists for the parasexual modification of the genetic content of plant cells. One basic prerequisite for the potential use of protoplasts in such studies is the ability to isolate them readily and in large numbers and to culture them *in vitro* to form cell colonies and whole plants.

1

II. Isolation of Protoplasts

A. MECHANICAL ISOLATION

All of the early attempts to isolate plant protoplasts relied entirely on mechanical methods and were essentially limited to tissues containing large and vacuolated cells or to filamentous structures made up of elongated cells. Induced plasmolysis of such cells caused the plasmalemma to retract away from the cell wall, resulting in the formation of a rounded protoplast in the center of the cell. Tissues containing such plasmolyzed cells were cut into thin strips, and often the end walls were removed without damaging the protoplasts, which lay safely away from the cell wall. Protoplasts from such cut cells were easily released by osmotic swelling when the tissue was placed in solutions causing rapid uptake of water.

Mechanical methods of protoplast isolation are suitable only for a limited variety of higher plant tissues, such as leaf, bulb scale, fruit epidermises, or storage tissues. One of the most serious disadvantages of this technique is that it yields a very small number of protoplasts after a rather tedious procedure. Nevertheless, many of the pioneering and even some later studies on the expansion capacity of protoplasts, the effects of plant hormones on the plasmalemma, the osmotic relations of plant cells, the effect of fungal toxins on the plasmalemma, and the fusion of protoplasts were carried out with mechanically isolated protoplasts (Klercker, 1892; Seifriz, 1928; Chambers and Höfler, 1931; Plowe, 1931; Levitt *et al.*, 1936; Michel, 1937; Törnäva, 1939; Köningsberger, 1947; Tribe, 1955; Whatley, 1956; Vreugdenhill, 1957; Pilet, 1971, 1973; Pilet *et al.*, 1972). More recently, a combination of mechanical and enzymatic methods has also been shown to be useful (Bui-Dang-Ha and Mackenzie, 1973; Harada, 1973).

B. ENZYMATIC ISOLATION

The disadvantages inherent in the mechanical methods of protoplast isolation were largely overcome by the development of an enzymatic procedure that used a crude cellulase preparation from the fungus *Myrothecium verrucaria* to isolate protoplasts from tomato roots (Cocking, 1960). Similar enzyme preparations were later used to isolate protoplasts from a variety of tissues (Gregory and Cocking, 1963, 1965; Ruesink and Thimann, 1965, 1966). It therefore became possible not only to obtain comparatively large populations of active and viable protoplasts, but also to avoid their exposure to the deleterious effects of excessive and sometimes prolonged plasmolysis and to obtain protoplasts from relatively nonvacuolated meristematic cells, which do not plasmolyze easily. Nevertheless, the enzymatically isolated protoplasts are exposed to a variety of exogenous as well as endogenous (from burst or damaged cells) enzymes for considerable periods of time, whereas the mechanically isolated protoplasts are exposed only to endogenous enzymes for very brief periods of time.

Protoplast preparations, whether obtained by mechanical or by enzymatic isolation, almost always contain considerable amounts of cellular debris in the form of undigested cuticle, vascular elements, cell walls, tissue fragments, membranes, nuclei, plastids, other cell organelles, etc., all of which must be removed rapidly and adequately. In most instances much of the debris can be satisfactorily removed by filtration through Miracloth and stainless steel or nylon filters, followed by washing on Millipore filters or by centrifugation at low speed. Protoplasts often suffer mechanical damage and show reduced viability as a result of extensive centrifugation and washing procedures. Two-phase density gradients have proved useful in some cases for the purification of protoplast preparations (Kanai and Edwards, 1973; Larkin, 1976), although they are not totally efficient in removing the cell wall-degrading enzymes. The cleanest protoplast preparations, particularly from cell cultures, are obtained by floating the protoplasts on sucrose solutions (Gregory and Cocking, 1965; Chupeau and Morel, 1970; Evans *et al.*, 1972; Pilet *et al.*, 1972; Vasil and Vasil, 1979, 1980; White and Vasil, 1979). A discontinuous density gradient system for separation of protoplasts based on differences in their buoyant densities also has been described (Harms and Potrykus, 1978; Hughes *et al.*, 1978).

During the last decade a number of potent but crude enzyme preparations have become available commercially, with a fair degree of quality control. The most commonly used enzyme preparations are: cellulase from *Trichoderma viride*, Driselase from a basidiomycete rich in cellulase and pectinase, Macerase from *Rhizopus* rich in pectinase, and various other pectinases and hemicellulases (Vasil, 1976; Potrykus *et al.*, 1976; Eriksson, 1977). Many toxic substances and impurities are present in these enzyme preparations, including nucleases (especially ribonucleases), lipases, peroxidases, proteolytic and various other enzymes, and phenolics. Partial purification of the commercial enzymes by elution through Sephadex G-25 or Biogel is sometimes useful (Schenk and Hildebrandt, 1969a; Kao *et al.*, 1971; Gamborg and Wetter, 1975; Vasil *et al.*, 1975). However, most workers routinely use commercial enzyme preparations without any purification to obtain high yields of viable protoplasts that are capable of normal growth and development.

Very highly purified and crystalline enzymes are not only prohibitively expensive but are also relatively useless for protoplast isolation, as these are unable to break down the chemically and structurally complex plant cell walls, which contain hemicellulose, pectin, lipids, and proteins in addition to cellulose. Complex enzyme mixtures, cleaned of toxic substances and impurities, therefore are most useful for the complete breakdown of cell walls and the release of protoplasts.

Two principal methods have been used for the isolation of protoplasts. In the sequential method, plasmolyzed leaves are cut into small pieces, after the lower epidermis is removed, and incubated in Macerozyme, a preparation that is rich in pectinase (Takebe *et al.*, 1968). This causes the maceration of tissues and the release of mesophyll cells. The enzyme mixture and some of the debris is then removed by filtration, and the mesophyll cells are suspended in a cellulase solution to digest the cell walls and release the protoplasts.

More commonly, a mixture of pectinase and cellulase is used that not only macerates the tissues by attacking the middle lamella but also releases protoplasts by digesting cell walls (Power and Cocking, 1970). For some plants, such as cereals, mixtures of cellulase, pectinase, Driselase, and Rhozyme are necessary to obtain the optimum yield of viable protoplasts (Vasil and Vasil, 1979, 1980).

A brief period of incubation in vacuum helps to remove the air that is trapped in the intercellular spaces and facilitates the penetration of enzymes resulting in improved yields. Protoplasts are generally isolated at 25–30°C, but in some cases, particularly in cereal species, prolonged incubation at low temperature (14°C) followed by a short time at about 30°C have proved most useful (Potrykus *et al.*, 1977; Vasil and Vasil, 1979, 1980).

The protoplast must be protected against osmotic swelling and bursting during and after the removal of the cell wall. This is easily accomplished by the inclusion of osmotic stabilizers, such as mannitol, sorbitol, glucose, or sucrose, in the enzyme solution or with a combination of ionic and nonionic osmotica. Various facets of cellular metabolism, including cell wall formation, are influenced by the nature of the osmoticum and the osmotic pressure. Reduced osmotic pressures appear to be helpful in the isolation and culture of protoplasts. This is achieved by growing plants or cells in the dark for a few days before isolation of protoplasts. The use of metabolically active osmotic stabilizers, such as glucose, sorbitol, and sucrose, along with metabolically inert osmotic substances such as mannitol is recommended, because the sugars are gradually used by the protoplasts during their early growth and cell wall formation, resulting in a gradual reduction of osmoticum in the nutrient solutions. This helps eliminate sudden changes in the osmotic conditions when regenerated cell colonies must be transferred to nutrient media without mannitol for continued growth.

The determination of the viability of protoplasts is an important aspect of protoplast culture. Of course, the best test for viability is the regeneration of cell wall and the induction of cell divisions. The presence of active cytoplasmic streaming in freshly isolated protoplasts is a good sign of viability. Vital staining, with Evans blue or flouorescein diacetate, has also been used successfully to differentiate between viable and nonviable protoplasts (Widholm, 1972; Kanai and Edwards, 1973; Glimelius *et al.*, 1974; Larkin, 1976).

C. SOURCES OF PROTOPLASTS

Root tips of seedlings and the locular tissue of solanaceous fruits were used in the early work on the isolation of protoplasts with the aid of cell wall-digesting enzymes (Cocking, 1960; Gregory and Cocking, 1965). With the availability of better and more efficient enzyme preparations, protoplasts can now be isolated from leaves (Takebe *et al.*, 1968; Otsuki and Takebe, 1969; Ohyama and Nitsch, 1972; Evans *et al.*, 1972; Durand *et al.*, 1973; Kartha *et al.*, 1974; Vasil and Vasil, 1974; Watts *et al.*, 1974; Vasil, 1976; Bourgin *et al.*, 1979), cladodes (Bui-Dang Ha and Mackenzie, 1973), shoot apices (Gamborg *et al.*, 1975), fruits (Raj and Herr, 1970), roots (Ruesink and Thimann, 1966; Power *et al.*, 1970; Bawa and Torrey, 1971; Landgren and Torrey, 1973; Vasil and Vasil, 1974), legume root nodules (Davey *et al.*, 1973; Vasil *et al.*, 1975, 1977), coleoptiles (Ruesink and Thimann, 1965; Hall and Cocking, 1971), the aleurone layer of cereal grains (Taiz and Jones, 1971), microspore mother cells (Ito, 1973), microspore tetrads (Bhojwani and Cocking, 1972; Rajasekhar, 1973; Wakasa, 1973), and pollen tubes (Condeelis, 1974). Mesophyll tissue of leaves is the most commonly used source of protoplasts. Young leaves obtained from shoot cultures are preferable to leaves from field or greenhouse plants.

Cells and tissues grown *in vitro,* particularly suspension cultures, have been shown to be an important source of protoplasts (Ruesink and Thimann, 1965; Motoyoshi, 1971; Wakasa, 1973; Vasil and Vasil, 1974, 1979, 1980; Vardi *et al.*, 1975; Eriksson and Jonasson, 1969; Schenk and Hildebrandt, 1969a,b; Kao *et al.*, 1971; Grambow *et al.*, 1972; Maretzki and Nickell, 1973; Gamborg *et al.*, 1974; Wallin *et al.*, 1977; White and Vasil, 1979). Cultured cells and tissues are grown under aseptic and controlled physiological and environmental conditions and provide a least variable source of cells at any given time. They are also conditioned to growth in culture, and in many cases the requirements for their growth and organogenesis *in vitro* are already known.

The yield as well as the viability of protoplasts is closely related to the conditions under which the source material has been grown and maintained. Cell wall modifications favorable to good protoplast production in cultured cells can be achieved sometimes by the addition of auxins, sulfur amino acids, reducing agents, or heavy metal ions (Eriksson *et al.*, 1978). Low concentrations of sucrose also improve the yield of protoplasts. Perhaps the best source of protoplasts are cell suspension cultures that have rapid growth rates and are subcultured every 2–4 days.

Plants grown under field conditions or in the greenhouse provide a most inconsistent source material. The growth conditions of these plants must be closely studied and maintained to obtain the maximum yield of protoplasts that will grow in culture. Light intensity (low light intensities are favorable) and duration (day/night cycles are required) and inorganic nutrition (nitrogen fertilizer is essential) are the most critical factors that affect protoplast yield and viability. In only a few instances have the various parameters that control protoplast yield been studied methodically (Watts *et al.*, 1974; Shepard and Totten, 1975; Cassells and Barlass, 1978). Similar detailed studies with other species are most desirable.

III. Culture of Protoplasts

A. NUTRIENT MEDIA

The nutrient requirements of isolated protoplasts are very similar to those of cultured cells and tissues. In the absence of the cell wall, protoplasts tend to be very efficient in the uptake of nutrients from the medium. Therefore, the nutrient media used for the culture of protoplasts generally are modified to contain reduced levels of inorganic substances. Leakage of some metabolites may also take place from protoplasts during the initial stages of culture in the absence of the cell wall. This may necessitate modification of nutrient media used for protoplast culture as compared to those found suitable for cell culture. Various modifications of the nutrient media developed by Murashige and Skoog (1962) and Gamborg *et al.* (1968) have been successfully used for the growth of protoplasts isolated from a variety of cells, tissues, and species. Kao and Michayluk (1975) used an extremely complex medium for the culture of single protoplasts of *Vicia hajastana*. Their medium contains, in addition to the normal complement of inorganic substances, 14 vitamins, auxins and cytokinins, various organic acids, 10 sugars and sugar alcohols, 21 amino acids, 6 nucleic acid bases, casein hydrolyzate, and coconut water. The addition of auxins to nutrient media for the culture of protoplasts should be carefully studied, as auxins may cause rapid bursting of protoplasts either by increasing the volume of existing vacuoles or by the formation of numerous new vacuoles.

B. Methods of Culture

The techniques for the culture of protoplasts are adapted, with only minor modifications, from the well-established procedures developed for the culture of single cells or cell suspensions. Various methods available for the culture of protoplasts are described below.

1. Suspension and Drop Cultures

Protoplasts are suspended in a liquid medium at a density of about 10^5/ml and cultured in very shallow layers in 25-ml Erlenmeyer flasks, with or without any shaking.

The procedure for drop cultures developed by Kao et al. (1971), has been used successfully and extensively (Grambow et al., 1972; Gamborg et al., 1973; Kartha et al., 1974; Gamborg and Wetter, 1975; Vasil and Vasil, 1980). Protoplast suspensions at a density of 10^4 or 10^5/ml are placed in 50-μl drops in plastic Petri dishes, sealed with Parafilm, and incubated. In order to maintain adequate humidity in the cultures, drops of water are placed in the middle of the Petri dish. Regular monitoring of such cultures is possible with the aid of an inverted microscope.

2. Plating

Protoplasts suspended in the liquid medium are mixed gently but rapidly with an equal volume of the medium prepared in agar (1–2%) and kept at about 45°C in a water bath (Nagata and Takebe, 1971). Small volumes (5 ml) of the protoplast-containing agar medium are then poured into Petri dishes, which are sealed with Parafilm and incubated. In order to provide adequate aeration to the protoplasts, it is important that the poured agar medium remain in a very thin layer. Protoplasts can also be suspended in a soft agar medium (0.4% agar) and overlaid on already plated nutrient medium gelled with 0.8% agar. Examination of such cultures is not as easy as that of drop cultures, but plated cultures have the advantage that the protoplasts remain in a fixed position, and the development of individual protoplasts can be followed with ease.

3. Microculture Chambers

Development of individual protoplasts can be followed with much more precision in microculture chambers, as demonstrated by Vasil and Vasil (1973, 1974) in tabacco and Durand et al. (1973) in *Petunia*.

Either single or only a very small number of protoplasts can be cultured by this method, which is particularly useful for following the development of mechanically isolated heterokaryons following protoplast fusion. Microculture chambers are prepared by placing a droplet of ca. 30 μl of the nutrient medium, containing one or more protoplasts, and enclosing it by a cover glass resting on two cover glasses placed on either side of the nutrient medium. The cultures are sealed with sterile paraffin oil. The formation of cell masses of 75 or more cells can be followed in this manner, after which the cell colonies can be transferred to agar media.

The practice of culturing single protoplasts in very small volumes of nutrient medium has been used to great advantage by Kao (1977) and Gleba and Hoffmann (1978) to obtain intergeneric hybrid cell lines of soybean and *Nicotiana glauca* and of *Arabidopsis thaliana* and *Brassica compestris,* respectively, and by Gleba (1978) to obtain tobacco plants from single mesophyll protoplasts. These authors used Cuprak dishes containing numbered small wells that could accommodate anywhere from 0.25 to 25 μl of nutrient medium. It has been shown that single protoplasts can be successfully cultured in as little as 0.25 μl of nutrient medium (Gleba, 1978).

4. *Feeder Layers and Nurse Cultures*

Nondividing but metabolically active, X-irradiated protoplasts embedded in nutrient agar support the growth of protoplasts plated at very low densities (5–50 protoplasts per milliliter) above them (Raveh and Galun, 1975). "Nurse cultures" also have been used by Menczel *et al.* (1978) to clone somatic hybrids from *Nicotiana* heterokaryons. There is some indication that the nurse effect is nonspecific, and nurse layers of one species may be able to support the growth of other species.

In the absence of generally applicable selection systems for the preferential growth of hybrid cells over parental types, the development of techniques for the growth of single protoplasts can become a major factor in the isolation of somatic hybrids, particularly after intergeneric fusions, where the growth of hybrid cells is very difficult and/or rare.

C. CELL WALL REGENERATION

Freshly isolated protoplasts cultured in suitable nutrient media rapidly regenerate cell walls. The newly synthesized wall can be visualized at the light microscope level by Calcofluor staining (Nagata and Takebe, 1970) or, more decisively, by the ultrastructural techniques of silver hexamine staining (Fowke *et al.*, 1974), freeze etching (Willison and Grout, 1978), platinum–palladium replicas (Williamson *et al.*, 1977), negative staining (Herth and Meyer, 1977), and the scanning microscope (Burgess *et al.*, 1978). Protoplasts isolated from cell cultures regenerate cell walls more rapidly than those from differentiated tissues, such as the mesophyll. Cell wall regeneration takes place by the deposition of cellulose microfibrils on the surface of the plasma membrane, which can be seen as early as 10–20 minutes after the culture of *Vicia* protoplasts (Williamson *et al.,* 1977). In general the first microfibrils are observed after 8–24 hours of culture (Burgess *et al.,* 1978; Fowke, 1978). It is conceivable that microfibril deposition starts soon after the removal of the protoplasts from the enzyme solutions, but the actual visualization of the microfibrils is not possible at the earliest stages owing to the loss of the matrix components into the surrounding nutrient medium. The wide range of differences observed in the appearance of the first microfibrils probably reflects the metabolic and physiological status of the isolated protoplasts. Exogenously supplied plant growth substances do not appear to directly influence cell wall formation.

The newly deposited cell wall is composed of loosely organized cellulose microfibrils, which later become more organized to form a typical plant cell wall (Fowke *et al.,* 1974; Burgess *et al.,* 1977; Fowke, 1978). The formation of microfibrils takes place at the surface of the plasma membrane rather than within the cytoplasm. It has been suggested that preformed microfibrils of indeterminate length rise out of the plasma membrane (Willison and Cocking, 1975; Willison, 1976) or by the addition of new material at both ends of very short microfibrils (Burgess and Linstead, 1976). The role of specific cell organelles in cell wall regeneration in protoplasts is not clear, although Golgi bodies and endoplasmic reticulum are said to be involved in the synthesis and deposition of the primary cell walls of higher plants. Cell wall regeneration precedes cell division in cultured protoplasts and is considered to be a prerequisite for mitosis (Vasil, 1976; Schilde-Rentschler, 1977).

D. CELL DIVISION

Protoplasts increase in size and numerous cytoplasmic strands are formed during and before the early stages of cell wall regeneration. There is also an increase in the number of cell organelles, cytoplasmic streaming, respiration, and synthesis of RNA, protein, and polysaccharides, all indicative of increased metabolic activity. Most of the cell organelles aggregate around the nucleus. With the completion of the new cell wall, protoplasts lose their characteristic spherical shape. Mitosis and cytokinesis are normal and occur after 2–7 days of culture. Protoplasts isolated from differentiated cells, such as the mesophyll cells of leaves, which do not divide in nature, take longer to undergo the first division than those isolated from cells dividing rapidly in culture (Vasil and Vasil, 1974). Multicellular clumps or colonies are formed within 1–3 weeks, and these can be either further subcultured as callus tissues or transferred to liquid nutrient media to obtain suspension cultures. The number and variety of species in which sustained cell divisions can be obtained from isolated protoplasts is quite limited. Most of the species listed in Table I belong to the Solanaceae, and important families of crop plants, such as the Gramineae and the Leguminosae, are poorly represented.

Although single, isolated protoplasts can be cultured in complex nutri-

TABLE I

SPECIES IN WHICH SUSTAINED CELL DIVISIONS HAVE BEEN REPORTED IN CULTURED PROTOPLASTS

Species	References
Ammi visnaga	Gamborg *et al.* (1974)
Antirrhinum majus	Poirier-Hamon *et al.* (1974)
Arabidopsis thaliana	Gamborg and Miller (1973)
Brassica oleracea var. *acephala*	Gatenby and Cocking (1977)
Catharanthus roseus	Koblitz (1975)
Cicer arietinum	Gamborg *et al.* (1974)
Citrus sinensis	Vardi *et al.* (1975)
Cucumis sativus	Coutts and Wood (1975)
Geranium	Abo-El-Nil and Hildebrandt, 1976
Glycine max	Kao *et al.* (1970)
Gossypium hirsutum	Bhojwani *et al.* (1977)
Haplopappus gracilis	Kao *et al.* (1971)
Hordeum vulgare	Koblitz (1976)
Hyoscyamus albus	Lorz *et al.* (1979)
Hyoscyamus niger	Kohlenbach and Bohnke (1975)
Linum usitatissimum	Gamborg *et al.* (1974)
Lycopersicon esculentum	Zapata *et al.* (1977)
Lycopersicon peruvianum	Zapata *et al.* (1977)
Medicago sativa	Gamborg *et al.* (1974)
Melilotus alba	Gamborg *et al.* (1974)
Nicotiana glutinosa	Bourgin *et al.* (1979)
Oryza sativa	Anonymous (1975); Deka and Sen (1976); Cai *et al.* (1978)
Pennisetum americanum	Vasil and Vasil (1979)
Pharbitis nil	Messerschmidt (1974)
Phaseolus vulgaris	Pelcher *et al.* (1974)
Pisum sativum	Constabel *et al.* (1973); Gamborg *et al.* (1975)
Saccharum	Maretzki and Nickell (1973)
Solanum tuberosum	Upadhya (1975)
Triticum monococcum	Nemet and Dudits (1977)
Vicia faba	Binding and Nehls (1978)
Vicia hajastana	Gamborg *et al.* (1974)
Vicia narbonensis	Donn (1978)
Vigna sinensis	Davey *et al.* (1974); Gamborg *et al.* (1974)
Zea mays	Potrykus *et al.* (1977, 1979)

ent media under certain conditions (see Section III, B), maintenance of a minimum cell density is generally necessary for inducing cell wall regeneration and cell division in most instances. Experience has shown that a population density of 10^3/ml to 10^5/ml gives the best results.

E. REGENERATION OF PLANTS

Cell colonies formed from protoplasts can be transferred to fresh nutrient media for further growth and multiplication. In many species shoot and root differentiation and plantlet formation can be induced in protoplast-derived callus tissues with the help of auxins and cytokinins (Table II). In some species embryoids are formed from protoplast-derived colonies, as in *Daucus carota* (Grambow *et al.*, 1972; Kameya and Uchimiya, 1972), *Citrus sinensis* (Vardi *et al.*, 1975), *Nicotiana tabacum* (Lorz *et al.*, 1977), *Atropa belladonna* (Gosch *et al.*, 1975), and *Pennisetum americanum* (Vasil and Vasil, 1980). Totipotency of higher plant protoplasts was first demonstrated in *N. tabacum* (Takebe *et al.*, 1971; Nagata and Takebe, 1971). A perusal of the species listed in Table II shows that plant regeneration from protoplasts has been achieved, with a few exceptions, only in members of the Solanaceae. Furthermore, this list includes only two species of major economic importance, namely *N. tabacum* and *Solanum tuberosum,* and only one representative of the cereals (*Pennisetum*). Plant regeneration from protoplasts is a prerequisite for the utilization of protoplast technology in somatic hybridization and genetic manipulation. In order for this technology to be used for crop improvement it will be essential that plant regeneration from protoplasts of legumes and cereals—the two most important groups of plants for man and his domesticated animals—be achieved soon.

F. CEREAL PROTOPLAST CULTURE—A SPECIAL PROBLEM

Large populations of viable protoplasts can be isolated from leaves or cell cultures of cereals, but the induction of cell wall regeneration and cell divisions in such protoplasts has become a most frustrating problem. Despite the use of innumerable variations in culture conditions, nutrient media, and the source of plant material, attempts to induce callus formation from cereal protoplasts have failed in most cases (Potrykus *et al.*, 1976; Cocking, 1978; Galston, 1978; King *et al.*, 1978; Pental and Gunckel, 1979; Thomas *et al.*, 1979). Instances of sporadic cell wall formation and cell division have been reported for several cereal species, but in the absence of sustained cell divisions and callus formation these reports need further confirmation. There are only five examples of callus formation from cereal protoplasts: *Hordeum vulgare* (Koblitz, 1976),

TABLE II

**Species in Which Plant Regeneration Has Been Achieved from
Cultured Protoplasts**

Species	References
Asparagus officinalis	Bui-Dang-Ha and Mackenzie (1973)
Atropa belladonna	Gosch *et al.* (1975)
Brassica napus	Kartha *et al.* (1974)
Brassica napus (haploid)	Thomas *et al.* (1976)
Bromus inermis	Kao *et al.* (1973)
Datura metel (haploid and diploid)	Schieder (1977)
Datura meteloides (haploid and diploid)	Schieder (1977)
Datura innoxia (haploid and diploid)	Schieder (1975)
Daucus carota	Grambow *et al.* (1972); Dudits *et al.* (1976)
Hyoscyamus muticus	Lorz *et al.* (1979)
Nicotiana acuminata	Bourgin *et al.* (1979)
Nicotiana alata (haploid)	Bourgin and Missonier (1978)
Nicotiana alata	Bourgin *et al.* (1979)
Nicotiana debneyi	H. H. Smith (personal communication)
Nicotiana glauca	Bourgin *et al.* (1979)
Nicotiana langsdorffi	Bourgin *et al.* (1979)
Nicotiana longiflora	Bourgin *et al.* (1979)
Nicotiana otophora	Banks and Evans (1976); Bourgin *et al.* (1979)
Nicotiana paniculata	Bourgin *et al.* (1979)
Nicotiana plumbaginifolia	Bourgin *et al.* (1979)
Nicotiana suaveolens	Bourgin *et al.* (1979)
Nicotiana sylvestris	Bourgin *et al.* (1976, 1979); Nagy and Maliga (1976); Banks and Evans (1976)
Nicotiana sylvestris × *Nicotiana otophora* (F₁ hybrid)	Banks and Evans (1976)
Nicotiana tabacum	Takebe *et al.* (1971); Nagata and Tabeke (1971)
Nicotiana tabacum × *Nicotiana otophora* (F₁ hybrid)	Banks and Evans (1976)
Nicotiana tabacum (haploid)	Ohyama and Nitsch (1972)
Pennisetum americanum	Vasil and Vasil (1980)
Petunia axillaris	Power *et al.* (1976a)
Petunia hybrida	Durand *et al.* (1973); Frearson *et al.* (1973); Vasil and Vasil (1974)
Petunia hybrida (haploid)	Binding (1974)
Petunia hybrida (cytoplasmic male sterile)	Vasil and Vasil (1974)
Petunia hybrida × *Petunia parodii* (hybrid)	Power *et al.* (1976b)
Petunia inflata	Power *et al.* (1976a)
Petunia parodii	Hayward and Power (1975)
Petunia parviflora	Sink and Power (1977)
Petunia violacea	Power *et al.* (1976a)
Ranunculus sceleratus	Dorion *et al.* (1975)
Solanum dulcamara	Binding and Nehls (1977)
Solanum tuberosum	Shepard and Totten (1977)

12

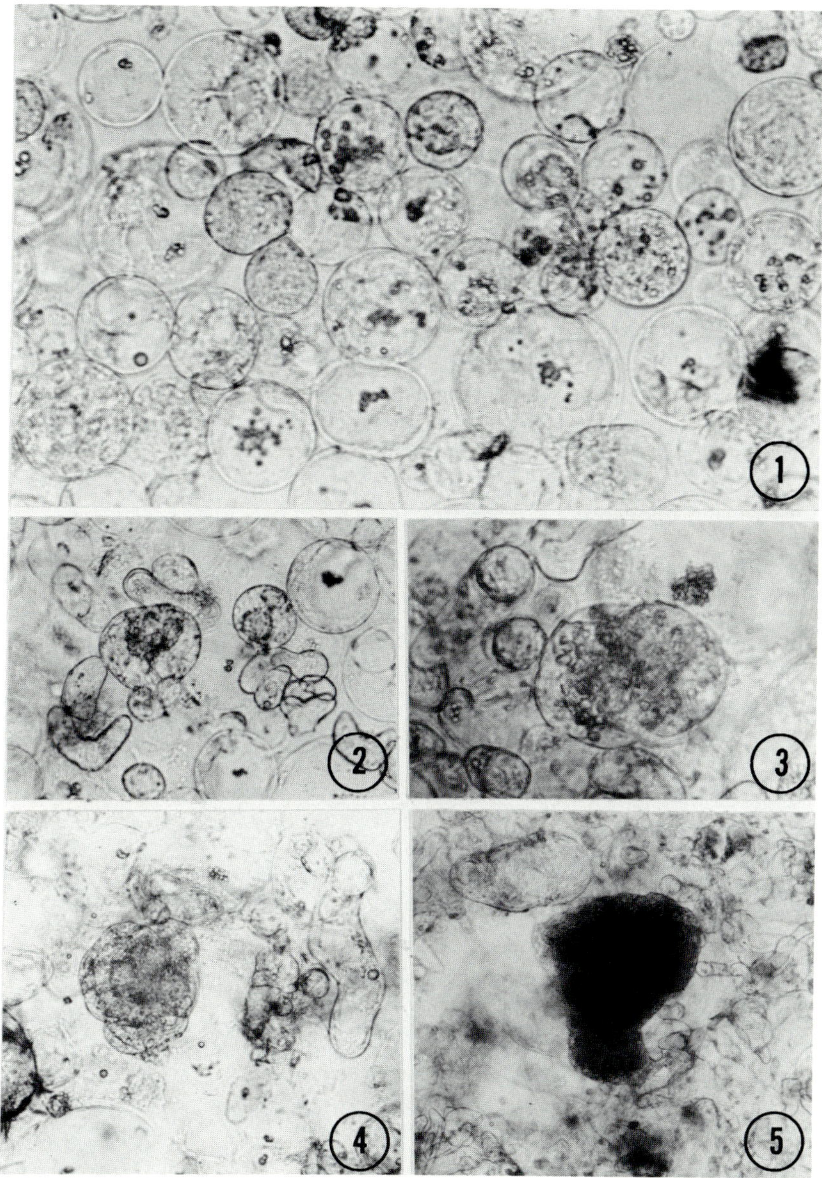

Figs. 1–5.   Culture of *Pennisetum americanum* protoplasts.
Fig. 1.   Vacuolated and densely cytoplasmic embryogenic protoplasts isolated from embryogenic suspension cultures.
Figs. 2–4.   Cell wall regeneration and cell division in embryogenic protoplasts.
Fig. 5.   Embryoid formation from embryogenic protoplasts.

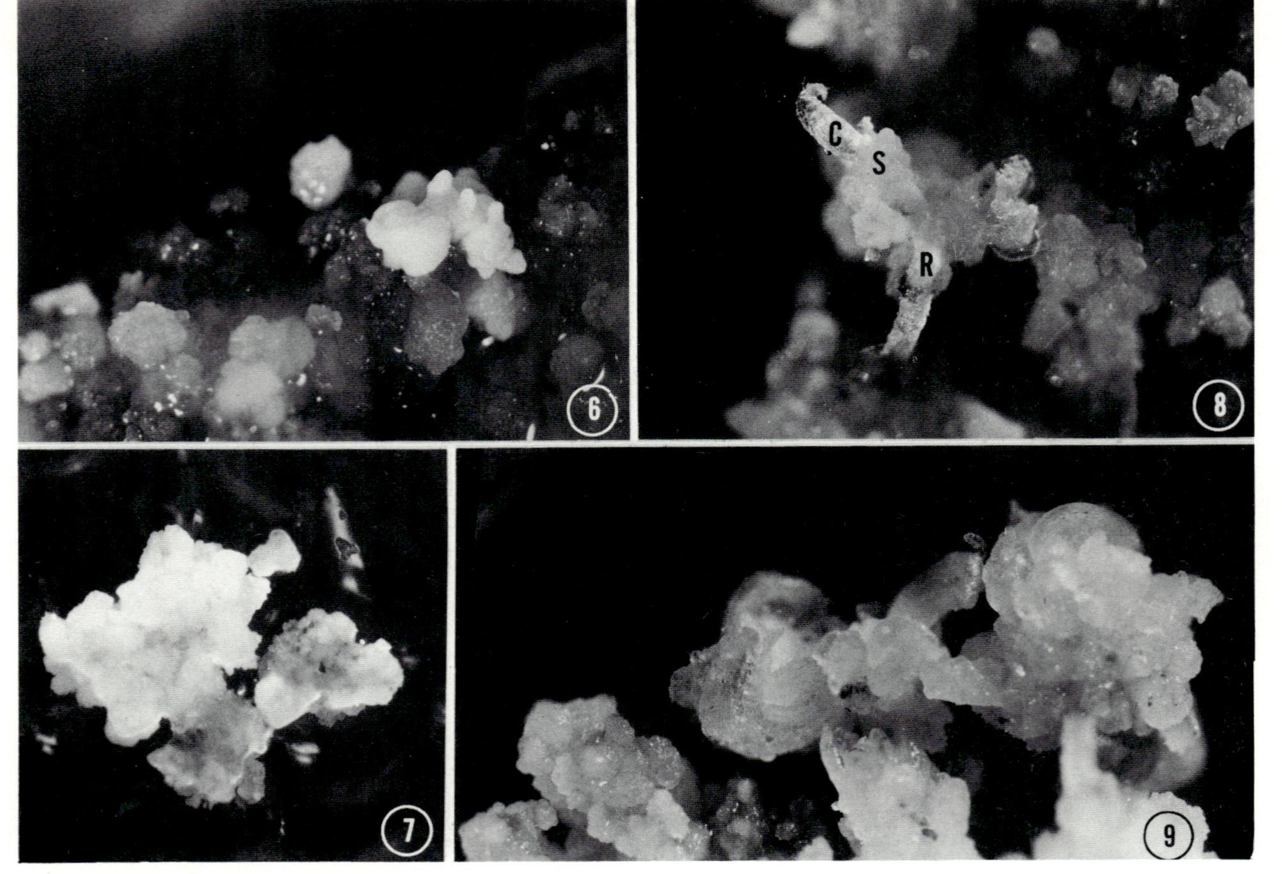

FIGS. 6–9.   Culture of *Pennisetum americanum* protoplasts.
FIG. 6.   Mature embryoids formed from embryogenic protoplasts. Note well-formed scutellum.
FIG. 7.   Several embryoids are seen in the cell mass resulting from the division of embryogenic protoplasts.
FIG. 8.   Elongation of the coleoptile (C) and coleorhiza (R) during ''germination'' of the embryoids and plantlet formation. Secondary proliferation of the scutellum(s) can be seen.
FIG. 9.   Development of shoots from several embryoids.

Oryza sativa (Deka and Sen, 1976; Cai *et al.*, 1978), *Pennisetum americanum* (I. K. Vasil and Vasil, 1980; V. Vasil and Vasil, 1979), *Triticum monococcum* (Nemet and Dudits, 1977), and *Zea mays* (Potrykus *et al.*, 1977, 1979). In all of these cases protoplasts were isolated from cell cultures, and in the only instance where protoplasts were isolated from young stem tissue of *Z. mays* and successfully cultured (Potrykus *et al.*, 1977), it has subsequently not been possible to repeat the results (Potrykus, personal communication). The report of plantlet formation from protoplasts of *P. americanum* (I. K. Vasil and Vasil, 1980; V. Vasil and Vasil, 1980) is therefore most encouraging. Immature embryos of *P. americanum* were induced to give rise to callus tissue, which in turn was placed in a liquid nutrient medium to obtain an embryogenic suspension culture. Such suspension cultures yield two types of protoplasts, one large and vacuolated and the other small, nonvacuolated, and richly cytoplasmic. The latter are derived from embryogenic cell masses in suspension culture, and under appropriate conditions give rise to plantlets via somatic embryogenesis (Figs. 1–9). In future work on cereal protoplasts more attention should be given to the nature of the cells from which protoplasts are isolated than to nutrient media and culture conditions.

IV. Conclusions

Serious attempts to isolate and culture protoplasts of flowering plants started in 1970. During the past decade much progress has been made, and it is now possible to obtain large populations of living protoplasts from a wide variety of plants. However, sustained cell divisions leading to callus formation have been achieved only in a limited number of species. Plant regeneration from protoplasts is even rarer, having been reported only in 12 genera, more than half of which belong to the Solanaceae.

The use of protoplasts in somatic hybridization and genetic manipulation experiments is dependent on our ability to regenerate plants from isolated protoplasts, particularly in species of agronomic and economic importance. It is disappointing, therefore, to realize that plant regeneration from protoplasts of the two most important angiosperm families, the Leguminosae and the Gramineae, has not been possible, except the recent success in *P. americanum*. Extensive as well as intensive efforts must be undertaken immediately to bridge this serious gap in protoplast technology.

Owing to the early successes with the culture of mesophyll protoplasts isolated from young leaves, most of the later attempts to culture protoplasts have also been limited to mesophyll protoplasts. Only in a few cases have protoplasts been isolated from other tissues or from cell cultures. In many cases, particularly in the legumes, cell masses and callus tissues can be obtained from protoplasts but these fail to undergo organogenesis. Experience indicates that the physiological as well as the morphogenetic status of the source material, whether plants or cultured cells, is of utmost value in determining not only the fate of cultured protoplasts but also the regeneration of plants from the resulting cell masses. Embryogenic tissue cultures obtained from immature embryos or other sources appear to be the most promising source of protoplasts for plant regeneration. This proved to be the key factor in the successful regeneration of plantlets, through somatic embryogenesis, in protoplast-derived cell masses of the cereal *P. americanum*. However, it appears from the available literature that minor changes in the composition of the nutrient medium are not a profitable approach to resolving the problem of the culture and regeneration of protoplasts in species that have proved to be recalcitrant.

REFERENCES

Abo-El-Nil, M. M., and Hildebrandt, A. C. (1976). *Can. J. Bot.* **54**, 1530–1534.
Anonymous (1975). *Sci. Sin.* **18**, 779–784.
Banks, M. S., and Evans, P. K. (1976). *Plant Sci. Lett.* **7**, 409–416.
Bawa, S. B., and Torrey, J. G. (1971). *Bot. Gaz. (Chicago)* **132**, 240–245.
Bhojwani, S. S., and Cocking, E. C. (1972). *Nature (London), New Biol.* **239**, 29–30.
Bhojwani, S. S., Power, J. B., and Cocking, E. C. (1977). *Plant Sci. Lett.* **8**, 85–89.
Binding, H. (1974). *Z. Pflanzenphysiol.* **74**, 327–356.
Binding, H., and Nehls, R. (1977). *Z. Pflanzenphysiol.* **85**, 279–280.
Binding, H., and Nehls, R. (1978). *Z. Pflanzenphysiol.* **88**, 327–332.
Bourgin, J. P., and Missonier, C. (1978). *Z. Pflanzenphysiol.* **87**, 55–64.
Bourgin, J. P., Missonier, C., and Chupeau, Y. (1976). *C. R. Acad. Sci., Ser. D* **282**, 1853–1856.
Bourgin, J. P., Chupeau, Y., and Missonier, C. (1979). *Physiol. Plant.* **45**, 288–292.
Bui-Dang-Ha, D., and Mackenzie, I. A. (1973). *Protoplasma* **78**, 215–221.
Burgess, J., and Linstead, P. J. (1976). *Planta* **131**, 173–178.
Burgess, J., Linstead, P. J., and Fisher, V. E. L. (1977). *Micron* **8**, 113–122.
Burgess, J., Linstead, P. J., and Bonsall, V. E. (1978). *Planta* **139**, 85–91.
Cai, Q., Qian, Y., Zhou, Y., and Wu, S. (1978). *Chih Wu Hsuch Pao* **20**, 97–102.
Cassells, A. C., and Barlass, M. (1978). *Physiol. Plant.* **42**, 236–242.
Chambers, R., and Höfler, K. (1931). *Protoplasma* **12**, 338–355.
Chupeau, Y., and Morel, G. (1970). *C. R. Acad. Sci., Ser. D* **270**, 2659–2662.
Cocking, E. C. (1960). *Nature (London)* **187**, 927–929.
Cocking, E. C. (1978). *Proc. Symp. Plant Tissue Cult Peking*, pp. 225–263.
Condeelis, J. S. (1974). *Exp. Cell Res.* **88**, 435–439.
Constabel, F., Kirkpatrick, J. W., and Gamborg, O. L. (1973). *Can. J. Bot.* **51**, 2105–2106.
Coutts, R. H. A., and Wood, K. R. (1975). *Plant Sci. Lett.* **4**, 189–193.
Davey, M. R., Cocking, E. C., and Bush, E. (1973). *Nature (London)* **244**, 460–461.
Davey, M. R., Bush, E., and Power, J. B. (1974). *Plant Sci. Lett.* **3**, 127–133.

Deka, P. C., and Sen, S. K. (1976). *Mol. Gen. Genet.* **145,** 239–243.

Donn, G. (1978). *Z. Pflanzenphysiol.* **86,** 65–75.

Dorion, N., Chupeau, Y., and Bourgin, J. P. (1975). *Plant Sci. Lett.* **5,** 325–331.

Dudits, D., Kao, K. N., Constabel, F., and Gamborg, O. L. (1976). *Can. J. Bot.* **54,** 1063–1067.

Durand, J., Potrykus, I., and Donn, G. (1973). *Z. Pflanzenphysiol.* **69,** 26–34.

Eriksson, T. (1977). *In* "Plant Tissue Culture and its Bio-technological Application" (W. Barz, E. Reinhard, and M. H. Zenk, eds.), pp. 313–322. Springer-Verlag, Berlin and New York.

Eriksson, T., and Jonasson, K. (1969). *Planta* **89,** 85–89.

Eriksson, T., Glimelius, K., and Wallin, A. (1978). *In* "Frontiers of Plant Tissue Culture, 1978" (T. A. Thorpe, ed.), pp. 131–139. Univ. of Calgary Press, Calgary, Alberta.

Evans, P. K., Keates, A. G., and Cocking, E. C. (1972). *Planta* **104,** 178–181.

Fowke, L. C. (1978). *In* "Frontiers of Plant Tissue Culture, 1978" (T. A. Thorpe, ed.), pp. 223–233. Univ. of Calgary Press, Calgary, Alberta.

Fowke, L. C., Bech-Hansen, C. W., and Gamborg, O. L. (1974). *Protoplasma* **79,** 325–248.

Frearson, E. M., Power, J. B., and Cocking, E. C. (1973). *Dev. Biol.* **33,** 130–137.

Galston, A. W. (1978). *In* "Propagation of Higher Plants Through Tissue Culture" (K. W. Hughes, R. Henke, and M. Constantin, eds.), pp. 200–212. U.S. Dep. Energy, Oak Ridge, Tennessee.

Gamborg, O. L., and Miller, R. A. (1973). *Can. J. Bot.* **51,** 1795–1799.

Gamborg, O. L., and Wetter, L. R., eds. (1975). "Plant Tissue Culture Methods." Natl. Res. Counc. Can., Ottawa.

Gamborg, O. L., Miller, R. A., and Ojima, K. (1968). *Exp. Cell Res.* **50,** 151–158.

Gamborg, O. L., Kao, K. N., Miller, R. A., Fowke, L. C., and Constabel, F. (1973). *Colloq. Int. C.N.R.S.* **212,** 155–173.

Gamborg, O. L., Constabel, F., Fowke, L. C., Kao, K. N., Ohyama, K., Kartha, K., and Pelcher, L. (1974). *Can. J. Genet. Cytol.* **16,** 737–750.

Gamborg, O. L., Shyluk, J. P., and Kartha, K. K. (1975). *Plant Sci. Lett.* **4,** 285–292.

Gatenby, A. A., and Cocking, E. C. (1977). *Plant Sci. Lett.* **8,** 275–280.

Gleba, Y. Y. (1978). *Naturwissenschaften* **65,** 158.

Gleba, Y. Y., and Hoffmann, F. (1978). *Mol. Gen. Genet.* **165,** 257–264.

Glimelius, K., Wallin, A., and Eriksson, T. (1974). *Physiol. Plant.* **31,** 225–230.

Gosch, G., Bajaj, Y. P. S., and Reinert, J. (1975). *Protoplasma* **86,** 405–410.

Grambow, H. J., Kao, K. N., Miller, R. A., and Gamborg, O. L. (1972). *Planta* **103,** 348–355.

Gregory, D. W., and Cocking, E. C. (1963). *Biochem. J.* **88,** 40.

Gregory, D. W., and Cocking, E. C. (1965). *J. Cell Biol.* **24,** 143–146.

Hall, M. D., and Cocking, E. C. (1971). *Biochem. J.* **124,** 33*p.*

Harada, H. (1973). *Z. Pflanzenphysiol.* **69,** 77–80.

Harms, C. T., and Potrykus, I. (1978). *Theor. Appl. Genet.* **53,** 57–63.

Hayward, C., and Power, J. B. (1975). *Plant Sci. Lett.* **4,** 407–410.

Herth, W., and Meyer, Y. (1977). *Biol. Cell.* **30,** 33–40.

Hughes, B. G., White, F. G., and Smith, M. A. (1978). *Biochem. Physiol. Pflanz.* **172,** 223–231.

Ito, M. (1973). *Bot. Mag.* **86,** 133–141.

Kameya, T., and Uchimiya, H. (1972). *Planta* **103,** 356–360.

Kanai, R., and Edwards, G. E. (1973). *Plant Physiol.* **52,** 484–490.

Kao, K. N. (1977). *Mol. Gen. Genet.* **150,** 225–230.

Kao, K. N., and Michayluk, M. R. (1975). *Planta* **126,** 105–110.

Kao, K. N., Keller, W. A., and Miller, R. A. (1970). *Exp. Cell Res.* **62,** 338–340.

Kao, K. N., Gamborg, O. L., Miller, R. A., and Keller, W. A. (1971). *Nature (London)*, *New Biol.* **232**, 124.

Kao, K. N., Gamborg, O. L., Michayluk, M. R., Keller, W. A., and Miller, R. A. (1973). *Colloq. Int. C.N.R.S.* **212**, 207–213.

Kartha, K. K., Michayluk, M. R., Kao, K. N., and Gamborg, O. L. (1974). *Plant Sci. Lett.* **3**, 265–271.

King, P. J., Potrykus, I., and Thomas, E. (1978). *Physiol. Veg.* **16**, 381–399.

Klercker, J. A. F. (1892). *Oefvers Vetensk. Akad. Foerh.*, Stockholm **9**, 463–471.

Koblitz, H. (1975). *Biochem. Physiol. Pflanz.* **167**, 489–499.

Koblitz, H. (1976). *Biochem. Physiol. Pflanz.* **170**, 287–293.

Köningsberger, V. J. (1947). *Meded. K. Vlaam. Acad. Wet., Lett. Schone Kunsten Kl. Wet. Belg.*, **9**, 5–28.

Kohlenbach, H. W., and Bohnke, E. (1975). *Experientia* **31**, 1281–1283.

Landgren, C. R., and Torrey, J. G. (1973). *Colloq. Int. C.N.R.S.* **212**, 281–289.

Larkin, P. J. (1976). *Planta* **128**, 213–216.

Levitt, J., Scarth, G. W., and Gibbs, R. D. (1936). *Protoplasma* **26**, 237–248.

Lorz, H., Potrykus, I., and Thomas, E. (1977). *Naturwissenschaften* **64**, 439.

Lorz, H., Wernicke, W., and Potrykus, I. (1979). *Planta Med.* **36**, 21–29.

Maretzki, A., and Nickell, L. G. (1973). *Colloq. Int. C.N.R.S.* **212**, 51–63.

Menczel, L., Lazar, G., and Maliga, P. (1978). *Planta* **143**, 29–32.

Messerschmidt, M. (1974). *Z. Pflanzenphysiol.* **74**, 175–178.

Michel, W. (1937). *Arch. Exp. Zellforsch, Besonders Gewebezuecht.* **20**, 230–252.

Motoyoshi, F. (1971). *Exp. Cell Res.* **68**, 452–56.

Murashige, T., and Skoog, F. (1962). *Physiol. Plant.* **15**, 473–497.

Nagata, T., and Takebe, I. (1970). *Planta* **92**, 301–308.

Nagata, T., and Takebe, I. (1971). *Planta* **99**, 12–20.

Nagy, J. I., and Maliga, P. (1976). *Z. Pflanzenphysiol.* **78**, 453–455.

Nemet, G., and Dudits, D. (1977). *In* "Use of Tissue Cultures in Plant Breeding" (J. Novak, ed.), pp. 145–163. Czech. Acad. Sci. Prague.

Ohyama, K., and Nitsch, J. P. (1972). *Plant Cell Physiol.* **13**, 229–236.

Otsuki, Y., and Takebe, I. (1969). *Plant Cell Physiol.* **10**, 917–921.

Pelcher, L. E., Gamborg, O. L., and Kao, K. N. (1974). *Plant Sci. Lett.* **3**, 107–111.

Pentel, D., and Gunckel, J. E. (1979). *In* "Plant Cell and Tissue Culture" (W. R. Sharp, P. O. Larsen, E. F. Paddock, and V. Raghavan, eds.), pp. 633–709. Ohio State Univ. Press, Columbus.

Pilet, P. E. (1971). *C. R. Acad. Sci., Ser. D* **272**, 2253–2256.

Pilet, P. E. (1973). *Colloq. Int. C.N.R.S.* **212**, 99–107.

Pilet, P. E., Prat, R., and Roland, J. C. (1972). *Plant Cell Physiol.* **13**, 297–309.

Plowe, J. Q. (1931). *Protoplasma* **12**, 196–220.

Poirier-Hamon, S., Rao, P. S., and Harada, H. (1974). *J. Exp. Bot.* **25**, 752–760.

Potrykus, I., Harms, C. T., and Lorz, H. (1976). *In* "Cell Genetics in Higher Plants" (D. Dudits, G. L. Farkas, and P. Maliga, eds.), pp. 129–140. Akadémiai Kiadó, Budapest.

Potrykus, I., Harms, C. T., Lorz, H., and Thomas, E. (1977). *Mol. Gen. Genet.* **156**, 347–350.

Potrykus, I., Harms, C. T., and Lorz, H. (1979). *Theor. Appl. Genet.* **54**, 209–214.

Power, J. B., and Cocking, E. C. (1970). *J. Exp. Bot.* **21**, 64–70.

Power, J. B., Cummins, S. E., and Cocking, E. C. (1970). *Nature (London)* **255**, 1016–1018.

Power, J. B., Frearson, E. M., George, D., Evans, P. K., Berry, S. F., Hayward, C., and Cocking, E. C. (1976a). *Plant Sci. Lett.* **7**, 51–56.

Power, J. B., Frearson, E. M., Hayward, C., George, D., Evans, P. K., Berry, S. F., and Cocking, E. C. (1976b). *Nature (London)* **263**, 500–502.

Raj, B., and Herr, J. M., Jr. (1970). *Protoplasma,* **69,** 291–300.

Rajasekhar, E. W. (1973). *Nature (London), New Biol.* **246,** 223–224.

Raveh, D., and Galun, E. (1975). *Z. Pflanzenphysiol.* **76,** 76–79.

Ruesink, A. W., and Thimann, K. V. (1965). *Proc. Natl. Acad. Sci. U.S.A.* **54,** 56–64.

Ruesink, A. W., and Thimann, K. V. (1966). *Science* **154,** 280–281.

Schenk, R. U., and Hildebrandt, A. C. (1969a). *Crop Sci.* **9,** 629–631.

Schenk, R. U., and Hildebrandt, A. C. (1969b). *Phyton (Buenos Aires)* **26,** 155–166.

Schieder, O. (1975). *Z. Pflanzenphysiol.* **76,** 462–466.

Schieder, O. (1977). *Z. Pflanzenphysiol.* **84,** 275–281.

Schilde-Rentschler, L. (1977). *Planta* **135,** 177–181.

Seifriz, W. (1928). *Protoplasma* **3,** 191–196.

Shepard, J. F., and Totten, R. E. (1975). *Plant Physiol.* **55,** 689–694.

Shepard, J. F., and Totten, R. E. (1977). *Plant Physiol.* **60,** 313–316.

Sink, K. C., and Power, J. B. (1977). *Plant Sci. Lett.* **10,** 335–340.

Taiz, L., and Jones, R. L. (1971). *Planta* **101,** 95–100.

Takebe, I., Otsuki, Y., and Aoki, S. (1968). *Plant Cell Physiol.* **9,** 115–124.

Takebe, I., Labib, G., and Melchers, G. (1971). *Naturwissenschaften* **58,** 318–320.

Thomas, E., Hoffmann, F., Potrykus, I., and Wenzel, G. (1976). *Mol. Gen. Genet.* **145,** 245–248.

Thomas, E., King, P. J., and Potrykus, I. (1979). *Z. Pflanzenzuecht.* **82,** 1–30.

Törnäva, S. R. (1939). *Protoplasma* **32,** 329–341.

Tribe, H. T. (1955). *Ann. Bot. (London)* **19,** 351–368.

Upadhya, M. D. (1975). *Potato Res.* **18,** 438–445.

Vardi, A., Spiegel-Roy, P., and Galun, E. (1975). *Plant Sci. Lett.* **4,** 231–236.

Vasil, I. K. (1976). *Adv. Argon.* **28,** 119–160.

Vasil, I. K., and Vasil, V. (1980). *Proc. Int. Symp. Plant Protoplasts, Szeged, Hungary.*

Vasil, I. K., Vasil, V., Sutton, W. D., and Giles, K. L. (1975). *Proc. Int. Symp. Yeast Other Protoplasts, Univ. Nottingham, 1974* pp. 82.

Vasil, I. K., Vasil, V., and Hubbell, D. H. (1977). *In* "Genetic Engineering for Nitrogen Fixation" (A. Hollaender *et al.*, eds.), pp. 197–211. Plenum, New York.

Vasil, V., and Vasil, I. K. (1973). *Colloq. Int. C.N.R.S.* **212,** 139–149.

Vasil, V., and Vasil, I. K. (1974). *In Vitro* **10,** 83–96.

Vasil, V., and Vasil, I. K. (1979). *Z. Pflanzenphysiol.* **92,** 379–383.

Vasil, V., and Vasil, I. K. (1980). *Theor. Appl. Genet* **56,** 97–99.

Vreugdenhill, D. (1957). *Acta Bot. Neerl.* **6,** 472–542.

Wakasa, K. (1973). *Jpn. J. Genet.* **48,** 279–289.

Wallin, A., Glimelius, K., and Eriksson, T. (1977). *Physiol. Plant.* **40,** 307–311.

Watts, J. W., Motoyoshi, F., and King, J. M. (1974). *Ann. Bot. (London)* **38,** 667–671.

Whatley, F. R. (1956). *In* "Moderne Methoden der Pflanzenanalyse" (K. Paech and M. V. Tracy, eds.), Vol. I, pp. 452. Springer-Verlag, Berlin and New York.

White, D. W. R., and Vasil, I. K. (1979). *Theor. Appl. Genet.* **55,** 107–112.

Widholm, J. M. (1972). *Stain Technol.* **47,** 189–194.

Williamson, F. A., Fowke, L. C., Weber, G., Constabel, F., and Gamborg, O. L. (1977). *Protoplasma* **91,** 213–219.

Willison, J. H. M. (1976). *In* "Microbial and Plant Protoplasts" (J. F. Peberdy, A. H. Rose, H. J. Rogers, and E. C. Cocking, eds.), pp. 283–298. Academic Press, New York.

Willison, J. H. M., and Cocking, E. C. (1975). *Protoplasma* **84,** 147–159.

Willison, J. H. M., and Grout, B. W. W. (1978). *Planta* **140,** 53–58.

Zapata, F. J., Evans, P. K., Power, J. B., and Cocking, E. C. (1977). *Plant Sci. Lett.* **8,** 119–124.

Chapter 11

Protoplast Fusion and Somatic Hybridization

OTTO SCHIEDER AND INDRA K. VASIL

*Max-Planck-Institut für Züchtungsforschung (Erwin-Baur-Institut), Cologne,
Federal Republic of Germany, and
Department of Botany, University of Florida, Gainesville, Florida*

I. Introduction

The crossing of plants possessing different genetic characteristics often leads to hybrids with better growth patterns than shown by the parental plants. Therefore, the production of hybrids is an important tool for the improvement of crop plants. Some hybrids contain the genome of two or three different plant species, even different genera. Important crop plants, such as rape or wheat, are interspecific or intergeneric hybrids that have arisen by spontaneous crosses or been produced by the plant breeder. However, the production of such hybrids is possible only in those cases where a cross leads to the production of viable seeds. It is known that the formation of viable embryos in most interspecific or intergeneric crosses is prevented because of specific inhibition or prevention of key steps in pollination, pollen tube growth, fertilization, and embryo or endosperm development.

Somatic or parasexual hybrids between species that are sexually incompatible can now be achieved through the induced fusion of protoplasts (Vasil *et al.*, 1979). This technique is of interest not only to the plant breeder but also to the geneticist. Studies of fusion products can give information about compatibility and incompatibility of the nuclei or cytoplasm. Chromosome elimination in such fusion products during culture can be used for gene mapping as in fusion products of animal cells (Bernhard, 1976). This review describes the most recent results of somatic hybridization in plants and provides an outlook, especially for the agronomist, as to whether and as to how this new technique may be useful for the improvement of crop plants.

The idea of producing hybrid plants between sexually incompatible species was first conceived at the beginning of this century. In a series of experiments, Winkler (1907, 1908, 1910, 1934, 1938) attempted to produce asexual hybrids between *Solanum lycopersicum* (*Lycopersicon esculentum*) and *Solanum nigrum* by grafting. He expected spontaneous cellular and nuclear fusion to occur in the graft area. For a long time (up to 1938) he tried to find such somatic hybrids, which he called "burdonen" (late Latin synonym for *mulus* = mule). During these experiments he found interesting periclinal chimeras of *Solanum tubingense* and *Solanum koelreutianum*, and many others. He also claimed to have found some possible "burdonen." These plants were later shown to be heteroploids with fluctuating chromosome numbers (Brabec, 1949, 1954, 1965). Similar claims by other scientists (Glutschenko, 1950) for "vegetative hybrids" also could not withstand critical investigations (Böhme, 1954).

Küster (1909, 1910) was the first to point out the possibility of producing somatic hybrids by protoplast fusion and made the early attempts in this direction. At the time no methods were available to provide large numbers of protoplasts. Therefore, his experiments to establish the fusion of protoplasts as a technique for the production of hybrids from somatic cells remained unsuccessful. Yet he remained very optimistic as he was convinced that in the future hybrids by fusion of protoplasts would become possible (Küster, 1935). He was interested especially in interspecific and intergeneric fusions and their fate during further development.

The methods of producing large amounts of protoplasts were developed in England by Cocking (1960), and in Japan by Takebe *et al.* (1968) and Otsuki and Takebe (1969) by employing crude fungal enzymes. Soon first reports appeared demonstrating the regeneration of plants from protoplasts of tobacco (Takebe *et al.*, 1971; Nagata and Takebe, 1971). A detailed account of the isolation and culture of protoplasts is provided in Chapter 10, this volume.

II. Fusion of Protoplasts

A. SPONTANEOUS FUSION

Spontaneous fusion of two or more adjoining somatic protoplasts takes place through the expansion either of plasmodesmatal connections during cell wall degradation in enzyme solutions or of meiotic protoplasts through simple contact and without the help of any inducing agents (Vasil, 1976). Such homokaryotic protoplasts can be stimulated to divide and to form plants. The triploid plants obtained from haploid protoplasts of *Datura innoxia* (Schieder, 1976a) may very well have arisen from spontaneous fusion events. The frequency of spontaneous fusions differs and seems to be dependent on the plant species. For example, no spontaneous fusion bodies could be observed after enzymatic isolation of protoplasts from the liverwort *Sphaerocarpos donnellii* (Schieder, 1974a). It should be noted that plasmodesmata are not present in mosses and liverworts. After isolation of *D. innoxia* mesophyll protoplasts normally about 8% of bi- and multinuclear protoplasts were found (Schieder, 1976a). In tobacco up to 30% multinuclear protoplasts, some containing up to 14 nuclei, have been reported (Power and Frearson, 1973).

The formation of spontaneous fusion bodies is of no practical use, but they may be useful in studies of the nature and function of pasmodesmata, the physiology and control of mitosis in multinucleated cells, and nuclear fusion. Perhaps spontaneous fusion has some practical importance for chromosome doubling.

Enzymatically isolated meiotic protoplasts of *Lilium longiflorum* and *Trillium kamtschaticum* show one or more nuclei (Ito, 1973; Ito and Maeda, 1973). These protoplasts fuse spontaneously even after isolation by simple contact and without the help of any inducing agent. Up to 90% of the meiotic protoplasts have been observed to be bi- or multinuclear. The ability of such protoplasts to fuse spontaneously at high frequencies could be observed only for about 10 minutes after isolation. Longer treatment or increased concentration of the enzymes dramatically decreased their ability to fuse spontaneously. The mixing of isolated meiotic protoplasts from *L. longiflorum* and *T. kamtschaticum* resulted in the formation of about 11% interspecific "hybrid" protoplasts (Ito and Maeda, 1973).

B. INDUCED FUSION

Apart from those occurring in the meiotic protoplasts of some Liliaceae, spontaneous fusions lead only to the formation of homokaryotic fusion bodies. Heterokaryotic fusion bodies from two different lines or species must be induced by chemical treatment. Fusion of mechanically isolated protoplasts was reported as early as 1909 by Küster (1909). He tested several chemicals for their ability to induce fusions. From a large spectrum of chemicals tested, he found calcium nitrate to be the most effective fusion agent, even though the frequency was low (Küster, 1910). There are several reports of the ability of sea water to induce protoplast fusions (Hofmeister, 1954; Binding, 1966, 1974; Eriksson, 1971). Others have claimed that sodium nitrate possesses a good ability to form fusion bodies (Michel, 1937; Power *et al.*, 1970; Carlson *et al.*, 1972). This seems to be limited to meristematic or nonvacuolated protoplasts. Sodium nitrate has a deleterious effect on protoplast viability and at best produces low fusion frequencies (Potrykus, 1973; Burgess and Fleming, 1974; Keller and Melchers, 1973). Several other techniques therefore, have been tried to obtain high fusion frequencies. Some of these effectively agglutinate protoplasts, as for example lectins, but do not cause fusions (Hartmann *et al.*, 1973; Withers, 1973; Burgess and Fleming, 1974; Glimelius *et al.*, 1974) or produce only low frequencies of fusion bodies and often cause reduced protoplast viability (Eriksson, 1971; Schenk and Hildebrandt, 1971; Kameya and Takahashi, 1972; Keller *et al.*, 1973; Potrykus, 1973).

Keller and Melchers (1973) introduced an effective fusion technique based on the treatment of protoplasts with Ca^{2+} ions. The good ability of Ca^{2+} to induce fusions, which was first recognized by Küster (1910), could be increased by incubating the protoplasts in media containing Ca^{2+} ions at high temperature (37°C) and at the highly alkaline pH of 10.5. Fusion frequencies of more than 25% could be obtained after low-speed (50 g) centrifugation of tobacco protoplast suspensions. Similar results were obtained with protoplasts of *Petunia hybrida* and the liverwort *S. donnellii* (Binding, 1974; Schieder, 1974a,b).

Another very successful and more popular method for the fusion of protoplasts was developed by Kao and associates (Kao and Michayluk, 1974; Constabel and Kao, 1974) and by Wallin *et al.* (1974); it also was based on the use of Ca^{2+} ions but with lower concentrations. A key step in this method involves the agglutination of protoplasts with the aid of high molecular weight (MW) polyethylene glycol (PEG, MW ca. 6000). Protoplasts treated with PEG solutions containing Ca^{2+} ions fuse during the elution and/or dilution of PEG with the protoplast culture medium. Similar observations have been made with the polymer polyvenyl alcohol (Nagata, 1978). Good fusion frequencies have also been reported following treatment with a positively charged phospholipid, 1,2-*O*-dipentadecyl-methylidene-glycerol-3-phosphoryl-(*N*-ethylamino)-ethanolamine (Nagata *et al.*, 1979). In this case also additional treatment with Ca^{2+} ions was useful. Polyethylene glycol-induced fusions are nonspecific and are there-fore useful for intra- and interspecific fusions, for fusions of animal cells, and for interkingdom fusions of mammalian cells and plant protoplasts (Pontecorvo, 1975; Jones *et al.*, 1976; Dudits *et al.*, 1976a; Vasil, 1976; Vasil *et al.*, 1979). Fusion frequencies of up to 100% have been demonstrated in plant protoplasts with the aid of PEG and Ca^{2+} (Kao *et al.*, 1974; Vasil *et al.*, 1975). Further improvement in the induction of fusions and in the survival of protoplasts is attained by treating protoplasts with PEG in the presence of, or by eluting with solutions containing, high Ca^{2+} at high pH and high temperature (Burgess and Fleming, 1974; Kao *et al.*, 1974; Wallin *et al.*, 1974; Binding, 1976; Schieder, 1977a). The actual mechanism of fusion is not clearly understood, although some theories have been proposed (Keller and Melchers, 1973; Kao and Michayluk, 1974; Constabel and Kao, 1974; Grout and Coutts, 1974; Wallin *et al.*, 1974).

Protoplasts are known to carry negative surface charges (Grout *et al.*, 1972). The ζ-potential of normal tobacco protoplasts is -25 to -35 mV, whereas protoplasts suspended in 100 mM $CaCl_2$ have a potential of 0 mV (Nagata and Melchers, 1978). This decrease is accompanied by the aggregation of protoplasts. Such aggregation is essential for fusions to take place. The ζ-potential can be reversed by the addition of positively charged polymers.

III. Somatic Hybridization by Complementation Selection

Induced fusions of protoplasts from two genetically different cell lines or species must necessarily result in a variety of homo- as well as heterokaryotic fusion products (Fig. 1). The selection of the few true hybrid cells formed from the mixed population of regenerating protoplasts is a key factor in successful somatic hybridization. Two successful selection procedures for the isolation of somatic hybrids have been established. In the first, selection of somatic hybrids depends on the appearance of a different phenotype or growth pattern in comparison with the parental plant material (complementation). In the second, direct mechanical selection/isolation of fusion bodies is made after fusion of the protoplasts or after a short period of culture for culturing individual fusion products in microdroplets or in nurse culture. Both methods aspire to isolate the somatic hybrids as early as possible. The benefit of an early isolation of the somatic hybrids is simple: Large numbers of developed individuals can be screened for their possible hybrid nature.

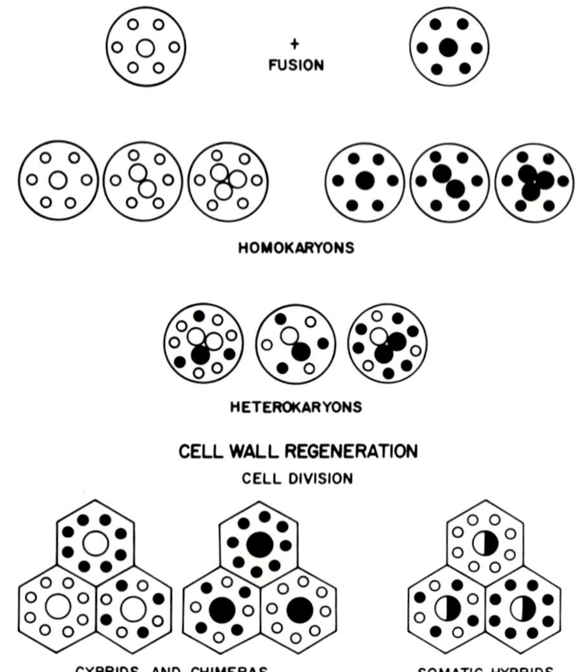

FIG. 1. The range of fusion products possible following the fusion of protoplasts from two cell lines or species, illustrated here with open and closed circles depicting nuclei and cell organelles. Cybrids, chimeras, and true hybrid cells may result as a result of heterokaryotic fusions.

A. COMPLEMENTATION TO THE WILD TYPE

The fusion of protoplasts from two nonallelic chlorophyll-deficient or auxotrophic mutants leads to somatic hybrids expressing the wild-type phenotype:

$$\text{Albino aB} + \text{albino Ab} \rightarrow \text{green AaBb}$$
$$\text{Auxotroph aB} + \text{auxotroph Ab} \rightarrow \text{autotroph AaBb}$$

A similar procedure is useful if protoplasts of a chlorophyll-deficient or auxotrophic mutant are combined with green wild-type protoplasts that under the chosen conditions are unable to regenerate:

$$\text{Albino aR} + \text{wild-type Ar} \rightarrow \text{green AaRr}$$
$$\text{Auxotroph aR} + \text{wild-type Ar} \rightarrow \text{autotroph AaRr}$$

The first experiments to select somatic hybrids with the aid of albino mutants were performed on haploid tobacco (Melchers and Labib, 1974; Melchers and Sacristán, 1977), and with the aid of auxotrophic mutants on the liverwort *S. donnellii* (Schieder, 1974b). In the albino complementation selection in tobacco, the light sensitivity of both mutants used facilitated the selection of somatic hybrids. Similar results were obtained in the interspecific combination of nonallelic light-sensitive albino mutants of *Nicotiana tabacum* and *Nicotiana sylvestris* (Melchers, 1977a). The additional marker in light sensitivity is not indispensable, as demonstrated in tobacco with other chlorophyll-deficient mutants (Gleba *et al.*, 1975; Gleba, 1979). Nonallelic albino mutants were also used for complementation selection of intraspecific somatic hybrids of *D. innoxia* (Schieder, 1977a), of interspecific hybrids of *P. parodii* + *Petunia hybrida* (Cocking, 1978), and of *N. tabacum* + *Nicotiana glauca* (Evans *et al.*, 1978).

It is not always necessary to fuse two nonallelic chlorophyll-deficient mutants or auxotrophic mutants for the production of somatic hybrids. As illustrated earlier, the combination of wild-type protoplasts of one line or species with protoplasts of a chlorophyll-deficient mutant of another line or species can also be used for complementation selection of somatic hybrids in cases where the wild-type protoplasts under the chosen culture conditions are not regenerable. Such a system was first reported by Cocking *et al.* (1977). They fused protoplasts of a chlorophyll-deficient mutant of *P. hybrida,* which can be regenerated in defined medium to shoots, with the wild-type protoplasts of *P. parodii,* which under the same culture conditions develop into cell aggregates of at best 50 cells. All developed green lines could be recognized by morphological traits and by chromosome analysis as somatic hybrids. The same method was used for the selection of somatic hybrids of *P. parodii* and *Petunia inflata* (Power *et al.*, 1979).

A similar system was employed for the production of somatic hybrids of sexually incompatible species of the genus *Datura*. The protoplasts of some chlorophyll-deficient mutants of *D. innoxia* can be regenerated in defined media to shoots, whereas the wild-type protoplasts of *Datura stramonium* and *Datura discolor* are unable to develop under the same culture conditions (Schieder, 1977b). After fusion of protoplasts from a chlorophyll-deficient mutant of *D. innoxia* with the wild-type protoplasts of *D. stramonium* and *D. discolor,* respectively, several green lines could be isolated; these were confirmed as hybrids by chromosome analysis; by their morphology as compared with the parental plants (Schieder, 1978a), and by isoenzyme analysis (Lönnendonker and Schieder, 1980). This method was also used successfully in hybridization experiments in which protoplasts of a chlorophyll-deficient mutant of *D. innoxia* were fused with wild-type protoplasts of the tree *Datura, Datura sanguinea*

(Schieder, 1978b, 1980), and *Datura candida* (Schieder, 1980), respectively. Sexual crosses between these species are incompatible. Isoenzyme (Lönnendonker and Schieder, 1980) and chromosome analyses and the intermediate morphologies clearly demonstrated the hybrid nature of these selected green lines. The hybrids of *D. innoxia* + *D. sanguinea* show some abnormalities. They had lost the ability to produce roots, and their grafted shoots formed stem tumors (Schieder, 1980) similar to those described for several sexual hybrids of the genus *Nicotiana* (Smith, 1962).

After fusion of protoplasts of a chlorophyll-deficient line of *N. tabacum* and of a cell line from *Nicotiana knightiana* that is unable to produce shoots, three somatic hybrids could be isolated (Maliga *et al.*, 1978). The potential for shoot production of tobacco and the potential for chlorophyll synthesis of *N. knightiana* were present in the hybrids.

One remarkable somatic hybrid has been obtained by using chlorophyll-deficient mutants for complementation selection. This result indicates the possibility that after fusion of protoplasts of two distantly related species, some genetic information from one species can be incorporated into the genome of the other species. Dudits *et al.* (1979) fused protoplasts of an albino mutant of *Dancus carota* with the green wild-type protoplasts of *Aegopodium podagraria*. The protoplasts of *A. podagraria* are not able to divide under the chosen culture conditions. They selected 10 green calluses from which three could be regenerated to plants. In control experiments no green calluses could be obtained. The green plants showed an abnormal morphology and possessed only the chromosome complement of *D. carota* ($2n = 18$). Based on ribonucleic acid (RNA) hybridization experiments, on root morphology, and on the production of root carotenoids normally present in the roots of *A. podagraria*, the authors discuss the possibility of some genes of *A. podagraria* being present in the selected plants. Such transfer could happen during development of the fusion bodies, where chromosome elimination showed incompatibility between the genomes.

Morphological markers in combination with albino mutations have been shown to be practicable for the selection of somatic hybrids in some instances. Dudits *et al.* (1977) fused protoplasts of an albino mutant from *D. carota* with the wild-type protoplasts of *Daucus capillifolius*, but the protoplasts of both species were able to divide under the chosen culture conditions. Therefore, selection of the somatic hybrids was undertaken by their green color and by the different morphology of their leaves compared with the simultaneously regenerated green plantlets of *D. capillifolius*. Chromosome and isoenzyme analyses confirmed the hybrid nature of the selected lines. The isolation of somatic hybrid cell lines of *Datura in-*

noxia + *Atropa belladonna* was undertaken in a similar manner, but the selection of the hybrids was possible at an earlier stage of development (Krumbiegel and Schieder, 1979). Protoplasts of an albino mutant of *D. innoxia* were fused with the green wild-type protoplasts of *A. belladonna*. In this case also the protoplasts of both species are able to develop under the culture conditions used. Calluses developed from protoplasts of *D. innoxia* normally produce numerous hairs as a first step in morphogenesis, whereas *A. belladonna* does not show this marker at any stage of development. Selected green and hair-producing calluses were demonstrated by chromosome analysis as real somatic hybrids, because the chromosomes of *A. belladonna* show only a punctiform size, whereas the chromosomes of *D. innoxia* can be recognized by their larger dimensions. These somatic hybrids produce only fleshy leaves and sometimes incomplete shoots. Root formation could not be induced.

The complementation selection of somatic hybrids with the aid of auxotrophic mutants, as compared to the use of chlorophyll-deficient mutants, would be more efficient because only the hybrid lines would survive. Unfortunately, in higher plants the production of such mutants is very limited (Schieder, 1976b, 1978b; see also Chapter 9 in Part A). In lower plants, such as liverworts and mosses, auxotrophic mutants have been described, for example, *S. donnellii* (Schieder, 1976b) and *Physcomitrella patens* (Ashton and Cove, 1977). For the complementation selection of somatic hybrids of *S. donnellii,* a female nicotinic acid-deficient mutant and a male pale-green and glucose-deficient mutant were used. The pale-green mutant was chosen to allow determination of the fusion rate immediately after the fusion treatment. The different sex of the mutants served as an additional marker. Female plants of *S. donnellii* possess seven autosomes plus a large X chromosome, whereas males possess the autosomes plus a very small Y chromosome. Somatic hybrids therefore, should be autotrophic and have both the sex chromosomes X and Y. On unsupplemented medium, one autotrophic strain was isolated showing the double number of autosomes (14 chromosomes) and both sex chromosomes X and Y (Schieder, 1974b).

In 1977 Grimsley *et al.* reported the recovery of numerous somatic hybrids of *P. patens* after fusion of protoplasts from different auxotrophic mutants. Culture on minimal medium facilitated the selection of the hybrids. With the aid of somatic hybridization they could also detect three different genetic groups of six independently isolated nicotinic acid-deficient mutants. A normal classification of these mutants via crosses was not possible, because most of the mutants showed various degrees of sexual sterility (Grimsley *et al.*, 1977a,b).

There is only one report describing complementation selection of somatic hybrids in higher plants with the aid of auxotrophic mutants. Glimelius *et al.* (1978) succeeded in selecting numerous somatic hybrids using the nitrate reductase-deficient and chlorate-resistant mutants of tobacco produced by Müller and Grafe (1978). Protoplasts of two such genetically different mutants were fused and cultured in a medium containing only nitrate as nitrogen source. Control experiments excluded back mutations and cross feeding. A total of 1061 somatic hybrids could be selected, from which some were further characterized in respect to nitrate reductase activity, chlorate sensitivity, chromosome number, and shoot formation. Wallin *et al.* (1979) have succeeded in producing somatic hybrids using the same mutants isolated by Müller and Grafe (1978), but they fused either normal protoplasts of one with miniprotoplasts of the other mutant or only the miniprotoplasts of both the mutants. The miniprotoplasts, containing only small amounts of cytoplasm but possessing a plasmamembrane, were prepared with cytochalasin B (Wallin *et al.*, 1979), as also described for the production of miniprotoplasts of animal and mammalian cells (Ege *et al.*, 1974). This procedure is important because it allows the transfer of nuclei without the transfer of much of the other genetic material contained in the mitochondria and plastids.

B. DIFFERENT GROWTH PATTERN

The first somatic hybrid plants were reported by Carlson *et al.* (1972) after fusion of *N. glauca* (2n = 24) and *Nicotiana langsdorffii* (2n = 18) protoplasts with the help of sodium nitrate. They used the differential growth characteristics and nutritional requirements—and a bit of "good fortune" (Carlson, 1973; Melchers and Labib, 1974)—of unfused as well as hybrid protoplasts and cells of both parents for selection of the somatic hybrids. Protoplasts of the two parental species do not regenerate into callus in the medium used, whereas the hybrids give rise to callus. They isolated 33 calluses, from which they regenerated three plants showing the amphidiploid chromosome number (2n = 42). It has been reported that sodium nitrate does not induce the high degree of fusion (25%) in tobacco mesophyll protoplasts (Melchers and Labib, 1974) as claimed by Carlson *et al.* (1972). Smith *et al.* (1976) produced somatic hybrids from the same species, but the fusion of protoplasts was induced by PEG. Preselection was performed by picking up the best growing calluses on complete medium. They regenerated 23 somatic hybrid plants that all possessed either the hexaploid chromosome number or a nearly hexaploid aneuploid number. Isoenzyme investigations pointed out the hybrid nature of the regenerated plants (Wetter and Kao, 1976). Smith *et al.* (1976)

discuss the possibility of the triple fusion products being more vigorous in growth in comparison to the amphidiploids, and so being picked as the best growing calluses. That this may be the case also has been discussed by Chupeau *et al.* (1978). They also isolated somatic hybrids of the same species of *Nicotiana,* but they did not undertake a preselection based on good growth. Out of 48 colonies selected, six regenerated flowering plants. Two of these plants showed 42 chromosomes and were morphologically identical to the sexual amphidiploid *N. glauca* + *N. langsdorffii.* This demonstrated that preselection of best growing calluses by Smith *et al.* (1976) indeed prefers the triple fusion products.

A similar system for selecting somatic hybrids of *P. hybrida* and *P. parodii* was reported by Power *et al.* (1977). In the protoplast regeneration medium used, only the hybrid protoplasts can grow, whereas the protoplasts of *P. parodii* develop only small cell clusters and the protoplasts of *P. hybrida* do not divide at all.

Such systems, in which the different growth patterns of the parents and the hybrids in defined media are employed for the selection of somatic hybrids, are useful only in those instances where sexual hybrids are available. Knowledge of the growth pattern of the hybrids in defined media is necessary for selection. Obviously, the use of such a system is not practicable in the production of somatic hybrids between species that are sexually incompatible.

C. RESISTANT FACTORS

The fusion of protoplasts of two different lines that are resistant to two different drugs, respectively, should result in the formation of double-resistant somatic hybrids, provided the resistances are expressed as dominant factors:

Resistance Ab + resistance aB → double resistance AaBb

It is irrelevant whether the resistance is based on an induced mutation or is a natural variant of the species. Selection of the somatic hybrids, following protoplast fusion, can be carried out on media containing both drugs.

The protoplasts of *P. parodii* do not form cell aggregates of more than 50 cells in a defined medium. This limited development is not suppressed by the addition of 1 mg/liter actinomycin D. The protoplasts of *P. hybrida,* which normally can be regenerated to plants, do not divide in the presence of this antibiotic. After fusion of the *P. parodii* and *P. hybrida* protoplasts, Power *et al.* (1976) isolated in selective medium several actinomycin D-resistant calluses that were regenerated to plants. Isoenzyme investigations and the comparison of these plants with the sexually produced hybrids confirmed their hybrid nature.

A quite similar method was used for the selection of somatic hybrids of *N. sylvestris + N. knightiana*. Protoplasts of a kanamycin-resistant mutant cell line of *N. sylvestris*, showing additionally a loss of the ability to regenerate shoots, were fused with the wild-type protoplasts of *N. knightiana*, which do not form shoots *in vitro*. Somatic hybrid selection was based on kanamycin resistance and on the restoration of the ability to form shoots. Isoenzyme investigations confirmed the hybrid nature of the isolated plants (Maliga *et al.*, 1977).

An interesting hybridization experiment has been undertaken by Wullems *et al.* (1979). Protoplasts of a tumorous and phytohormone autotrophic cell line of *N. tabacum* (induced by *Agrobacterium tumefaciens*) deficient in shoot regeneration, in which the enzyme lysopin dehydrogenase, responsible for the synthesis of octopine, was present, were fused with mesophyll protoplasts of a streptomycin-resistant line of the same species. Selection of somatic hybrids was based on screening for streptomycin resistance as a first step, and for phytohormone autotrophy and greening as a second step. The hybrid nature of three independent, isolated cell strains was shown by studies on lysopine dehydrogenase activity and by greening and shoot formation.

Further evidence of the validity of this approach has been provided by White and Vasil (1979). *Nicotiana sylvestris* cell lines resistant to S-2-aminoethylcysteine (AEC[R]) or 5-methyltryptophan (5MT[R]) were isolated in suspension culture and were shown to require in excess of 100 (5MT[R]) or 1000 (AEC[R]) times as much of their respective analogs as the parental wild-type culture for total growth inhibition. Eight calluses were selected on double-analog medium after heterokaryotic fusion of AEC[R] and 5MT[R] protoplasts. Prior to selection, protoplasts were grown to a stage at which calluses were capable of independent growth. The 1.8×10^4 control calluses obtained from mixed but unfused AEC[R] and 5MT[R] protoplasts and from AEC[R] and 5 MT[R] homokaryotic fusions were placed on the double-analog selection, but none survived. There was no evidence for cross feeding, high-frequency appearance of double resistance within the parental cell lines, or increased resistance after homokaryotic fusion. Double-resistant AEC[R] + 5MT[R] calluses appeared only after fusion of parental protoplasts. These hybrid cell lines were stable, retaining double resistance after 6 months in the absence of the analog. Analysis of the hybrid calli showed the AEC resistance to be dominant and the 5MT resistance to be semidominant.

D. VIGOROUS GROWTH PATTERN

It is well known that sexually produced amphidiploids often show a more vigorous growth pattern than the parental plants (Gottschalk, 1976). Such a heterosis effect has been demonstrated even at the early callus phase in plant somatic hybrids. For example, Smith *et al.* (1976) made a preselection of the best growing calluses after fusion of protoplasts from *N. glauca* and *N. langsdorffii* before they employed the selective medium lacking in phytohormones. Most of all, interspecific somatic hybrid calli of the genus *Datura* produced by protoplast fusion (Schieder, 1978a,b, 1980) showed a much better growth in comparison to the calluses developed from nonhybrid protoplasts (Fig. 2A,B). A similar result was observed in fusion products of *D. innoxia* + *A. belladonna* (Krumbiegel and Schieder, 1979) (Fig. 2C). This phenomenon is not the result of the higher ploidy level, because intraspecific somatic hybrid cell lines of *D. innoxia* showed, if at all, only a slightly better growth than the diploid nonhybrid calluses. This well-known effect on amphidiploid plants seems to be a good marker for the isolation of somatic hybrids at an early stage of development. Selection based on the vigorous growth pattern of the somatic hybrids has the advantage of being independent of mutants or of different growth patterns of the parental cell lines as well as of the somatic hybrids. It is obvious that heterosis cannot be expected in all cases. In those cases where heterosis is expressed, such a selection system is more practicable, especially for the plant breeder. No somatic hybrid has been isolated by the use of heterosis as the sole selection criterion, but the results with *Datura* demonstrate clearly that such a method can be very useful. Enhanced growth as a selective marker is not useful if only hexaploid somatic hybrids can be selected as described by Smith *et al.* (1976) for *N. glauca* + *N. langsdorffii* hybrids. It is hoped that this will not happen in all combinations.

E. MORPHOLOGY

Some somatic hybrid cell lines and plants have been isolated only by their abnormal morphology. Melchers *et al.* (1978) fused mesophyll protoplasts of a yellow mutant of *Lycopersicon esculentum* with protoplasts derived from cell cultures of potato. Regenerated calluses produced shoots of potato and the presumed somatic hybrids. Regenerated plants that showed abnormal morphology were shown to be somatic hybrids by the analysis of chromosomes and fraction-1 protein, ribulose-1,5-biphosphate carboxylase. The plants recognized as somatic hybrids have slightly

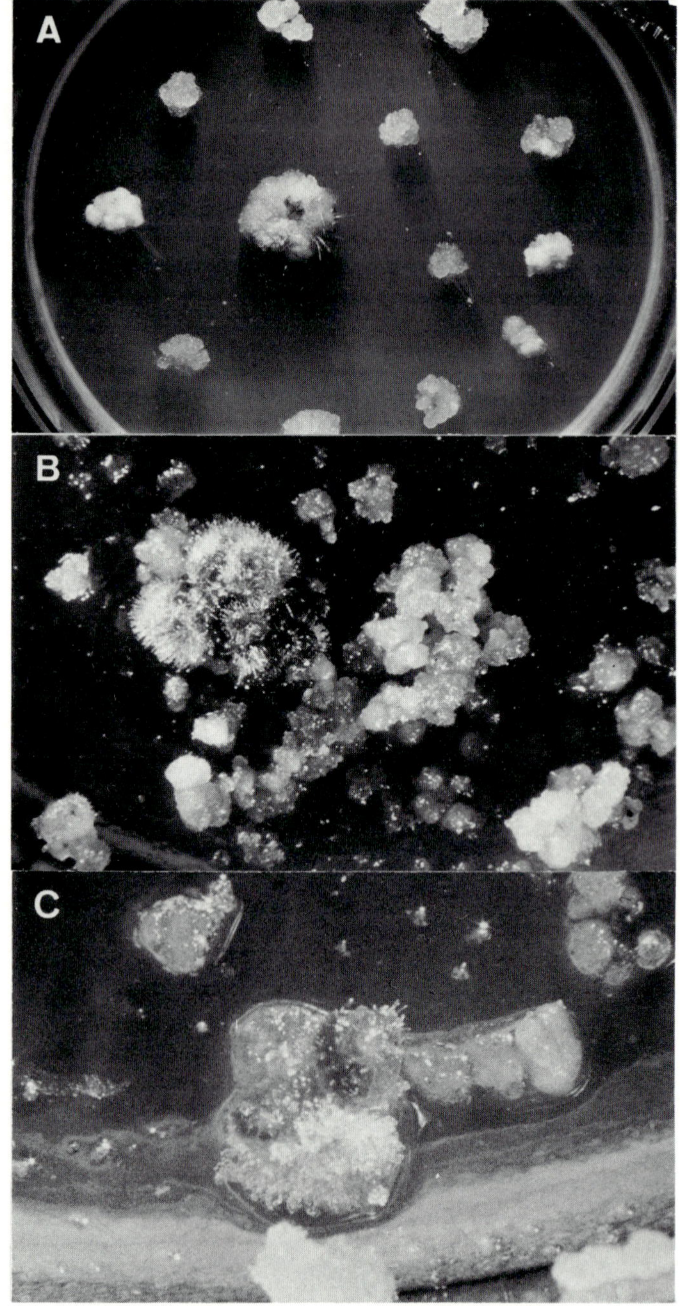

FIG. 2. Somatic hybrid calluses of (A) *Datura innoxia* + *Datura candida*, (B) *Datura innoxia* + *Datura sanguinea*, and (C) *Datura innoxia* + *Atropa belladonna* surrounded by smaller nonhybrid calluses. (Schieder, 1978b, 1980; Krumbiegel and Schieder, 1979.)

thickened rhizomes, filled flowers, and a different leaf shape from potato and tomato plants.

Binding and Nehls (1978) have isolated somatic hybrid cell lines of *Vicia faba* and *P. hybrida.* Detection of the somatic hybrids was possible by the intermediate appearance of their calluses. The larger dimension of the *V. faba* chromosomes in comparison to those of *P. hybrida* gave clear evidence for the hybrid nature of the isolates.

As shown by Gosch and Reinert (1976, 1978), the nuclei of intergeneric fusion products of *P. hybrida* and *A. belladonna* can fuse and subsequently undergo mitosis. Several regenerated shootlike structures showing fleshy leaves could not be confirmed as shoots of either *P. hybrida* or *A. belladonna* (Gosch and Reinert, 1978), but chromosome analysis suggested their hybrid nature (G. Gosch, personal communication). The larger chromosomes of *P. hybrida* could be clearly distinguished from the punctiform chromosomes of *A. belladonna.*

IV. Mechanical Isolation by Visual Means

Kao (1977) introduced a successful method of mechanically isolating heterokaryocytes and cultivating them individually. Fusion of colorless protoplasts of *Glycine max* derived from a cell culture with the green mesophyll protoplasts of *N. glauca* facilitated the isolation with micropipets of heterokaryotic fusion products after 24–48 hours. The transfer of the heterokaryons into Cuprak dishes containing many individual small wells allowed the culture and observation of individual cells. With these manipulations 20 hybrid cell lines were obtained. Chromosome analysis confirmed the hybrid nature of the isolates. Additional evidence of their hybrid nature was obtained from a study of isoenzyme patterns (Wetter, 1977).

Similar results were obtained by the isolation of heterokaryons of *Arabidopsis thaliana* and *Brassica campestris* with micropipets (Gleba and Hoffmann, 1978). Again, a study of the chromosomes and isoenzyme patterns of the cell lines showed that they were somatic hybrids. Root and shoot formation could also be induced (Gleba *et al.,* 1978; Gleba and Hoffmann, 1979).

An alternative method to the single cultivation of heterokaryocytes is their "nurse culture" in suspensions of phenotypically different cells. Such a system is independent of the specific and complex media that are necessary for single-protoplast culture (Kao, 1977; Gleba, 1978). Menczel *et al.* (1978) fused protoplasts of a kanamycin-resistant cell line of *N. knightiana,* deficient in inducible shoot redifferentiation and chlorophyll synthesis, with green wild-type protoplasts of *N. sylvestris*. After 48–72 hours interspecific fusion products were transferred to protoplast cultures of the kanamycin-resistant and chlorophyll-deficient cell line of *N. knightiana*. Thus, all green and shoot-producing calluses could be selected as somatic hybrids.

V. Cybrids

All experiments discussed up to now were fusion experiments in which either two genetically different nuclei were combined or nuclear genes of one species were possibly incorporated into the genome of another species. Gleba and co-workers (Gleba *et al.*, 1975; Gleba, 1979) conducted several fusion experiments on tobacco using plastome mutants in relation to chlorophyll synthesis and cytoplasmic male sterility (CMS); he studied the fate of these plastome markers in the resulting intraspecific and interspecific somatic hybrids. Segregation of the mixed cytoplasm into mutant and wild-type plastome was observed. An interesting experiment on tobacco was described by Zelcer *et al.* (1978) using the protoplast fusion technique for combining only the cytoplasmic genes of one line with the nuclear genes of another line. The nuclei of protoplasts from plants of a cytoplasmic male-sterile cultivar of *N. tabacum* were inactivated by X irradiation and fused with protoplasts of *N. sylvestris*. The selection of fusion products was based on the suppression of cell division in the X-irradiated protoplasts of the CMS parent and on the use of a defined medium that was unfavorable to the protoplasts of *N. sylvestris*. A feeder-layer technique (Raveh *et al.*, 1973) was utilized in order to rescue the low frequency of fusion products that withstood the selection procedure. Several mature plants, regenerated from seven calluses were obtained. Of these, 21 plants had the morphology of *N. sylvestris*, but their anthers resembled the anthers of the CMS *N. tabacum* parent and were in fact male sterile. Seven of these male-sterile plants (cybrids) possessed the diploid chromosome number ($2 n = 24$), whereas the others were tetraploid.

The inactivation of the nuclei does not appear to be absolutely necessary for the transfer of CMS in tobacco (Belliard *et al.*, 1977, 1978) and *Petunia* (Izhar and Power, 1979). The results obtained in tobacco were based on the abnormal flower morphology expressed by the CMS cytoplasm and on nuclear markers concerning the leaf shape. Fusion of protoplasts of two varieties of tobacco of which one contained the CMS cytoplasm leads to somatic hybrids as well as to cybrids. The cybrids possessed only the chromosome number of diploid tobacco ($2 n = 48$) but showed the genetic markers of the variety that possessed male fertility and the abnormal flower morphology expressed by the CMS cytoplasm of the other variety. Such a transfer of CMS cytoplasm without the nuclear genetic material is also possible in interspecific fusion experiments. Protoplasts of a cytoplasmic male-sterile line of *P. hybrida* were fused with the wild-type protoplasts of *Petunia axillaris* (Izhar and Power, 1979). The selection system was designed to allow only the growth of protoplasts possessing the genome of the fertile line. The majority of plants regenerated were phenotypically similar to the fertile species. Some of these plants were cytoplasmic hybrids (cybrids) combining the cytoplasm from the male-sterile line with the genome of the fertile line.

These results demonstrate that with the method of protoplast fusion the cytoplasmic genetic material of one line or species can be combined, separately from the nuclei, with the nuclear genetic material of another line or species. This procedure may be of general applicability in male-sterility-based hybrid seed breeding.

VI. Observed Divisions in Heterokaryocytes

The fusion of green mesophyll protoplasts with colorless protoplasts derived from cell cultures has in many cases allowed the observation of divisions in the fusion products. The identification of small multicellular aggregates of heterokaryocytes of distantly related species was made possible by the presence of green and colorless plastids (e.g., Kartha *et al.*, 1974; Constabel *et al.*, 1975a,b, 1976; Binding, 1976; Reinert and Gosch, 1976; Brar *et al.*, 1979). However, these observations provide no evidence of nuclear fusion and the formation of common metaphases, because in such heterokaryocytes only one of the nuclei may have undergone divisions. Cytological evidence for the formation of common metaphases in fusion products of distantly related species, based on differences in the chromosome size of the parents, has been provided by several authors (Kao *et al.*, 1974; Constabel *et al.*, 1975b; Gosch and Reinert, 1978; Dudits *et al.*, 1976b). These reports provide no information concerning the fate of such somatic hybrid aggregates during further development, because the observations normally have been undertaken only for a short time after fusion. The experiments establish only that in fusion products of distantly related species nuclear fusion and subsequent divisions can occur, but there is no evidence that such hybrid aggregates are stable. Nevertheless, somatic hybrid fusion bodies of *G. max* and *Vicia hajastana* maintained their hybrid character for 4 weeks of culture, that is, five to six generations, without losing chromosomes (Constabel *et al.*, 1977). Many of the somatic hybrid aggregates of distantly related species can be expected to be unstable during further development, similar to those reported for somatic hybrid cell lines of distantly related animal species (Bernhard, 1976), where chromosome elimination frequently has been observed. Some evidence for the occurrence of chromosome elimination in somatic hybrid plant cell lines has been reported (see Section VII).

VII. Cytology and Genetics of Somatic Hybrids

A. Cytology

In general, diploid protoplasts have been used for the production of somatic hybrids in higher plants. Therefore, predominantly tetraploid somatic hybrids should be expected. Indeed, the double-diploid or amphidiploid chromosome set in intraspecific as well as interspecific somatic hybrid plants could be found in *Nicotiana* (Melchers and Labib, 1974; Melchers, 1977a; Melchers and Sacristán, 1977; Gleba *et al.*, 1975; Chupeau *et al.*, 1978; Carlson *et al.*, 1972), *Datura* (Schieder, 1977a, 1978b), *Petunia* (Power *et al.*, 1977; Cocking *et al.*, 1977), and *Daucus* (Dudits *et al.*, 1977), together with somatic hybrid lines that showed either higher ploidy levels or aneuploid chromosome numbers.

Somatic hybrids with ploidy levels higher than tetraploid may be the result either of fusion events in which more than two protoplasts have fused or of endomitosis during postfusion development. Hexaploid somatic hybrids, which are often selected (Smith *et al.*, 1976; Schieder, 1977a, 1978a; Dudits *et al.*, 1977), must originate predominantly by triple fusion events. The aneuploid somatic hybrids with various chromosome numbers indicate irregular mitosis during culture or they may be attributed to the use of cell cultures for protoplast isolation (Melchers and Labib, 1974; Melchers and Sacristán, 1977; Melchers *et al.*, 1978; Schieder, 1977a, 1978a; Maliga *et al.*, 1978; Glimelius *et al.*, 1978), because in cell cultures instabilities of chromosome numbers are quite common.

The cytological data so far discussed have concerned only the intraspecific somatic hybrids or hybrids between species within one genus, except those that have been produced between very closely related genera *Solanum* and *Lycorpersicon* (*Solanum tuberosum* + *Lycopersicon esculentum*, Melchers *et al.*, 1978). In all these somatic hybrids chromosome elimination was not observed. A somewhat different situation is encountered in somatic hybrid cell lines of distantly related species. The somatic hybrid cell lines of *N. glauca* and *G. max* produced by Kao (1977), in

TABLE I

SOMATIC HYBRID PLANTS OR CELL LINES SELECTED
AFTER FUSION OF PROTOPLASTS

Hybrid plant or cell line	References
Plants regenerated from intraspecific fusion experiments	
Sphaerocarpos donnellii "nic2" + "pal2"	Schieder (1974b)
Physcomitrella patens aux. mutants	Grimsley *et al.* (1977a,b)
Nicotiana tabacum "v" + "s"	Melchers and Labib (1974)
Nicotiana tabacum "albino + "susu"	Gleba *et al.* (1975)
Nicotiana tabacum "CMS" + "fertile"	Belliard *et al.* (1977, 1978)
Nicotiana tabacum "albino" + "Ws1Ws2"	Kameya (1975)
Nicotiana tabacum "albino" + "CMS"	Gleba (1979)
Nicotiana tabacum "nia" + "cnx"	Glimelius *et al.* (1978)
	Wallin *et al.* (1979)
Nicotiana tabacum "SR1" + "tumor"	Wullems *et al.* (1979)
Datura innoxia "A1/5a" + "A7/1s"	Schieder (1977a)
Plants regenerated from interspecific fusion experiments	
Nicotiana glauca + *Nicotiana langsdorffii*	Carlson *et al.* (1972)
	Smith *et al.* (1976)
	Chupeau *et al.* (1978)
Nicotiana tabacum + *Nicotiana sylvestris*	Melchers (1977a)
Nicotiana tabacum "CMS" + *Nicotiana sylvestris*	Zelcer *et al.* (1978)
Nicotiana tabacum + *Nicotiana knightiana*	Maliga *et al.* (1978)
Nicotiana tabacum + *Nicotiana glauca*	Evans *et al.* (1978)
Nicotiana sylvestris + *Nicotiana knightiana*	Maliga *et al.* (1977)
Petunia hybrida + *Petunia parodii*	Power *et al.* (1976, 1977); Cocking *et al.* (1977)
Petunia hybrida "CMS" + *Petunia axillaris*	Izhar and Power (1979)
Petunia parodii + *Petunia inflata*	Cocking (1978)
Daucus carota + *Daucus capillifolius*	Dudits *et al.* (1977)
Daucus carota + *Aegopodium podagraria*	Dudits *et al.* (1979)
Datura innoxia + *Datura stramonium*	Schieder (1978a)
Datura innoxia + *Datura discolor*	Schieder (1978a)
Datura innoxia + *Datura sanguinea*	Schieder (1978b)
Datura innoxia + *Datura candida*	Schieder (1980)
Solanum tuberosum + *Lycopersicon esculentum*	Melchers *et al.* (1978)
Arabidopsis thaliana + *Brassica campestris*	Gleba and Hoffmann (1978)
Cell lines developed from intra- and interspecific fusion experiments	
Nicotiana sylvestris "AEC^R" + *Nicotiana sylvestris* "5MT^R"	White and Vasil (1979)
Nicotiana glauca + *Glycine max*	Kao (1977)
Datura innoxia + *Atropa belladonna*	Krumbiegel and Schieder (1979)
Petunia hybrida + *Atropa belladonna*	Gosch (personal communication)
Petunia hybrida + *Parthenocissus tricuspidata*	Power *et al.* (1975)
Petunia hybrida + *Vicia faba*	Binding and Nehls (1978)

which the chromosomes of both parents could be easily distinguished by their different sizes, were found to be unstable. The number of chromosomes of *N. glauca* decreased during culture, and after 6 months of culture only a few *N. glauca* chromosomes could be detected. The number of isoenzymes expressed by the *N. glauca* genome decreased simultaneously with the decrease of the chromosomes (Wetter, 1977). There was no apparent change in the number of soybean chromosomes. Similar observations were made on the somatic hybrid cell lines of *V. faba* and *P. hybrida* (Binding and Nehls, 1978). In these somatic hybrids, chromosome elimination was random, and chromosomes of one or the other partner were shown to have been lost in three investigated hybrid calluses. The somatic hybrid cell lines of *A. thaliana* and *B. campestris* (Gleba and Hoffmann, 1978) were more stable, because no evidence for chromosome elimination by chromosome counts and isoenzyme investigations was found after a culture period of 7 months. Shoot-forming calluses, however, showed loss of the chromosomes of *A. thaliana* (Y. Y. Gleba and F. Hoffmann, personal communication). No indication of chromosome elimination during 8 months of culture was observed in somatic hybrid cell lines of *D. innoxia* and *A. belladonna* (Krumbiegel and Schieder, 1979).

In animal somatic cell hybrids chromosome elimination occurs frequently. This phenomenon has been used for chromosome mapping of mammalian genes (Bernhard, 1976). Similar studies in somatic hybrid cell lines of distantly related plant species should also be possible. The observation of the successive disappearance of isoenzymes coded from the *N. glauca* genome in somatic hybrid cell lines of *G. max* and *N. glauca*, simultaneously with the successive elimination of the *N. glauca* chromosomes, demonstrates the possibility of such studies in plants. Chromosome mapping of genes may in the future play an important role in experiments wherein the transfer of single chromosomes or even single genes may have to be undertaken.

B. Genetics

The hybrid nature of most of the selected somatic hybrids was demonstrated only by chromosome analysis, isoenzyme investigations, morphological comparison with the sexual hybrids, or their growth behavior. Genetic analysis was employed in only a few cases to demonstrate additionally their hybrid nature. The diploid intraspecific somatic hybrids of tobacco that have been produced after fusion of haploid protoplasts of two genetically different chlorophyll-deficient mutants (Melchers and Labib, 1974) showed after self-crossing an offspring in green and mutated seedlings comparable to that of the diploid sexual hybrids (Melchers, 1977b). This fact demonstrates clearly the hybrid nature of the selected green lines. The hybrid nature of the tetraploid intraspecific somatic hybrids of two albino mutants of *D. innoxia* was demonstrated by the use of anther culture (Schieder, 1978c). Out of anthers from four somatic hybrids, green and chlorophyll-deficient androgenetic plantlets showing the phenotype of the albino mutants segregated in a high percentage.

Power *et al.* (1978) compared the flower color segregation after self-crosses of sexual tetraploid hybrids of *P. hybrida* + *P. parodii* with that of two somatic hybrids. Minor variations were detected. This flower color segregation is the result of a high percentage of quadrivalents found in meiosis I. In the progeny of amphidiploid somatic hybrids of *D. innoxia* + *D. stramonium* or *D. discolor*, respectively, a small portion (< 1%) of albino seedlings was found. Similar results, but with higher percentages of albinos, could be obtained in androgenetic plantlets produced via anther culture of tetraploid somatic hybrids of the same species. For the hybridization experiments one albino mutant of *D. innoxia* was employed. The appearance of chlorophyll-deficient seedlings and androgenetic plantlets, respectively, may be the result of the formation of a small percentage of tri- and quadrivalents in meiosis I (O. Schieder, unpublished observations).

VIII. Conclusions

The results in somatic hybridization presented here are limited only to model plants predominantly from the genera *Nicotiana, Petunia, Daucus,* and *Datura*. Somatic hybridization experiments have not yet been done with important crop plants. This is because of the limited number of species from which protoplasts can be regenerated to plants, and among these there are no important crop plant species (see Chapter 10, this volume). Nevertheless, the model systems described here show that intra- and interspecific flowering somatic hybrids can be obtained by protoplast fusion. Such flowering somatic hybrids, in some cases, show the same genetic behavior as sexual hybrids. Furthermore, it has been demonstrated that prezygotic (*P. parodii* + *P. inflata*) as well as embryo–endosperm (e.g., *D. innoxia* + *D. stramonium*) incompatibility can be overcome by protoplast fusion. These facts indicate that the technique of protoplast fusion and somatic hybridization may in the future become an excellent complement to the classical methods of plant breeding.

The regeneration of dihaploid as well as tetraploid protoplasts of *S. tuberosum* has been reported (Binding *et al.*, 1978; Melchers, 1978; Shepard

FIG. 3. Potato ($2n = 4x = 48$) breeding scheme, according to Wenzel *et al.* (1979), that combines classical breeding methods with parthenogenetic and androgenetic chromosome reduction, respectively, and somatic hybridization. α, potato genome; an asterisk indicates favorable genes with qualitative inheritance (e.g., resistance).

and Totten, 1977). Potato is a crop plant that is extensively cultivated in the northern hemisphere. Successful fusion experiments have been done with potato and tomato (Melchers *et al.*, 1978), which demonstrates the applicability of this technique for potato. Wenzel *et al.* (1979) have suggested a breeding scheme for potato (Fig. 3) that combines classical breeding methods with parthenogenetic and androgenetic induction of dihaploids and somatic hybridization. The scheme indicates that intraspecific somatic hybridization can also be of benefit, especially for plant species in which seed propagation normally is neither necessary nor practicable. The selection of the somatic hybrids in this case can perhaps be undertaken by the more vigorous growth of the hybrids as demonstrated in *Datura* protoplast fusion experiments (Schieder, 1978a).

The transfer of CMS cytoplasm by protoplast fusion to somatic hybrids should also be of interest to the plant breeder. This may be of critical importance in male-sterility-based hybrid seed production.

In conclusion, the fundamental techniques for somatic hybridization experiments are available. There exist very successful protoplast fusion methods and several techniques that allow the early detection of somatic hybrids. For further successful somatic hybridization in the direction of plant improvement, it is essential to increase the number of species from which protoplast regeneration into plants is possible. This is especially necessary for important crop plants, such as the cereals and the legumes. The development of more efficient procedures for the preferential growth of hybrid cells in selective media should also be undertaken on an urgent basis.

REFERENCES

Ashton, N. W., and Cove, D. J. (1977). *Mol. Gen. Genet.* **154,** 87–95.
Belliard, G., Pelletier, G., and Ferault, M. (1977). *C. R. Acad. Sci., Ser. D* **248,** 749–752.
Belliard, G., Pelletier, G., Vedel, F., and Quetier, F. (1978). *Mol. Gen. Genet.* **165,** 231–237.

Bernhard, H. P. (1976). *Int. Rev. Cytol.* **47**, 289–325.

Binding, H. (1966). *Z. Pflanzenphysiol.* **55**, 305–321.

Binding, H. (1974). *Z. Pflanzenphysiol.* **72**, 422–426.

Binding, H. (1976). *Mol. Gen. Genet.* **144**, 171–175.

Binding, H., and Nehls, R. (1978). *Mol. Gen. Genet.* **164**, 137–143.

Binding, H., Nehls, R., Schieder, O., Sopory, S. K., and Wenzel, G. (1978). *Physiol. Plant.* **43**, 52–54.

Bŏhme, H. (1954). *Z. Pflanzenzuecht.* **33**, 367–418.

Brabec, F. (1949). *Planta* **37**, 57–95.

Brabec, F. (1954). *Planta* **44**, 562–606.

Brabec, F. (1965). *In* "Handbuch fŭr Pflanzenphysiologie" (W. Ruhland, ed.), Vol. XV, pp. 388–494. Springer-Verlag, Berlin and New York.

Brar, D. S., Rambold, S., Constabel, F., and Gamborg, O. L. (1978). *Abstr., Int. Congr. Plant Tissue Cell Cult., 4th Calgary,* p. 64.

Burgess, J., and Fleming, E. N. (1974). *Planta* **118**, 183–193.

Carlson, P. S. (1973). *Colloq. Int. C.N.R.S.* **212**, 497–505.

Carlson, P. S., Smith, H. H., and Dearing, R. D. (1972). *Proc. Natl. Acad. Sci. U.S.A.* **69**, 2292–2294.

Chupeau, Y., Missonier, C., Hommel, M.-C., and Goujaud, J. (1978). *Mol. Gen. Genet.* **165**, 239–245.

Cocking, E. C. (1960). *Nature (London)* **187**, 962–963.

Cocking, E. C. (1978). *In* "Frontiers of Plant Tissue Culture" (T. A. Thorpe, ed.), pp. 151–158. Univ. of Calgary Press, Calgary, Alberta.

Cocking, E. C., George, D., Price-Jones, M. J., and Power, J. B. (1977). *Plant Sci. Lett.* **10**, 7–12.

Constabel, F., and Kao, K. N. (1974). *Can. J. Bot.* **52**, 1603–1606.

Constabel, F., Kirkpatrik, J. W., Kao, K. N., and Kartha, K. K. (1975a). *Biochem. Physiol. Pflanz.* **168**, 319–325.

Constabel, F., Dudits, D., Gamborg, O. L., and Kao, K. N. (1975b). *Can. J. Bot.* **53**, 2092–2095.

Constabel, F., Weber, G., Kirkpatrik, J. W., and Pahl, K. (1976). *Z. Pflanzenphysiol.* **79**, 1–7.

Constabel, F., Weber, G., and Kirkpatrik, J. W. (1977). *C. R. Acad. Sci. D.* **285**, 319–322.

Dudits, D., Rasko, I., Hadlaczky, G., and Lima-de-Faria, A. (1976a). *Hereditas* **82**, 121–124.

Dudits, D., Kao, K. N., Constabel, F., and Gamborg, O. L. (1976b). *Can. J. Genet. Cytol.* **18**, 263–269.

Dudits, D., Hadlaczky, G., Lévi, E., Fejér, O., Haydu, Z., and Lázár, G. (1977). *Theor. Appl. Genet.* **51**, 127–132.

Dudits, D., Hadlaczky, G., Koncz, C., Lázár, G., and Horváth, G. (1979). *Plant Sci. Lett.* **15**, 101–112.

Ege, T., Hamberg, H., Krondahl, U., Eriksson, J., and Ringertz, N. R. (1974). *Exp. Cell Res.* **87**, 365–377.

Eriksson, T. (1971). *Colloq. Int. C.N.R.S.* **193**, 297–301.

Evans, D. A., Gamborg, O. L., Shyluk, J. P., and Wetter, L. R. (1978). *Abstr., Int. Congr. Plant Tissue Cell Cult., 4th, Calgary* p. 70.

Gleba, Y. Y. (1978). *Naturwissenschaften* **65**, 158–159.

Gleba, Y. Y. (1979). *In* "Plant Cell and Tissue Culture—Principles and Applications" (W. R. Sharp, P. O. Larsen, E. F. Paddock, and V. Raghavan, eds.), pp. 774–788. Ohio State Univ. Press, Columbus.

Gleba, Y. Y., and Hoffmann, F. (1978). *Mol. Gen. Genet.* **165**, 257–264.

Gleba, Y. Y., and Hoffmann, F. (1979). *Naturwissenschaften* **66**, 547–554.

Gleba, Y. Y., Butenko, R. G., and Sytnik, K. M. (1975). *Dokl. Akad. Nauk SSSR* **221**, 1196–1199.

Gleba, Y. Y., Kohlenbach, H. W., and Hoffmann, F. (1978). *Naturwissenschaften* **65**, 655–657.

Glimelius, K., Wallin, A., and Eriksson, T. (1974). *Physiol. Plant.* **31**, 225–230.

Glimelius, K., Eriksson, T., Grafe, R., and Müller, A. J. (1978). *Physiol. Plant.* **44**, 273–277.

Glutschenko, I. J. (1950). "Die vegetative Hybridisation." Verlag Kult. Fortschr., Berlin.

Gosch, G., and Reinert, J. (1976). *Naturwissenschaften* **63**, 534–535.

Gosch, G., and Reinert, J. (1978). *Protoplasma* **86**, 23–38.

Gottschalk, W. (1976). "Die Bedeutung der Polyploidie für die Evolution der Pflanzen," Fischer, Stuttgart.

Grimsley, N. H., Ashton, N. W., and Cove, D. H. (1977a). *Mol. Gen. Genet.* **154**, 97–100.

Grimsley, N. H., Ashton, N. W., and Cove, D. J. (1977b). *Mol. Gen. Genet.* **155**, 103–107.

Grout, B. W. W., and Coutts, R. H. A. (1974). *Plant Sci. Lett.* **2**, 397–403.

Grout, B. W. W., Willison, J. H. M., and Cocking, E. C. (1972). *J. Bioenerg.* **4**, 311–328.

Hartmann, J. X., Kao, K. N., Gamborg, O. L., and Miller, R. A. (1973). *Planta* **112**, 45–56.

Hofmeister, L. (1954). *Protoplasma* **43**, 278–326.

Ito, M. (1973). *Plant Cell Physiol.* **14**, 865–872.

Ito, M., and Maeda, M. (1973). *Exp. Cell Res.* **80**, 453–456.

Izhar, S., and Power, J. B. (1979). *Plant Sci. Lett.* **14**, 49–55.

Jones, C. W., Mastrangelo, I. A., Smith, H. H., and Liu, H. Z. (1976). *Science* **193**, 401–403.

Kameya, T. (1975). *Jpn. J. Genet.* **50**, 235–246.

Kameya, T., and Takahashi, N. (1972). *Jpn. J. Genet.* **3**, 215–217.

Kao, K. N. and (1977). *Mol. Gen. Genet.* **150**, 225–230.

Kao, K. N., and Michayluk, M. R. (1974). *Planta* **115**, 355–367.

Kao, K. N., Constabel, F., Michayluk, M. R., and Gamborg, O. L. (1974). *Planta* **120**, 215–227.

Kartha, K. K., Gamborg, O. L., Constabel, F., and Kao, K. N. (1974). *Can. J. Bot.* **52**, 2435–2436.

Keller, W. A., and Melchers, G. (1973). *Z. Naturforsch., Teil C* **28**, 737–741.

Keller, W. A., Harvey, B. L., Kao, K. N., Miller, R. A., and Gamborg, O. L. (1973). *Colloq. Int. C.N.R.S.* **212**, 455–463.

Krumbiegel, G., and Schieder, O. (1979). *Planta* **145**, 371–375.

Küster, E. (1909). *Ber. Dtsch. Bot. Ges.* **27**, 589–598.

Küster, E. (1910). *Arch. Entwicklungsmech. Org.* **30**, 351–355.

Küster, E. (1935). "Die Pflanzenzelle." Fischer, Jena.

Lönnendonker, N., and Schieder, O. (1980). *Plant Sci. Lett.* **17**, 135–139.

Maliga, P., Lázár, G., Joó, F., Nagy, A. H., and Menczel, L. (1977). *Mol. Gen. Genet.* **157**, 291–296.

Maliga, P., Kiss, Z. R., Nagy, A. H., and Lázár, G. (1978). *Mol. Gen. Genet.* **163**, 145–151.

Melchers, G. (1977a). *In* "International Cell Biology" (B. R. Brinkley and K. R. Porter, eds.), pp. 207–215. Rockefeller Univ. Press, New York.

Melchers, G. (1977b). *Naturwissenschaften* **64**, 184–194.

Melchers, G. (1978). *In* "Production of Natural Compounds by Cell Culture Methods" (A. W. Alfermann and E. Reinhard, eds.), BPT-Rep. 1/78, pp. 306–311. Ges. Strahlen Umweltforsch., Munich.

Melchers, G., and Labib, G. (1974). *Mol. Gen. Genet.* **135**, 277–294.

Melchers, G., and Sacristán, M. D. (1977). *In* "La Culture des Tissues et de Cellules des Végétaux, Travaux dédiés à la mémoire de Georges Morel" (R. Gautheret, ed.), pp. 169–177. Masson, Paris.

Melchers, G., Sacristán, M. D., and Holder, A. A. (1978). *Carlsberg Res. Commun.* **43**, 203–218.

Menczel, L., Lázár, G., and Maliga, P. (1978). *Planta* **143**, 29–32.

Michel, W. (1937). *Arch. Exp. Zellforsch.* **20**, 230–252.

Müller, A. J., and Grafe, R. (1978). *Mol. Gen. Genet.* **161**, 67–76.

Nagata, T. (1978). *Naturwissenschaften* **65**, 263–264.

Nagata, T., and Melchers, G. (1978). *Planta* **142**, 235–238.

Nagata, T., and Takebe, I. (1971). *Planta* **99**, 12–20.

Nagata, T., Eibel, H., and Melchers, G. (1979). *Z. Naturforsch.* **34**, Section C, 460–462.

Otsuki, Y., and Takebe, I. (1969). *Plant Cell Physiol.* **10**, 917–921.

Pontecorvo, G. (1975). *Somat. Cell Genet.* **1**, 397–400.

Potrykus, I. (1973). *In* "Yeast, Mould and Plant Protoplasts" (J. R. Villanueva, J. Garcia-Acha, S. Gascón, and S. Uruburu, eds.), pp. 319–332. Academic Press, New York.

Power, J. B., and Frearson, E. M. (1973). *Colloq. Int. C.N.R.S.* 212, 409–414.

Power, J. B., Cummins, S. E., and Cocking, E. C. (1970). *Nature (London)* **225**, 1016–1018.

Power, J. B., Frearson, E. M., Hayward, C., and Cocking, E. C. (1975). *Plant Sci. Lett.* **5**, 197–207.

Power, J. B., Frearson, E. M., Hayward, C., George, D., Evans, P. K., Berry, S. F., and Cocking, E. C. (1976). *Nature (London)* **263**, 500–502.

Power, J. B., Berry, S. F., Frearson, E. M., and Cocking, E. C. (1977). *Plant Sci. Lett.* **10**, 1–6.

Power, J. B., Sink, K. C., Berry, S. F., Burns, S. F., and Cocking, E. C. (1978). *J. Hered.* **69**, 373–376.

Power, J. B., Berry, S. F., Chupman, J. V., Sink, K. C., and Cocking, E. C. (1979). *Theor. Appl. Genet.* **F69**, 373–376.

Raveh, D., Hubermann, E., and Galun, E. (1973). *In Vitro* **9**, 216–222.

Reinert, J., and Gosch, G. (1976). *Naturwissenschaften* **63**, 534.

Schenk, R. U., and Hildebrandt, A. C. (1971). *Colloq. Int. C.N.R.S.* **193**, 321–331.

Schieder, O. (1974a). *Biochem. Physiol. Pflanz.* **165**, 433–435.

Schieder, O. (1974b). *Z. Pflanzenphysiol.* **74**, 357–365.

Schieder, O. (1976a). *Mol. Gen. Genet.* **149**, 251–254.

Schieder, O. (1976b). *Mol. Gen. Genet.* **144**, 63–66.

Schieder, O. (1977a). *Planta* **137**, 253–257.

Schieder, O. (1977b). *Z. Pflanzenphysiol.* **84**, 275–281.

Schieder, O. (1978a). *Mol. Gen. Genet.* **162**, 113–119.

Schieder, O. (1978b). *In* "Frontiers of Plant Tissue Culture, 1978" (T. A. Thorpe, ed.), pp. 393–401. Univ. of Calgary Press, Calgary, Alberta.

Schieder, O. (1978c). *Planta* **141**, 333–334.

Schieder, O. (1980). *Z. Pflantenphysiol.*, in press.

Shepard, J. F., and Totten, R. E. (1977). *Plant Physiol.* **60**, 313–316.

Smith, H. H. (1962). *Trans. N.Y. Acad. Sci.* **24**, 741–746.

Smith, H. H., Kao, K. N., and Combatti, N. C. (1976). *J. Hered.* **67**, 123–128.

Takebe, I., Otsuki, Y., and Aoki, S. (1968). *Plant Cell Physiol.* **9**, 115–124.

Takebe, I., Labib, G., and Melchers, G. (1971). *Naturwissenschaften* **58**, 318.

Vasil, I. K. (1976). *Adv. Agron.* **28**, 121–160.

Vasil, I. K., Vasil, V., Sutton, W. D., and Giles, K. L. (1975). *Abstr., Int. Symp. Yeast Other Protoplasts, 4th, Nottingham,* p. 82.

Vasil, I. K., Ahuja, M. R., and Vasil, V. (1979). *Adv. Genet.* **20**, 127–217.

Wallin, A., Glimelius, K. G., and Eriksson, T. (1974). *Z. Pflanzenphysiol.* **74**, 64–80.

Wallin, A., Glimelius, K., and Eriksson, T. (1979). *Z. Pflanzenphysiol.* **91**, 89–94.

Wenzel, G., Schieder, O., Przewoźny, T., Sopory, S. K., and Melchers, G. (1979). *Theor. Appl. Genet.* **F55**, 49–55.

Wetter, L. R. (1977). *Mol. Gen. Genet.* **150**, 231–236.

Wetter, L. R., and Kao, K. N. (1976). *Z. Pflanzenphysiol.* **80**, 455–462.

White, D. W. R., and Vasil, I. K. (1979). *Theor. Appl. Genet.* **54**, 239–244.

Winkler, H. (1907). *Ber. Dtsch. Bot. Ges.* **25**, 568–576.

Winkler, H. (1908). *Ber. Dtsch. Bot. Ges.* **26**, 595–608.

Winkler, H. (1910). *Z. Bot.* **2**, 1–38.

Winkler, H. (1934). *Planta* **21**, 613–656.

Winkler, H. (1938). *Planta* **27**, 680–707.

Withers, L. A. (1973). *Colloq. Int. C.N.R.S.* **212**, 517–545.

Wullems, G. J., Molendijk, L., and Schilperoort, R. A. (1979). *Theor. Appl. Genet.* **56**, 203–208.

Zelcer, A., Aviv, D., and Galun, E. (1978). *Z. Pflanzenphysiol.* **90**, 397–407.

Chapter 12

Genetic Modification of Plant Cells Through Uptake of Foreign DNA

C. I. KADO

Department of Plant Pathology, University of California, Davis, California

A. KLEINHOFS

*Department of Agronomy and Soils, Washington State University,
Pullman, Washington*

I. Introduction

Genetic modification of plants by man has long been central to his activities in continually generating an efficient agriculture for food and fiber. In these modern times it is realized that the boundaries set by natural genetic barriers have placed limits on extending the gene pool and further governing new genetic variability to enhance agricultural efficiency. Endeavors to circumvent classical genetic technologies that are limited by natural genetic barriers are the approaches being examined in recent times. Technologies for efficient insertion of foreign genes into crop plants is one of these endeavors.

In this contribution attention is focused on the uptake, fate, and biologic expression of foreign genetic material in plants. Several major breakthroughs and new genetic technologies are discussed. Past claims of genetic expression of foreign genes in plants have been critically reviewed by several workers (Johnson and Grierson, 1974; Kado, 1976; Kleinhofs and Behki, 1977; Lurquin, 1977; Cocking, 1977; Chaleff and Polacco, 1977).

Presently, the evidence for biologic expression of foreign genes remains sketchy and very recent claims need confirmation. The new technological advances in the insertion of foreign deoxyribonucleic acid (DNA) into plant cells try to avoid the natural genetic barriers in plants, which can be formidable. It presently is realized by workers in the field that it will take a great deal more study to understand how exogenous genes can be made compatible in the foreign environment of a higher cell. Much enthusiasm has already been generated from the fact that various eukaryotic genes are expressed into biologically functional products in bacteria (Struhl *et al.*, 1976; Itakura *et al.*, 1977; Carbon *et al.*, 1977; Ratzkin and Carbon, 1977; Rambach and Hogness, 1977; Villa-Komaroff *et al.*, 1978; Walz *et al.*, 1978; Clarke and Carbon, 1978; Fraser and Bruce, 1978; Goeddel *et al.*, 1979). Furthermore, various foreign prokaryotic and eukaryotic genes are also transcribed and translated in the environment of *Xenopus* cells (Colman, 1975; Mertz and Gurdon, 1977; Brown and Gurdon, 1977; DeRobertis and Mertz, 1977; Kressman *et al.*, 1978; Trendelenberg *et al.*, 1978).

On the basis of these experimental results, it appears that the genetic barriers are less formidable than previously has been speculated. Therefore, the genetic engineering of plant cells may be a reality in the future.

II. Assessment of DNA Uptake, Integration, and Replication in Plants

The idea of feeding exogenous DNA to plants in order to modify the genetic constitution of the recipient has been around for a long time (early work reviewed in Ledoux, 1965). Deoxyribonucleic acid uptake experiments are carried out routinely in bacteria. In higher plants a considerable controversy exists about the fate of exogenously supplied DNA. Here we will attempt to describe the present state of knowledge in this field instead of chronicling the efforts of those who have studied it. Consequently, the references are often illustrative rather than comprehensive. Several comprehensive reviews dealing with this subject have been published (Kleinhofs and Behki, 1977; Lurquin, 1977).

The uptake of DNA by plants can be demonstrated in numerous ways. Typically, the recipient is exposed to donor DNA that is tagged by some physical marker, such as radioactivity, density, or both. After a period of incubation and washing or deoxyribonuclease (DNase) treatment to remove the external DNA, the recipient is ground up, the DNA extracted, and the presence of the donor DNA demonstrated. In the case of radioactivity this can be accomplished by counting the DNA preparation or by autoradiography of sections of the treated tissues. The early work using these techniques has been extensively reviewed (Ledoux, 1965). Degradation and reutilization of the donor DNA cannot be excluded in these experiments. Subsequently, these experiments were improved by choosing donor DNA with a widely different density from the host DNA. Deoxyribonucleic acid molecules with different densities (d) can be readily separated by CsC1 equilibrium density centrifugation. For example, if radiolabeled DNA from *Agrobacterium tumefaciens* ($d = 1.718$ gm/cm^3), *Sarcina flava* ($d = 1.730$ gm/cm^3), *Micrococcus lysodeikticus* ($d = 1.731$ gm/cm^3), or phage 2C ($d = 1.742$ gm/cm^3) is fed to higher plants (the characteristic DNA density of which ranges from 1.690 to 1.702 gm/cm^3), then the subsequent analysis of the DNA extracted from recipient plants will tell whether the donor DNA is present and if it has been preserved or reutilized by the host cell. Experiments of this type were first used by Ledoux and co-workers (Stroun *et al.*, 1966; Ledoux and Huart, 1968, 1969; Ledoux *et al.*, 1971; Ledoux and Charles, 1972) to demonstrate uptake of exogenous DNA by higher plants. Subsequently, others (Bendich and Filner, 1971; Hotta and Stern, 1971; Kleinhofs, 1977) have made similar observations. In general these experiments show that the exogenously applied DNA is recovered along with the recipient DNA even from extensively washed plant organs. The donor DNA shows variable degrees of degradation and reutilization, as observed by the broadening of the donor DNA density peak and the appearance of radioactivity in the recipient DNA density peak. The degree of donor DNA degradation observed is dependent upon the length of incubation of the recipient after the exposure to the donor DNA. In general the longer the time period, the less donor DNA is observed. In our experience, little or no donor DNA is found by these techniques after 4–5 days growth of the recipient plants. The failure to observe the donor DNA after this time is probably a reflection of the limited sensitivity of the CsC1 gradient analysis technique.

At this point it becomes difficult to determine the fate of the exogenous DNA. It can be argued that the exogenous DNA has been degraded and eliminated or that it has been simply diluted by cell division and replication with nonlabeled precursors to undetectably low levels. It is possible, of course, to turn to more sensitive techniques, such as DNA–DNA renaturation experiments. This has been done (Kleinhofs, 1975, 1977; Lurquin and Hotta, 1975; Kado and Lurquin, 1975, 1976). Such DNA–DNA reassociation experiments would also detect the foreign DNA if it were replicated. Nevertheless, similar arguments as before apply here except at greater sensitivity. Clearly, there are limits to techniques that permit the escape of biologically significant amounts of DNA.

If the synthesis and/or integration of the exogenous DNA in the plant cells could be demonstrated, then the arguments (i.e., is the DNA inside the cells, adsorbed to the cell walls, or in the intercellular spaces) about uptake would be moot.

In order to detect the synthesis of new donor DNA molecules, unlabeled heteropycnic DNA can be fed to the plants, which are subsequently labeled with suitable radioactive precursors of DNA synthesis. Again, theCsC1 density gradient analysis of the extracted DNA will tell whether any DNA molecules with a density corresponding to that of the donor DNA have incorporated label as the result of replication in the host cell. Experiments of this nature were performed by Ledoux and Huart (1969) and led to the startling observation that the radioactivity was incorporated into DNA molecules of density intermediate between the host and donor DNA molecules. A model was proposed in which the double-stranded donor DNA molecules were presumed to covalently join end-to-end with the host DNA. The new structure was presumed to be replicated in the host cell (Ledoux and Huart, 1969). Similarly designed experiments in other laboratories have failed to confirm these results (Hotta and Stern, 1971; Bendich and Filner, 1971; Kleinhofs, 1975; Kleinhofs *et al.*, 1975). In our experience, experiments of this type lead to labeling of the host DNA only. New density DNA peaks appear only when bacterial contamination of the samples occurs. Interestingly, many of the bacterial contaminants isolated from barley seedlings possess DNA densities approximately intermediate between the barley DNA and a commonly used donor DNA from *M. lysodeikticus* (Kleinhofs, 1975).

Unusual density DNA peaks have been observed by others in some experiments. Rebel *et al.* (1973) observed that *Mattiola incana* ($d = 1.698$ gm/cm^3) seedlings exposed to ^{32}P-labeled T4 DNA ($d = 1.694$ gm/cm^3) and incubated for 72 hours produced a radioactive DNA peak at $d = 1.724$ gm/cm^3. The authors explained this anomaly by postulating that the T4 phage DNA became associated with a very high-density plant DNA fraction to produce the intermediate-density 1.725 gm/cm^3 DNA band. Filter DNA–DNA hybridization revealed the presence of T4 DNA sequences in this high-density component (Gradmann-Rebel and Hemleben, 1976). The presence of a plant component in the high-density complex was not convincingly demonstrated (Gradmann-Rebel and Hemleben, 1976).

Treatment of tomato plants with *Agrobacterium tumefaciens* followed by [^{3}H]thymidine caused the appearance of a DNA fraction whose buoyant density was between that of the bacterial DNA ($d = 1.718$ gm/cm^3) and the plant main-band DNA ($d = 1.692$ gm/cm^3) (Hanson and Chilton, 1975). This DNA fraction was clearly shown to have no significant base sequence homology with *A. tumefaciens* DNA. Homology with tomato DNA was established (Hanson and Chilton, 1975). Therefore, this unusual density DNA fraction is clearly a plant satellite, perhaps induced by the nonphysiological treatment conditions.

In the absence of convincing evidence to the contrary, it must be concluded that the intermediate-density DNA peaks observed in some plant experiments either constitute unusual density plant satellite DNA or are artifacts. The fact that similar intermediate-density DNA peaks are not observed in the much cleaner protoplast experiments (see Section III) also supports this conclusion.

In summary, we conclude that the integration and replication of exogenous DNA in higher plant cells remains to be proved. This is not to say that it does not perhaps happen. If it has occurred, adequately sensitive assays simply have not been used to detect it. It is clear that even the most sensitive techniques used to date cannot have detected the transfer and stable maintenance of one or very few genes. The presence of such small quantities of DNA among the massive amounts of the host DNA can only be detected by specific gene probes or by biologic experiments (see Section IV and V).

III. DNA Uptake by Plant Protoplasts

In most instances studied so far, higher cells, unlike microbial systems, do not seem to require a special state of competence for DNA uptake, although cell cultures in an exponential growth phase are more suited for DNA uptake than stationary cultures (Farber *et al.*, 1975). Protoplasts of higher plants behave like mammalian cells in their uptake of DNA, presumably by pinocytosis (Burgess *et al.*, 1973; Cocking, 1975) or by cell fusion (Sarkar *et al.*, 1974; Lurquin, 1977). Successful transformation resulting in the genetic expression of inserted DNA has been accomplished with viral genomes inserted in human and mammalian cells (e.g., Crawford *et al.*, 1964; Bourgaux *et al.*, 1965; Aaronson and Martin, 1970; Merril *et al.*, 1971) and various plant protoplasts (e.g., Aoki and Takebe, 1969; Motoyoshi *et al.*, 1973; Motoyoshi and Hull, 1974; Sarkar *et al.*, 1974; Howell and Hull, 1977).

The success achieved with plant virus nucleic acids has engendered high hopes that transformation and genetic recombination may be also achieved through the insertion of other exogenous nucleic acids into plant protoplasts, thus circumventing the classical genetic methods, which are limited by sexual reproduction. Natural mechanical barriers that exist with whole plants, their seeds, and intact cells are essentially obviated through the use of protoplasts. The nuclease barrier (Kado, 1979) found mainly in suspension cell cultures (Bendich and Filner, 1971; Oleson *et al.*, 1974) is much reduced by the use of freshly isolated protoplasts. It has been observed that whole plant cells and callus tissues are highly refractory to DNA uptake and much of the DNA is degraded quickly (Hotta and Stern, 1971; Bendich and Filner, 1971; Kado and Lurquin, 1976). Although the use of isolated protoplasts reduces nucleases, the level of nuclease activity is dependent on the type of plant used as the source of the protoplasts. Carrot protoplasts from suspension cell cultures seem to possess relatively high endonuclease activity (Slavik and Widholm, 1978), whereas protoplasts obtained from the primary leaves of cowpea and from turnips are relatively low in nuclease activity (Fernandez *et al.*, 1978; Liu and Kado, unpublished results). Furthermore, it should be emphasized here that the cell wall-degrading enzymes can contribute to high levels of endonuclease activity and therefore the source and type of "cellulase" used is important (S.-T. Liu and C. I. Kado, unpublished observations).

The conformation of the DNA also diminishes the nuclease problem. Superhelical DNA is much more efficiently maintained in cowpea protoplasts than are nicked or linearized DNA molecules (Lurquin and Kado, 1979). Any DNA nicking activity taking place on the exterior of plant cells will therefore lower the efficiency of DNA maintenance following their uptake. Indeed, Ohyama *et al.* (1978) had observed exonuclease activity associated with their protoplast preparations and showed that plasmid DNA (pTi plasmid of *A. tumefaciens*) was degraded at slightly lower rates than bacterial chromosomal DNA.

Besides the nuclease barrier, foreign DNA is confronted by an apparent genetic barrier (Kado, 1979). In bacteria, there are specific restriction–modification systems that exclude foreign DNA (Boyer, 1971). Such a system has been postulated to be operating in higher cells as well (Sager and Kitchum, 1975). In the eukaryotic system, foreign DNA may undergo direct intranuclear destruction, may be lost by chromosome elimination or by crossing over and excision, or may be rendered inoperative by heterochromatization. These losses can take place at any time during the developmental stage of the cell. Although seemingly formidable, these barriers may not be as significant as formerly believed. Mertz and Gurdon (1977) have demonstrated that purified DNAs, such as plasmids Col E1 and cloned *Drosophila melanogaster* DNA, are readily transcribed after injection into the nucleus of *Xenopus laevis* oocytes. Some of the ribonucleic acid (RNA) synthesized is faithfully translated within these injected cells to produce the correct protein products (DeRobertis and Mertz, 1977). Although not all plasmids are transcribed in *Xenopus* oocytes (Trendelenburg and Gurdon, 1978), RNA transcripts that hybridize to Col E1 DNA have been reported (Mertz and Gurdon, 1977). Goebel and Schiess (1975) exposed baby hamster kidney cells (BHK21) to Col E1 DNA. They found that 2–8% of the input DNA was taken up by these cells (in the presence of DEAE–dextran) as measured by acid-precipitable counts remaining in the cells after extensive washing. Most of the radioactivity became concentrated in the nuclear fraction. After 16 hours of incubation most of the Col E1 DNA appeared as nicked circles, and after an additional 16 hours these nicked circles were cleaved into smaller 7–12S fragments. Hybridization experiments showed that 0.15% of the labeled Col E1-specific RNA was detected in the treated cells. This RNA gradually decreased with incubation time.

Cortese *et al.* (1978) have shown that tRNA genes of the nematode *Caenorhabditis elegans* ligated to the Col E1 plasmid are accurately transcribed in *X. laevis* oocytes. Transcription of yeast tyrosine tRNA genes has also been demonstrated in these oocytes (DeRobertis and Olson, 1979). This indicates that foreign eukaryotic DNA cloned on a bacterial plasmid can be functionally expressed in *Xenopus* cells and these cells can be useful for the expression of various foreign higher cell DNA.

Microinjection of foreign DNA into the nuclei of recipient cells may help reduce the effects of any nucleases present on cell surfaces. Cairns *et al.* (1978) used microinjections to insert adenovirus type-2 DNA into isolated nuclei of *Acetabularia major*. The injected nuclei were then implanted into dissected *Acetabularia*. Adenovirus protein was detected in the cytoplasm using an immunofluorescence assay. They therefore concluded that adenovirus was transcribed and translated in the plant cell. These studies suggest that foreign DNA can be transcribed in the nucleus of the higher cells.

A. Uptake of Complex DNA

Earlier studies involved the uptake of plant, bacterial, and phage DNAs. Ohyama *et al.* (1972) originally showed that protoplasts from *Amii visnaga*, carrot, and soybean, treated with *Escherichia coli*-labeled DNA, took up 0.05 to 1.0% of the total acid precipitable radioactivity. DEAE–dextran markedly stimulated incorporation of radioactivity. Polycations, such as poly-L-lysine and poly-L-ornithine, also enhanced incorporation. Additional work (Ohyama *et al.*, 1978) showed that donor *E. coli* DNA isolated from the protoplasts was highly degraded within 24 hours of incubation as judged by the appearance of a very broad band after density gradient centrifugation in CsCl.

Hoffman and Hess (1973) followed the uptake of radiolabeled petunia DNA in protoplasts that were obtained from a white-flowering pure line of petunia. Upon autoradiographic analysis of DNA-treated protoplasts, they found much of the radioactivity accumulated in the nucleus. Hoffmann (1973) further studied the fate of double-labeled DNA after incubating protoplasts for 12 hours. Nuclei were isolated and nuclear DNA was separated by CsCl equilibrium density gradient centrifugation. The ratio of ^{3}H to ^{14}C of the DNA banded in the gradient corresponded well with the ^{3}H/^{14}C ratio of the donor DNA. Under his experimental conditions 0.04% of the radioactivity was found in the nuclear DNA and 0.007% in cytoplasmic DNA that was also isolated by CsCl density gradient centrifugation. Because of the excessive damage caused by DEAE–dextran in different salt solutions, as judged by the release from the protoplasts of ultraviolet absorbing substances, it was concluded that DEAE–dextran was more deleterious than useful in DNA uptake studies using petunia protoplasts.

Studies undertaken by Uchimiya and Murashige (1977) with tobacco [^{3}H]DNA and tobacco (*Nicotiana tabacum* cv. "Xanthi-nc") protoplasts showed that there was a continued progressive uptake of the homologous DNA. Protoplasts of the tobacco mosaic virus-susceptible cultivar Xanthi were treated with DNA from the local lesion-forming cultivar "Xanthi-nc." The resulting plants regenerated from these protoplasts showed no phenotypic change (remained susceptible to the virus). When uptake experiments were performed with *E. coli* DNA using *Nicotiana glutinosa* protoplasts, the high molecular weight bacterial DNA was degraded within 5 hours of incubation. A significant portion of the exogenously added DNA was reutilized, but about 2.6% remained as *E. coli* DNA within 5–20 hours of incubation. It was found that 18–19% of the total radioactivity became associated with nuclei of tobacco protoplasts.

Liebke and Hess (1977) also observed a substantial association of *E. coli* DNA with nuclei of *Petunia* protoplasts if poly-L-lysine (10 μg/ml) was supplemented in the incubation medium. Without poly-L-lysine they observed much less *E. coli* DNA in the nuclear fraction. They estimated about 10^4 DNA molecules per nucleus in the control and 10^5 DNA molecules per nucleus in the poly-L-lysine experiments.

Other workers (Gleba *et al.*, 1974; Hughes *et al.*, 1978a) have essentially confirmed exogenous DNA uptake by various protoplasts and the association of significant portions of the DNA with protoplast nuclei. Hughes *et al.* (1978a) have reported that damaged protoplasts can become saturated with DNA and that certain divalent cations and polycations can increase protoplast damage. Using essentially the protoplast sucrose density shelf purification procedure of Shepard and Totten (1975), they were able to isolate undamaged protoplasts and concluded that, indeed, barley and tobacco protoplasts actively took up bacterial DNA. Suzuki and Takebe (1976) had earlier observed tobacco protoplast denaturation with Cu^{2+} but not with Zn^{2+}. Although these protoplasts aggregated in the presence of Zn^{2+}, they were readily redissociated by washing with 0.7 M mannitol. This was confirmed by Lurquin and Kado (1977a) with cowpea protoplasts. In fact, cowpea protoplasts treated with poly-L-ornithine and Zn^{2+} can be easily regenerated into calluses, demonstrating no apparent denaturation (Liu and Kado, unpublished observations).

The studies noted so far have involved mainly the use of DNA of relatively high complexity derived from bacteria or a plant homolog. Uptake studies employing homogenous DNA of lower complexity were first carried out by Suzuki and Takebe (1976). They used radioactive single-stranded DNA of phage fd, a flexible, rod-shaped virus containing 12.2% DNA of molecular weight (MW) 1.37×10^6 (Hoffmann-Berling et al., 1963). Their uptake studies with protoplasts, which were isolated from the palisade tissue of leaves of mature tobacco (*N. tabacum* cv. Xanthi), showed that about 30% of radioactivity remained with these cells up to 80 minutes after incubation with fd DNA. Because the protoplasts were rigorously treated with DNase (100 $\mu g/ml$), it is apparent that much of the fd DNA had penetrated into the protoplasts. This was substantiated by analysis of the subcellular fractions. About 70% of the radioactivity in protoplasts was present in the cytoplasmic fraction and 63% of this was in an acid-insoluble form. Therefore, about 27% of the radioactivity was associated with nuclei, chloroplasts, mitochondria, and membrane fragments. The fd DNA remained relatively intact as judged by velocity sedimentation analysis in a sucrose density gradient. Subsequent studies (Suzuki and Takebe, 1978) using phage λ DNA gave similar results.

Because protoplasts were derived from nonmeristematic tissues and because there were no significant differences in the radioactivities (total and acid insoluble) taken up by protoplasts that were treated with inhibitors of DNA synthesis, it was concluded that the radioactivity associated with protoplasts was not the result of plant DNA synthesized from degraded fd DNA. This conclusion is supported by observations by other workers (Uchimiya and Murashige, 1977; Liebke and Hess, 1977). Also, a high amount of reutilization is not expected based on the well-known observation that DNA replication in higher cells is confined to a specific portion of the cell cycle. Protoplasts generally have lengthy generation times coincident with DNA replication, which takes place well after reisolation of DNA in uptake experiments; therefore, any major DNA replication is expected during the S phase of the cell cycle, which occurs primarily when protoplasts are plated on proper energy-providing medium.

B. Uptake of DNA of Low Complexity (Plasmids)

Lurquin and Kado (1977a) first introduced the recombinant plasmid pBR313 (Bolivar *et al.*, 1977a) into cowpea leaf mesophyll protoplasts by using poly-L-ornithine and ZnSO$_4$. An alternative method was also used to prevent the deleterious action of any endonucleases that might accompany cowpea protoplast preparations. Kado and Lurquin (1978) reconstituted plasmid pBR313 and the pTi plasmid of *A. tumefaciens* with the capsid protein of tobacco mosaic virus. Rod-shaped particles resembling oversized tobacco mosaic virus were observed, and ultraviolet and circular dichroism analysis showed that the particles possessed nucleoprotein proteins (C. I. Kado and P. F. Lurquin, unpublished observations).

With either procedure of plasmid uptake much of the plasmid was cleaved in the cowpea protoplast cytoplasm, but a significant portion did become associated with the nuclei (Lurquin and Kado, 1977a). An estimated 20,000 copies of the plasmid pBR313 remained in each nucleus as nicked but intact molecules. Using radioactive pBR313 DNA, Hughes *et al.* (1977) showed that barley protoplasts could incorporate this molecule. They found that uptake kinetics were linear for both isolated protoplasts and isolated nuclei. This could not be attributed to uptake by damaged (nonspherical) protoplasts, because the number of damaged protoplasts and the extent to which they rapidly saturated with exogenous DNA were constant and approximately equal to the zero-time uptake value.

In agreement with Lurquin and Kado (1977a), Hughes *et al.* (1977) found a significant proportion of pBR313 DNA in the nuclear fraction. Although relatively intact molecules (as covalently closed circular DNA) were present in the nuclei, Lurquin and Kado (1977a) found that the plasmid DNA was not integrated with plant DNA. These experiments were based on network analysis of protoplast chromosomal DNA using radioactive pBR313 DNA as a molecular probe. The probe was not of sufficiently high specific radioactivity to detect very low levels of integration.

Fernandez *et al.* (1978) observed that plasmid pBR313 DNA was maintained for lengthy periods (45.5 hours) in turnip protoplasts. These workers observed that plasmid DNA uptake was inhibited substantially when protoplasts were exposed to KCN, suggesting that plasmid DNA uptake is an active process much like that observed with highly complex chromosomal DNAs (Uchimiya and Murashige, 1977; Hughes *et al.*, 1978a).

Based on the series of studies just described, it seems clear that protoplasts can take up foreign DNA. There is substantial evidence that foreign

DNA becomes associated with the nuclei of DNA-treated protoplasts. Localization of linear bacterial DNA in nuclei of protoplasts has also been observed (Hoffmann and Hess, 1973; Hoffmann, 1973; Uchimiya and Murashige, 1977; Liebke and Hess, 1977; Hughes *et al.*, 1978a,b). Hotta and Stern (1971) had shown with nuclei of broad bean (*V. faba*) embryos that the inner nuclear membrane had a fairly strong affinity for nicked or denatured DNA and a much lower affinity for double-stranded DNA. Ohyama *et al.* (1977) examined isolated nuclei from soybean protoplasts. They found that a mixture of adenosine triphosphate (ATP), phosphoenolpyruvate, and phosphoenolkinase strongly stimulated binding and uptake of radioactive soybean DNA, suggesting that an energy-dependent process might be operating. However, the binding of single-stranded DNA of phage fd was not enhanced by ATP (Ohyama, 1978). Pronase strongly inhibited binding of homologous soybean DNA, suggesting that there might be site-specific binding on soybean nuclei. Competition experiments with heterologous DNA showed that *Salmonella typhimurium* DNA and poly dAT·dAT were good competitors, whereas calf thymus DNA and *Micrococcus* DNA were poor competitors for the presumptive binding site of soybean DNA.

When these experiments are examined from an evolutionary point of view, it is difficult to generate a hypothesis on why nuclear-synthesized DNA would need to bind to the outer surface of the nucleus. It may be that adsorption of DNA to the outer surface is nonspecific binding. This is supported by the data of Ohyama (1978), who observed increased affinity of soybean nuclei for single-stranded fd DNA as the pH was lowered or in the presence of calcium or magnesium, divalent cations that are known to enhance binding of single-stranded DNA to negatively charged membrane filters (Gillespie, 1968). Strongly negatively charged membranes of various protoplasts have been shown to be nullified by calcium ions (Nagata and Melchers, 1978) and negative charges are associated with nuclear membranes. The rationale of investigating DNA uptake by isolated nuclei has been pointed out (Liebke *et al.*, 1977; Ohyama *et al.*, 1978). Such studies might culminate in the incorportion of genetically modified nuclei into protoplasts. Several reports have shown that plant protoplasts can be induced to take up plant organelles (Carlson, 1973; Potrykus, 1973; Potrykus and Hoffmann, 1973; Bonnett and Eriksson, 1974; Kung *et al.*, 1975) and microorganisms (Davey and Cocking, 1972; Giles and Whitehead, 1975; Davey and Power, 1975; Meeks *et al.*, 1978). The encapsidation of DNA with tobacco mosaic virus coat protein (Kado and Lurquin, 1978) or by using positively charged vesicles (Ostro *et al.*, 1977; Cassells, 1978) are also techniques for the uptake of pseudovirions (Lurquin and Kado, 1977b) and pseudoorganelles such as liposomes (Lurquin, 1979).

IV. Potential Gene Vector Systems for Plant Protoplasts

The technological advances made in the cloning of foreign genes using molecular cloning vehicles in *E. coli* have permitted broadening our horizon to seek similar technologies in higher cells. Before this technology could advance, it was imperative that a transformation system be developed in plants as it has now been accomplished with yeasts (Hinnen *et al.*, 1978; Beggs, 1978). So far the reported successes with plants have been limited and those genetic transformations reported using intact plants or their seeds have been controversial (see Section V).

The physical (mechanical), biochemical (molecular), and genetic barriers of the higher cell appear to be exceedingly difficult to obviate. The mechanical barrier of the cell wall is eliminated by the use of plant protoplasts. Yet other barriers are confronted that may prohibit the genetic expression of any inserted foreign gene. These barriers, shown in Table I, also have to be considered before rational approaches to achieving genetic modifications of plant cells by laboratory methods can be made.

It is nevertheless hoped that at least some of these barriers can be crossed because various molecular specificities may not be as stringent as sometimes envisioned. Indeed, it is well known that plant viruses possessing few genes are capable of genome replication and vegetative propagation in the host plant cell. Because very few genes are required for the virus' existence and yet it is capable of crossing the molecular barriers, it can only be concluded that the virus genome has evolved to mesh with the molecular machinery of the host cell without alteration of its capacity to replicate and propagate its own kind. Therefore, several potential vector

TABLE I

BARRIERS TO GENETIC EXPRESSION OF FOREIGN GENES IN PLANT PROTOPLASTS

Barrier	Mode	Means of eliminating the barrier
Cell wall	Mechanical	Use of protoplasts
Plasmalemma	Mechanical and biochemical (nucleases)	Use of chemical promoters for DNA uptake and protection of DNA
Compartmentation	Mechanical and biochemical	Unknown
Molecular		
Integration	DNA modifying enzyme specificity	Unknown
Transcription	Leader sequence, promoter and termination specificity	Unknown
Translation	Signal peptide sequence, ribosome binding, initiation and termination factor specificities	Unknown

systems for plants have been considered (Helinski, 1977; Kado, 1977, 1979). These include viruses, organelle DNA, plasmids, and plant bacteria.

A. VIRUSES

The obvious transducing vectors are plant viruses, particularly those with DNA genomes. A number of DNA plant viruses have been isolated and partially characterized. There are presently two and possibly three taxonomic groups of plant DNA viruses. These are the caulimoviruses (Harrison *et al.*, 1971), in which virions with double stranded DNA are grouped, cauliflower mosaic virus being the best known (Shepherd *et al.*, 1968); geminiviruses (Harrison *et al.*, 1977), in which virions with single-stranded DNA are grouped, bean golden mosaic virus being the representative of this group; and potato leafroll virus. The details of each member of these groups have been reviewed (Shepherd, 1979) and properties of their genomes are shown in Table II. Of these groups, the genome of cauliflower mosaic virus has been best characterized. Its double-stranded DNA is capable of directly infecting crucifers (Shepherd *et al.*, 1970) and turnip leaf protoplasts (Howell and Hull, 1977 and personal communications). Therefore, the mechanical barrier of the cell wall has not been limiting for the DNA of this virus. The potentials of cauliflower mosaic virus as a transducing vector appear temptingly attractive. However, certain properties of this, and of other viruses listed in Table II, make them less appealing than other transducers to be discussed. Although yields of cauliflower mosaic virus derived from infective tissues are sufficient to obtain DNA, the process of obtaining virions is time consuming and purification of the viral DNA is especially tedious and provides less than desirable amounts for use in gene cloning, relative to others (cf. Section IV,B). Also, the cauliflower mosaic virus host range is mainly restricted to crucifers (Shepherd, 1979). Cauliflower mosaic virus DNA possesses a single cleavage site for restriction endonuclease *Sal*I (Szeto *et al.*, 1977; Meagher *et al.*, 1977; Hull and Howell, 1978). Szeto *et al.* (1977) used a composite plasmid cloning vehicle pGM706 (Hamer and Thomas, 1976), which carries ampicillin and tetracycline resistance and a single *Sal*I site, to clone the cauliflower mosaic virus DNA in *E. coli* K12 $r_{\bar{B}}$ $m_{\bar{B}}$ recA$^-$. Infectivity tests on "Tendergreen" mustard with the hybrid plasmid were negative, even after excision and religation of the amplified DNA. Furthermore, cauliflower mosaic virus DNA cleaved by *Sal*I also showed no infectivity. The reason for this lack of infectivity may be the possibility that there are closely spaced *Sal*I sites and the small segment between them is lost. Alternatively, cauliflower mosaic virus may have a bipartite genome, in which two virus particles are required to establish infection.

TABLE II

PROPERTIES OF PLANT VIRUSES WITH DNA GENOMES

Group virus	Size of virion (μm)	Size of genome (MW)	Susceptible plant family	References
Caulimovirus				
Carnation etched ring virus	42	—	Caryophyllaceae	Hollings and Stone (1969); Fujisawa *et al.* (1972); Rubio-Huertos *et al.* (1972); Lawson and Civerolo (1976)
Cauliflower mosaic virus	50	$4.5-5 \times 10^6$	Cruciferae, Solanaceae (*Datura*)	Smith (1957); Pirone *et al.* (1961); Shepherd *et al.* (1970); Shepherd and Wakeman (1971); Lung and Pirone (1972)
Dahlia mosaic virus	45	—	Amaranthaceae, Chenopodiaceae, Compositae, Solanaceae	Brierly and Smith (1950) Brunt (1966, 1971)
Figwort mosaic virus	—	—	Scrophulariaceae, Chenopodiaceae	Duffus (cited in Shepherd, 1979)
Mirabilis mosaic virus	45–50	—	Nyctaginaceae	Brunt and Kitajima (1973)
Strawberry vein banding virus	45–50	—	Rosaceae	Frazier (1955); Kitajima *et al.* (1973)
Geminivirus				
Bean golden yellow mosaic virus	18	$0.66-0.95 \times 10^6$	Leguminosae	Meiners *et al.* (1975); Goodman (1977); Goodman *et al.* (1977)
Beet curly top virus	20		Some 44 families	Bennett (1971); Mumford (1974)
Cassava latent virus	15–20	0.8×10^6	Euphorbiaceae, Solanaceae	Lister (1959); Harrison *et al.* (1977)
Euphorbia mosaic virus			Euphorbiaceae, Solanaceae, Leguminosae	Bird and Maramorosch (1978)
Maize streak virus	15–20	0.71×10^6	Gramineae	Storey (1928); Bock (1974); Harrison *et al.* (1977)
Tobacco leaf curly virus	16–18		Solanaceae	Osaki and Inouye (1978)
Leafroll virus				
Potato leafroll virus	25	1.3×10^6	Amaranthaceae	Sarkar (1976, 1978)

The minor bands generated by *Sal*I digestions (Szeto *et al.*, 1977, Meagher *et al.*, 1977; Volovitch *et al.*, 1978; Hull and Howell, 1978) may be derived from a helper virus. A third possibility, suggested by Szeto *et al.* (1977), is that infectivity is perhaps lost because the native viral DNA has some crucial secondary structure that is lost on cleavage or some state of modification that is not present in the cloned material. Detailed studies of the cabbage B strain of cauliflower mosaic virus revealed variations in a *Sal*I site. Meagher *et al.* (1977) reported that cleavage with *Sal*I results in the conversion of less than 5% of circular DNA into linear molecules. This was put forth as evidence for nucleotide sequence heterogeneity in the native virus DNA populations, in which only a small portion of the DNA molecules possess uncommon restriction endonuclease sites. Hull and Howell (1978) suggested several possibilities. The minor DNA fragments generated by *Sal*I could be attributed either to heterogeneity with some of the viral strains to the presence of linear DNA molecules, which appear to arise from double-stranded breaks at specific points in the circular genome; or to the occurrence of a few "difficult" restriction sites in the genome of some virus strains. When the physical maps (determined by the use of single and coupled restriction endonucleases) of four different strains of cauliflower mosaic virus were compared, they found that one region of the viral genome is subject to change, whereas the rest is highly conserved.

These unresolved properties of the cauliflower mosaic virus genome add undesirable complexities for its use as a transducing vector. The loss of infectivity by a single cleavage of the genome, which remains unexplained, adds to the complication.

Another limiting feature of this virus (and perhaps other less well-studied viruses of the caulimoviruses group) is that they are likely to be replicated in the cytoplasm of the host cell instead of in the nucleus (Shepherd, 1976; Fujisawa and Matsui, 1972; Fujisawa *et al.*, 1967). Virions undergo vegetative reproduction and viral DNA replication in inclusion bodies that are situated in the cytoplasm of the host cell (Shepherd, 1979).

Cauliflower mosaic virus is limited mainly to crucifers. This would probably limit its use to these plants for the purpose of introducing foreign genes by means of this virus as the transducing vector. Dahlia mosaic virus, which has not been as well studied, may be somewhat better than cauliflower mosaic virus because its host range extends to four plant families (Table II).

Although these are some of the more obvious limiting factors concerning the use of caulimoviruses, further detailed studies of their genomes may reveal a replication origin region, a promotor region, and peptide leader sequences that may be useful in combination with other cloning vectors (for example plasmids).

Viruses of the gemini group possess single-stranded genomes (Table II). This by no means limits their use as possible cloning vectors because their replicative form (RF) can be employed. Indeed, this may be advantageous because genes cloned on a geminivirus will be naturally propagated in virions as single-strand progeny DNA. They are therefore readily amendable to the rapid single-stranded DNA sequencing technique of Sanger *et al.* (1977). A possible limiting factor of these geminiviruses is that they are usually limited to the phloem elements and so may not be conducive for invasive infection to other plant regions. Although mechanical transmission is possible with some of these viruses, such as beet curly top virus (Bennett, 1934; Fulton, 1955), it is not consistently reproducible for use in assaying genes on hybrid viral DNA. Moreover, because of their general confinement to phloem elements, the yields of virus and its RF DNA will always be small. This will limit accessibility of large quantities of virions and RF DNA, prerequisites for gene cloning.

Sufficient information is lacking for possible use of potato leafroll virus as a cloning vehicle. Its small double-stranded genome (MW 1.3×10^6) makes the virus attractive. However, like the geminiviruses, yields of potato leafroll virus are especially small, making it technically infeasible for use as a cloning vector.

Bacteriophages may prove to be an effective alternative cloning vector for plants. Examples of transfection of higher cells by some bacteriophages have been reported (Kleinhofs and Behki, 1977). Because there is no technical difficulty in easily obtaining limitless quantities of phage DNA, its use in plants should be further investigated. Those who view this alternative with skepticism should be aware of the awesome potentials of two phage systems already developed to investigate plant genes.

Thomas *et al.* (1974) have constructed a series of λ phage mutants, designated as λgt-λc, that carry two *Eco*RI restriction sites between which *Eco*RI-generated DNA segments of 1–14 kilobase pairs can be inserted. The mutants themselves are unable to produce plaques in the *E. coli* cloning host. However, plaques are produced if a new DNA segment within the above DNA size range is inserted between the two *Eco*RI sites because sufficient length of DNA is apparently necessary for packaging into viable phage particles (Thomas *et al.*, 1974). The formation of plaques indicates an insertion of a new DNA and new DNA clones therefore can be obtained by this powerful positive selection system. Furthermore, Davis and co-workers (1977) showed that λgt mutants containing a particular size of DNA can be selected either physically by its buoyant density (Kellenberger *et al.*, 1961) or genetically by growth on *pel*⁻ *E. coli* (Emmons *et al.*, 1975; Scandella and Arber, 1976), a host that is very stringent for phage containing DNA of wild-type size or larger. Therefore, specific sizes of DNA can be selected from a random mixture of DNA to be cloned in these λ mutants. These clones can be screened for a particular DNA sequence by employing one of several techniques, such as plaque hybridization (Jones and Murray, 1975; Kramer *et al.*, 1976), *in situ* hybridization (Grunstein and Hogness, 1975), or immunological procedures (Sanzey *et al.*, 1976).

One of the disadvantages of using λ vehicles is that there may be a possibility of integrating bacterial host genes. Although certain aspects of the fate of λ DNA in tobacco protoplasts have been studied (Suzuki and Takebe, 1978), evidence for λ DNA replication and possible transcription in plant hosts requires careful evaluation and consideration. Nevertheless, cloned plant genes may have a good chance of being expressed in plant cells. The analogous situation in animal cells has been experimentally demonstrated as described in Section III.

A second phage vehicle for use as a potential vector is the filamentous phage M13. The advantages of using M13 over using λ are that its genome is a single-stranded circular DNA molecule.Therefore genes cloned using its replicative form, RFI, can be separated naturally into single-stranded DNA during phage propagation. Hence, DNA sequencing of genes cloned in M13 can be accomplished readily by the sequenciing procedure outlined by Sanger *et al.* (1977). Indeed, Messing *et al.* (1977) and Messing and Gronenborn (1978) have contructed an ideal cloning vehicle using M13. A nonessential region of the M13 genome was located and a hybrid phage, M13 mp 1, was constructed that contained the Hind II fragment from the *lac* regulatory region of the *E. coli lac* operon (Gilbert *et al.*, 1975) and the part of the β-galactosidase gene that codes for the α-peptide. The synthesis of the α-peptide encoded by the *lac* structural part is governed by the lac controlling elements. Its ability to mediate α-complementation in an appropriate host can be monitored by a color reaction on agar medium containing isopropylthiogalactoside (an inducer of lactose operon) and 5-bromo-4-chloroindolyl-β-D-galactoside, a colorless compound that, when hydrolyzed by β-galactosidase, releases deep-blue 5-bromo-4-chloroindigo. Therefore, infection of *E. coli* K12 71–18 (Δ[lac, pro],F'lacI^{q}Z Δ M15 pro$^+$) by M13-mp 1 is reflected by the formation of blue plaques. *In vitro* insertion of foreign DNA in the *lac* region of this phage genome results in the loss of α-complementation and therefore gives rise to white plaques that are readily distinguishable from the blue ones, which contain phages without any insertions. Gronenborn and Messing (1978) improved the versatility of M13 mp 1 by introducing an *Eco*RI site by conversion of the *Eco*RI* sequences in the *lac* structural gene by alkylation using *N*-methyl-*N*-nitrosourea into sequences recognized by restriction enzyme *Eco*RI. The introduction of the *Eco*RI site into the *lac* structural part of M13 mp 1, which itself is resistant to cleavage by endonuclease *Eco*RI, creates an unique site to clone DNA fragments.

Besides the positive means of selection devised by this phage, there seems to be no upper limit of foreign DNA than can be inserted into M13 because of its filamentous structure. At least DNA up to 10×10^6 MW can be packaged as intact phage particles (H. Schaller, cited in Messing *et al.*, 1977). Therefore, plant DNA containing promotor sequences recognized by plant transcriptional systems may be cloned with M13 mp 1 derivatives, such as M13 mp 2. These, in turn, in RFI amplified form, can be used in uptake experiments with protoplasts.

Uptake studies of M13 mp 1 RFI in cowpea protoplasts have been performed. Using plaque assays with *E. coli* K12 71–18, Kado, Lurquin, and Messing (unpublished results) showed that the replicative form of M13 mp 1 is continually nicked into open circular molecules at the rate of about 2.5% nicking per hour. Therefore biologic activity, expressed as infectivity (transfection of the *E. coli*), is reduced to nearly undetectable levels after 24 hours in cowpea protoplasts. Nevertheless, nicked open circular molecules remain and the question remains of whether cloned genes within M13 mp 1 DNA would still function in protoplasts.

B. ORGANELLE DNA

Mitochondria and chloroplast DNA have been scrutinized by several laboratories as possible gene-cloning vehicles. Certainly they can be viewed as counterparts of bacterial plasmids, and therefore derivatives of mitochondrial or chloroplast DNA can be generated by procedures that are used for the isolation of low molecular weight derivatives of plasmids.

When introduced into protoplasts by technologies already discussed (see Section III), the fate of such reconstructed organelle DNAs remains unknown. It is clear that such a DNA must not only be capable of surviving in the cell but it must also be replicated and transcribed accurately. Compartmentation of the restructured organelle DNA is uncertain and depends on the extent of reconstruction of the molecule. Furthermore, it is unknown whether or not compartmentation affects the expression of foreign genes. These and a number of other uncertainties require many in-depth studies before endeavors in this direction can be undertaken. However, it should be stressed here that organelle DNA is attractive because it is a natural extrachromosomal ''resident'' of the higher cell and therefore possesses replication origins, promoter sites, and leader sequences for signal peptides of translational products. These can be isolated by the available technologies and perhaps constructed into a functional cloning vehicle.

C. PLASMIDS

A number of plasmids have been reconstructed to serve as vehicles for cloning genes in *E. coli* (for review, see Helinski, 1977; Sutcliffe and Ausubel, 1978). Reconstruction is generally necessary so that the plasmid has certain features that are desirable for cloning genes efficiently. The plasmid should carry genes for easy identification; generally one or more antibiotic resistance genes serve this purpose. It should possess uncomplicated insertion sequences. These are usually single sites for restriction endonucleases. The plasmid vector should also carry genes for high production (amplification under relaxed control) and should be of minimal size to reduce background noise.

Plasmids pBR313 and pBR322 (Bolivar *et al.*, 1977a,b) fit the above criteria and serve as examples here. There are a number of other plasmids suitable for cloning work (Helinski, 1977; Sutcliffe and Ausubel, 1978). Both plasmids carry genes for conferring ampicillin and tetracycline resistance. Single restriction endonuclease sites for *Eco*RI, *Hin*dIII, *Bam*HI, *Sal*I, *Hpa*I, and *Sma*I are present in pBR313 and an additional *Pst*I site in the ampicillin gene is on pBR322. This permits the cloning of foreign DNA with each of these restriction enzymes, and the probability of inserting a fully functional piece of foreign DNA therefore is increased. The substrate sequence for *Hin*dIII, *Bam*HI, and *Sal*I restriction endonuclease are in the gene that confers tetracycline resistance on *E. coli*. The insertion of foreign DNA into any one of these restriction sites inactivates the expression of tetracycline resistance. Therefore, *E. coli* cells haboring recombinant molecules can be first enriched by growth of only tetracycline-resistant *E. coli* in medium containing tetracycline. These cells are then killed by the addition of cycloserine. The remaining *E. coli* cells, harboring pBR313 with a gene insertion in the tetracycline locus, are grown in medium containing ampicillin. The plasmid is then amplified to 1000–2000 copies per cell by the addition of chloramphenicol to the medium (Hershfield *et al.*, 1974). Roberts *et al.* (1979) have described a method that allows the promoter of the *E. coli lac* operon to be placed at virtually any distance in front of a gene cloned in pBR322. In principle, such a method can be used to insert plant genes downstream from the *lac* promoter, thus permitting the expression of plant genes in *E. coli*. Furthermore, the plant gene itself can be produced in quantities more than sufficient for determining its DNA sequence by the method of Maxam and Gilbert (1977). These prospects are under current study and the technologies for the uptake of pBR313 and pBR322 by plant protoplasts have been described elsewhere (Lurquin and Kado, 1977a; Fernandez *et al.*, 1978; Kado and Lurquin, 1978) (also see Section III). The introduction of plant genes carried on promoter-sufficient cloning vehicles into protoplasts is one step toward gaining a means of transforming plant cells. Unfortunately, the lack of biochemical mutants of plants hinders the detection of expression by complementation methods. The pTi plasmid of *A. tumefaciens* has been cloned in pBR313 and pBR322 and inserted into cowpea protoplasts (S.-T. Liu, S. Sonoki, and C. I. Kado, unpublished observations). Although RNA transcripts of the plasmids have been detected in these protoplasts, biologic expression in the form of auxin prototrophy and the presence of nopaline were unconvincing (S.-T. Liu, P. F. Lurquin, and C. I. Kado, unpublished observations). Detection of a known gene product of the pTi plasmid, such as the Per-1 and Per-2 proteins (Sonoki and Kado, 1978), in these "transformed" cells would be convincing.

Other markers, such as the streptomycin- or valine-resistant lines of tobacco (Maliga *et al.*, 1975; Bourgin, 1978), might be useful in counterselection. They then could be used as markers in transformation experiments with plant protoplasts. Streptomycin resistance might be used as a means for positive selection of the transformed clones. However, it must be realized that streptomycin resistance is a recessive character and may not be adequately expressed following uptake of the cloned streptomycin gene.

The problem of replication-promoter recognition has been of concern to plant molecular biologists mainly because bacterial DNA replication origins and promoters may not function adequately in plant cells. In order to avoid this complication, plasmids of eukaryotes might be considered as potential gene-cloning vehicles. This is based on the assumption that DNA replication and promotor sites of eukaryotes would be better recognized than those from prokaryotes.

Besides organelle DNA, discussed in Section IV,B, studies of plasmids or plasmid-like molecules in eukaryotic cells should be considered. Such molecules have been observed in yeast (Billheimer and Avers, 1969; Petes and Williamson, 1975; Cameron *et al.*, 1977), *Neurospora* (Agsteribbe *et al.*, 1972), *Drosophila* (Sorsa, 1973; Stanfield and Helinski, 1976), HeLa cells (Smith and Vinograd, 1972), and tobacco (Wong and Wildman, 1972). Very little is known about these plasmid-like molecules in plants and an understanding regarding their role and function must await further studies.

D. BACTERIA

Certain groups of bacteria are known to exist in close association with plants. Of special interest is *Agrobacterium tumefaciens* because it causes the crown-gall disease of many different plants. A large plasmid, known as the pTi plasmid, confers the oncogenic phenotype on this bacterium (for review, see Lippincott and Lippincott, 1975; Kado, 1976; Braun, 1978). Deoxyribonucleic acid sequences of this plasmid have been detected (Chilton *et al.*, 1977) and are transcribed in bacteria-free crown-gall tissue (Ledeboer, 1978; Drummond *et al.*, 1977). Therefore, it is believed that the pTi plasmid or a portion thereof is the crown-gall-transforming genetic element that is transferred by *A. tumefaciens* into the host plant during infection. Smith and Hindley (1978) first developed an *in vitro* transformation system using *A. tumefaciens* cells mixed with suspension cultures of habituated tobacco cells. The tobacco cells were treated with rifampicin 48 hours after infection and allowed to grow as calluses for 4 weeks. Analysis of the callus tissue showed that nopaline, a pTi plasmid-mediated guanido amino acid, was indeed present in only those cells exposed to an oncogenic strain of *A. tumefaciens*. Márton *et al.* (1979) also demonstrated *in vitro* transformation of tobacco cell SR, a cell line resistant to streptomycin (Maliga *et al.*, 1975). Examination of the tobacco tranformants revealed heterogeneity in two tumor phenotypes, lysopine dehydrogenase activity and hormone independence, suggesting that genetic instability occurred among the transformants. Nevertheless, certain stable lines were able to regenerate shoots and showed extensive lateral shoot formation. Like the calluses, these shoots also showed lysopine dehydrogenase activity.

These *in vitro* transformation experiments also can be applied to plant protoplasts. The technologies for the uptake of pTi plasmid in protoplasts already have been developed (Lurquin and Kado, 1977a; Fernandez *et al.*, 1978; Kado and Lurquin, 1978) and reports of transformation of various protoplasts using pTi DNA are likely to be forthcoming (e.g., preliminary results expressed in Márton *et al.*, 1979). The pTi plasmid is transcribed in cowpea protoplasts (Liu *et al.*, unpublished observations).

Therefore, there are two possible approaches to the addition of new genetic information in the plant cell: first, by means of infection using *A. tumefaciens* that harbors a reconstructed pTi plasmid with a set of desired genes that is coupled to the pTi sequences in tumor cells (called the T-region); and second by direct introduction of reconstructed plasmid DNA into plant protoplasts, which in turn are regenerated into calluses and then into plants.

The first approach has some very obvious problems. If new genes are to be introduced in plants by the formation of a phenotypically detectable trait—the crown-gall tumor—then the cells of the tumor must be freed of bacteria and propagated in tissue culture, which in turn must be regenerated into plants. The *in vitro* transformation systems discussed above by-passes the need to infect intact plants. However, regeneration of transformants into whole plants from calluses is still limited to a few species. If plants of desired species were capable of regeneration, there would yet be the problem of whether or not the pTi genes remained stably maintained following sexual propagation of the plants. Deoxyribonucleic acid hybridization analysis of a mature tobacco plant regenerated from crown-gall callus tissue (in at least one tumor line, T37) showed that pTi DNA sequences were present; however, no pTi DNA sequences were detected in the progeny of this plant (E. W. Nester, M. P. Gordon, and M. D. Chilton, personal communications). This suggests that the pTi DNA is only stably maintained in tumor calluses and in plants regenerated from these calluses but not in plants propagated from the regenerated plants. Therefore, the use of *A. tumefaciens* as a vector has its limitations. Nevertheless, crop plants, such as the potato, that are vegetatively propagated may be genetically modified by the use of this bacterial gene vector.

V. Transformation in Plants

The introduction of exogenous DNA into cells can result in a stable and heritable change in phenotype. This process, known as transformation, is well established in several bacterial genera. In higher plants numerous transformation experiments have been reported. These have been extensively and critically reviewed by Kleinhofs and Behki (1977) and Lurquin (1977). Little actual progress has been made on clarifying the questions raised in these reviews and in resolving the problems of higher plant cell transformation. A glimmer of hope has been raised by the *in vitro* transformation of tobacco cells by *A. tumefaciens* cells (Márton *et al.*, 1979) and advances in the transformation of yeast (Hinnen *et al.*, 1978; Beggs, 1978) and mammalian cells (Bacchetti and Graham, 1977; Maitland and McDougall, 1977; Wigler *et al.*, 1977, 1978, 1979a,b; Minson *et al.*, 1978; Pellicer *et al.*, 1978).

The initial report of successful transformation in higher plants involved the *Petunia hybrida* anthocyanin synthesis system (Hess, 1969, 1972). Seedlings of a white-flowering strain were transformed with DNA from a red-flowering strain. These experiments, although extensively conducted over several years, suffer from the lack of a basic proof that DNA from the red-flowered strain has been transferred to the white-flowered strain and is responsible for the appearance of the few red-flowered plants among the white-flowered recipients. This lack of irrefutable evidence for transformation is common to all the DNA transformation systems studied in higher plants to date.

Transformation of the *waxy* (*glx glx*) endosperm character to the non-waxy (*Glx Glx*) state was reported in barley (Turbin *et al.*, 1975). The *waxy* mutant seed endosperm and starch grains contain amylopectin but not amylose. When they are stained with I_2-KI, a reddish-brown color is observed in place of the deep blue. This difference is easily detected in pollen preparations and consequently very large haploid cell populations can be easily screened. This is an obvious advantage in transformation studies but does not fully offset the disadvantage that the *waxy* locus in barley has not been characterized at the molecular level. Further genetic studies (Soyfer *et al.*, 1976) have shown that one plant having nearly 100% wild-type pollen in the initial generation segregated in subsequent generations. Of 28 self-fertilized seeds produced, 23 produced plants with nearly all wild-type pollen and five produced plants with nearly all *waxy* pollen. No heterozygotes were reported. The plants producing the wild-type pollen were two rowed (as the DNA donor) and plants producing the *waxy* pollen were six rowed (as the DNA recipient). This strict correlation of the two characters is unexpected because they are located on different chromosomes (Nilan, 1964). The genetic studies of the presumed transformants, although interesting, do not provide any further evidence on the transformation itself.

Transformation of genes for fruiting habit (*Fa* vs *fa*) and fruiting direction (*Up* vs *up*) was studied in red peppers, *Capsicum annuum* L. (Nawa *et al.*, 1975). Seeds of cultivar Tochigisantaka (T) (*up up*, *fa fa*, YY, $C_1 C_1$) were treated with DNA from cultivar Kiiro (K) (*Up Up*, *Fa Fa*, $y\,y$, $c_1 c_1$) and vice versa. In the self-progeny of T plants treated with K DNA, one plant out of 393 was found to carry the trait of the donor DNA. The change was from the recessive homozygous (*fa fa*) to the heterozygous or homozygous dominant (*Fa __*). The other characters remained unchanged, therefore ruling out possibilities of seed contamination or outcrossing. The use of multiple gene markers helps to strengthen the conclusions drawn in this study. The use of a nonbiochemical trait, however, prevents a more detailed analysis of the transformation at the gene product level.

In some very interesting experiments, Wardell (1976) reported that DNA from tobacco stems in a flowering state could induce flowering in tobacco buds maintained under conditions that did not permit flowering. Denaturation (heating followed by rapid cooling) enhanced the floral activity of the DNA solutions. Flowers developed from buds treated with native DNA were normal. Buds treated with denatured DNA developed many abnormal as well as normal flowers. Seeds produced by the abnormal flowers produced progeny with strikingly abnormal flowers. The floral activity of the DNA solution was completely eliminated by DNase treatment. A control using DNA from stems or leaves of vegetative tobacco plants was negative. In a subsequent report, Wardell (1977) presented evidence that the flowering stems contained large amounts of rapidly renaturing DNA and that this DNA was responsible for the flower induction. Rapid renaturation of the heated and then rapidly cooled DNA was demonstrated by CsCl equilibrium gradient and sucrose-velocity gradient centrifugation as well as by S1 nuclease digestion. These experiments suggest that exogenous DNA applied to tobacco buds can be at least transiently functional. Unfortunately, the current lack of understanding of the flowering induction process precludes more profound conclusions.

A different approach to gene transfer in higher plants has been reported by Pandey (1975). During experiments on the role of "mentor pollen" in overcoming intraspecific incompatibility in *Nicotiana* some unexpected observations were made. (Mentor pollen is defined as compatible, irradiated pollen that produces a pollen tube but fails to fertilize the ovule.) In experiments in which *Nicotiana forgetiana* was fertilized with mentor pollen from *Nicotiana alata* a few viable seeds were produced along with a large number of collapsed, nonviable ovules. These seeds gave rise to normal, fertile, diploid plants that resembled their maternal parent in most respects. Surprisingly, most of the plants showed flower color and an incompatibility allele characteristic of the mentor pollen source. About one-half of all the plants were examined and shown to be cytologically normal. Some of the plants showing the flower color or S allele typical of the mentor pollen were test crossed to plants of known genotypes. The most significant result of these crosses was the recovery of three S triallelic plants. The presence of the triallelic plants was argued as providing strong evidence for the occurrence of an unusual genetic transfer process.

These transformation experiments involve the use of homologous DNA. Several other studies have involved heterologous DNA transformation. The most extensively studied systems is that of *Arabidopsis thaliana* mutants blocked in the thiamine pathway (Ledoux *et al.*, 1974). Mutations controlling synthesis of thiamine occur at four loci in *Arabidopsis* (Redei, 1975). These have been designated as *py*, *tz*, *thi*$^{-1}$, and *thi*$^{-2}$. The *py* mutants require 4-amino-5-aminomethyl-2-methylpyrimidine, the *tz* mutants require 5-β-hydroxymethyl-4-methylthiazole, and the *thi* mutants require thiamine for normal growth and seed production (Li and Redei,

1969). Homozygous recessive seeds are produced on mutant seedlings supplemented with thiamine either in sterile culture or in soil. Homozygous mutant seeds treated with DNA from a large number of bacterial sources produced viable and fertile plants with variable frequency (Ledoux *et al.*, 1971, 1972, 1974). Controls using bacterial DNA extracted from a thiamine A mutant of *E. coli* or from bacteriophages T7 and 2C, which do not contain the thiamine genes, failed to produce "corrected" plants.

The experimental details and controversy surrounding these experiments have been reviewed in detail (Kleinhofs and Behki, 1977; Lurquin, 1977). Genetic studies by Redei *et al.* (1977) of some "corrected" mutants raise doubts about the participation of foreign genetic information in the "correction" phenomenon. In fact, Redei *et al.* concluded that the apparent "correction" was probably caused by mechanical seed contamination. Such experiments, however, cannot prove that all of the "corrected" types result from mechanical seed contamination. Ultimate proof of "correction" of these *Arabidopsis* mutants by bacterial DNA requires the demonstration of the presence of the bacterial DNA in the plant cells and sufficient refinement of the transformation techniques so that they can be reproduced routinely in independent laboratories.

Other heterologous DNA transformation attempts have been made with purified bacteriophages applied to plant cells in culture (Doy *et al.*, 1973a,b; Carlson, 1973; Johnson *et al.*, 1973). These experiments produced interesting and controversial results. The experimental approach, the results, and criticisms have been extensively reviewed (Kleinhofs and Behki, 1977; Lurquin, 1977). The data, if taken at face value, clearly suggest that the genes carried by the bacteriophage can, at least temporarily, be expressed. In no case was a stable, heritable change demonstrated. Consequently, this approach is not of primary interest to plant cell genetic engineering.

An interesting approach has been proposed and tested by Hess and co-workers (1974; Hess, 1975, 1978). The idea is to use pollen grains as a vehicle to transfer foreign genetic information to plant cells. Deoxyribonucleic acid and bacteriophages were apparently taken up by the swelling and germinating pollen grains. It was reported that progenies of *N. glauca* derived from *N. glauca* pollen treated with *N. langsdorffii* DNA produced tumors (Hess *et al.*, 1976). Moreover, *P. hybrida* pollen treated with bacteriophage carrying the *E. coli* β-galactosidase gene has been used for pollination of *Petunia* flowers. Controls included the use of pollen treated with bacteriophage lacking the β-galactosidase gene and homologous *Petunia* DNA. The progeny were tested for growth on lactose agar (Hess, 1978). The results were quantitatively but not qualitatively different; that is, seedlings from the treated as well as control batches grew on the lactose agar. The seedings from the batches treated with phage carrying the β-galactosidase gene grew better.Statistical significance was established. These results unfortunately do not seem to constitute clear-cut evidence for the functional transfer of the bacterial β-galactosidase gene.

Although there are numerous reports on transformation in higher plants, the approach so far has not yielded a great deal of information about the process. The experiments seem to be mostly random hit or miss propositions and the results are not helpful in designing future more successful experiments. Substantial improvement in gaining knowledge in future experiments would be realized by using strains with several markers, preferably closely linked ones, and by attempting transformation with traits that could be identified directly at the gene or gene product level.

It has been demonstrated that it is possible to recover hormone-independent calluses from cell wall-regenerating tobacco protoplasts treated with *A. tumefaciens* (Márton *et al.*, 1979). Because *A. tumefaciens* is able to transform numerous plant species *in vivo,* it appears to be a good candidate for transformation studies *in vitro*. Now that the transformation of tobacco cells in culture with the bacterium seems to be established, the next obvious step is to accomplish the same effect with the purified Ti plasmid DNA. Such experiments are underway in several laboratories.

Successes with transformation of animal cells suggest that the protoplast transformation approach should work. Transformation of human (Bacchetti and Graham, 1977) or mouse (Wigler *et al.*, 1977; Maitland and McDougall, 1977) cells from the thymidine kinase-negative (tk^-) to the thymidine kinase-positive (tk^+) phenotype was achieved using purified DNA or DNA fragments from herpes simplex virus. The presence of the virus-specific enzyme was demonstrated by biochemical and serological techniques. It is worthwhile to note that the tk^+ phenotype was stable over several hundred cell generations, variable stability in enzyme expression occurred among different clones, and a relatively high reversion rate was observed. Minson *et al.* (1978) have reported that DNA extracted from transformed cells is also able to transform tk^- mouse cells to the tk^+ phenotype.

Pellicer *et al.* (1978) examined the distribution of tk sequences in the DNA of several stably transformed mouse cell clones. Experiments involved reassociation kinetics in solution and annealing with tk DNA to restriction-cleaved cellular DNA following electrophoresis and transfer to filters. It was concluded that the viral tk gene is present in all cells at a frequency of one copy per chromosomal complement, the tk gene is stably integrated in the DNA of all transformants, and the integration is not site specific, occurring at different loci in the DNA of all transformants examined.

Wigler *et al.* (1978, 1979a,b) have demonstrated that the *tk* gene can be transfered by using DNA from tissues and cultured cells of a variety of animal species. The isolation of cells transformed with nonselectable biochemical markers was achieved by cotransformation with the *tk* gene and the nonselectable marker. The nonselectable marker, that is, ϕX174, plasmid pBR322, or cloned rabbit β-globin, was used in great excess and the transformants were selected by the tk^+ phenotype (Wigler *et al.*, 1979b). The yeast transformation work should also be of great interest as a model system to plant transformation workers. Hinnen *et al.* (1978) reported the transformation of a stable *leu*$^-$ yeast strain to *Leu*$^+$ by using a chimeric Col El plasmid carrying the yeast *leu* 2 gene (*leu* 2 codes for the leucine biosynthetic enzyme β-isopropylmalate dehydrogenase). The observed frequency was one in 10^7 viable cells. Hybridization and restriction endonuclease mapping techniques were used to demonstrate directly the presence of the transforming DNA in the yeast genome and to determine the arrangement of the sequences introduced. Three types of transformants were recovered. Type I, the most frequent class, appeared to involve the integration of the complete yeast–bacteria plasmid in the chromosome region adjacent to the *leu* 2 gene. Type II, the next most frequent class, appeared to result from the integration of the chimeric plasmid in chromosomal regions other than the one occupied by the *leu* 2 gene. Type III appeared to result from the integration of only the yeast portion of the chimeric plasmid at the site of the *leu* 2 gene. This type of transformant could not be differentiated from a revertant. It was pointed out, however, that no revertants were ever observed in the control experiments.

A highly efficient yeast transformation system has been developed by Beggs (1978). Chimeric plasmids were contructed from pMB9, a derivative of the Col El plasmid from *E. coli* and the 2 μM yeast plasmid. These chimeric plasmids can replicate in both *E. coli* and *Saccharomyces cerevisiae*. Yeast nuclear DNA fragments were linked to the chimeric plasmid and two plasmids were isolated that complemented *leu* B mutations in *E. coli*. These plasmids were used to transform *S. cerevisiae* from *leu* 2$^-$ to *Leu*$^+$ with a high frequency (5×10^{-4} to 3×10^{-3} transformants per viable cell). The yeast transformants contained multiple plasmid copies, which could be recovered by transformation in *E. coli*. The inheritance of the *Leu*$^+$ phenotype in the transformed clones was shown to be cytoplasmic.

In summary, it can be concluded that clear-cut proof of genetic transformation still remains to be established in higher plants. Advances in protoplast studies and in other eukaryotic systems, such as yeast and mammalian cells, suggest that transformation in higher plants should be demonstrated in the relatively near future.

VI. Conclusions

Dramatic advances in the area of cell and molecular biology have engendered new enthusiasm in the field of plant genetics. The selection, amplification, and isolation of desired genes and the technologies for sequencing the DNA of such genes are realities. The manipulation of plant protoplasts so that they take up recombinant DNA molecules and functionally express them at the molecular level has been demonstrated. Attention is now focused on solving the problems of stable maintenance of introduced DNA molecules in plant cells. Detailed studies are necessary to understand DNA replication, promoter, and peptide signal sequences that are recognized by plant cells. These are some of the major molecular and genetic barriers that require further understanding. Molecular cloning studies of plant genes in bacteria and the detailed analysis of their functions will certainly give insight to these questions.

Various technical limitations in the regeneration of protoplasts into whole plants require elucidation. So far this area has been empirically developed, indicating a great gap in the understanding of plant molecular biology.

Finally, a great deal of effort needs to be expended in characterizing the plant genes that it may be desirable to engineer for improved plant productivity. The detailed knowledge of their structure, function, and regulation will be urgently needed when gene transfer for practical plant breeding becomes a reality. In this review we express cautious optimism in the development of plant genetics through molecular approaches. We fully realize the diversity and possible awesome barriers that confront foreign DNA in plant cells. Nevertheless, knowledge accumulated through the use of recently developed molecular genetic tools will certainly pave the way in understanding gene function in plants and perhaps their direct manipulation in the future.

REFERENCES

Aaronson, S. A., and Martin, M. A. (1970). *Virology* **42,** 848–856.
Agsteribbe, E., Kroon, A. M., and Van Bruggen, E. F. J. (1972). *Biochim. Biophys. Acta* **269,** 299–303.
Aoki, S., and Takebe, I. (1969). *Virology* **39,** 439–448.
Bacchetti, S., and Graham, F. L. (1977). *Proc. Natl. Acad. Sci. U.S.A.* **74,** 1590–1594.
Beggs, J. D. (1978). *Nature (London)* **275,** 104–109.

Bendich, A. J., and Filner, P. (1971). *Mutat. Res.* **13,** 199–214.

Bennett, C. W. (1934). *J. Agric. Res.* **48,** 665–701.

Bennett, C. W. (1971). *Am. Phytopathol. Soc. Monogr.* No. 7.

Billheimer, F. E., and Avers, C. J. (1969). *Proc. Natl. Acad. Sci. U.S.A.* **64,** 739–746.

Bird, J., and Maramorosch, K. (1978). *Adv. Virus Res.* **22,** 55–110.

Bock, K. R. (1974). *Commonw. Mycol. Inst./Assoc. Appl. Biol. Descr. Plant Viruses* No. 133.

Bolivar, F., Rodriguez, R. L., Betlach, M. C., and Boyer, H. W. (1977a). *Gene* **2,** 75–93.

Bolivar, F., Rodriguez, R. L., Greene, P. J., Betlach, M. C., Hegneker, H. L., and Boyer, H. W. (1977b). *Gene* **2,** 95–113.

Bonnett, H. T., and Eriksson, T. (1974). *Planta* **120,** 71–79.

Bourgaux, P., Bourgaux-Ramoisy, D., and Stoker, M. (1965). *Virology* **25,** 364–371.

Bourgin, J.-P. (1978). *Mol. Gen. Genet.* **161,** 225–230.

Boyer, H. W. (1971). *Annu. Rev. Microbiol.* **25,** 153–176.

Braun, A. C. (1978). *Biochim. Biophys. Acta* **516,** 167–191.

Brierly, P., and Smith, F. F. (1950). *Plant Dis. Rep.* **34,** 363–370.

Brown, D. D., and Gurdon, J. B. (1977). *Proc. Natl. Acad. Sci. U.S.A.* **74,** 2064–2068.

Brunt, A. A. (1966). *Virology* **28,** 778–779.

Brunt, A. A. (1971). *Ann. Appl. Biol.* **67,** 357–368.

Brunt, A. A., and Kitajima, E. E. (1973). *Phytopathol. Z.* **76,** 265–275.

Burgess, J., Motoyoshi, F., and Fleming, E. N. (1973). *Planta* **111,** 199–208.

Cairns, E., Doerfler, W., and Schweiger, H. G. (1978). *FEBS Lett.* **96,** 295–297.

Cameron, J. R., Phillippsen, P., and Davis, R. W. (1977). *Nucleic Acids Res.* **4,** 1429–1448.

Carbon, J., Rutzkin, B., Clarke, L., and Richardson, D. (1977). *Brookhaven Symp. Biol.* **29,** 277–296.

Carlson, P. S. (1973). *Proc. Natl. Acad. Sci. U.S.A.* **70,** 598–602.

Cassells, A. C. (1978). *Nature (London)* **275,** 760.

Chaleff, R. S., and Polacco, J. (1977). *In* "Molecular Biology of Plant Cells" (H. Smith, ed)., Vol. 14, pp. 429–441. Univ. of California Press, Berkeley.

Chilton, M.-D., Drummond, M. H., Merlo, D. J., Sciaky, D., Montoya, A. L., Gordon, M. P., and Nester, E. W. (1977). *Cell* **11,** 263–271.

Clarke, L., and Carbon, J. (1978). *J. Mol. Biol.* **120,** 517–532.

Cocking, E. C. (1975). *In* "Genetic Manipulations with Plant Materials" (L. Ledoux, ed.), pp. 311–327. Plenum, New York.

Cocking, E. C. (1977). *Int. Rev. Cytol.* **48,** 323–344.

Colman, A. (1975). *Eur. J. Biochem.* **57,** 85–96.

Cortese, R., Melton, D., Tranquilla, T., and Smith, J. D. (1978). *Nucleic Acids Res.* **5,** 4593–4611.

Crawford, L., Dulbecco, R., Fried, M., Montagnier, L., and Stoker, M. (1964). *Proc. Natl. Acad. Sci. U.S.A.* **52,** 148–152.

Davey, M. R., and Cocking, E. C. (1972). *Nature (London)* **239,** 455–456.

Davey, M. R., and Power, J. B. (1975). *Plant Sci. Lett.* **5,** 269–274.

Davis, R. W., Thomas, M., Benton, D., Cameron, J., Phillippsen, P., Struhl, K., St. John, T., and Kramer, R. (1977). *In* "Molecular Cloning of Recombinant DNA" (W. A. Scott and R. Werner, eds.), pp. 155–159. Academic Press, New York.

DeRobertis, E. M., and Mertz, J. E. (1977). *Cell* **12,** 175–182.

DeRobertis, E. M., and Olson, M. V. (1979). *Nature (London)* **278,** 137–143.

Doy, C. H., Gresshoff, P. M., and Rolfe, B. G. (1973a). *Proc. Natl. Acad. Sci. U.S.A.* **70,** 723–726.

Doy, C. H., Gresshoff, P. M., and Rolfe, B. G. (1973b). *Nature (London), New Biol.* **244,** 90–91.

Drummond, M. H., Gordon, M. P., Nester, E. W., and Chilton, M.-D. (1977). *Nature (London)* **269**, 535–536.

Emmons, S. W., MacCosham, V., and Baldwin, R. I. (1975). *J. Mol. Biol.* **91**, 133–146.

Farber, F. E., Melnick, J. L., and Butel, J. S. (1975). *Biochem. Biophys. Acta* **39**, 298–311.

Fernandez, S. M., Lurquin, P. F., and Kado, C. I. (1978). *FEBS Lett.* **87**, 277–282.

Fraser, T. H., and Bruce, B. J. (1978). *Proc. Natl. Acad. Sci. U.S.A.* **75**, 5936–5940.

Frazier, N. W. (1955). *Phytopathology* **45**, 307–312.

Fujisawa, I., and Matsui, C. (1972). *Virus* **22**, 30–32.

Fujisawa, I., Rubio-Huertos, M., Matsui, C., and Yamaguchi, A. (1967). *Phytopathology* **57**, 1130–1132.

Fujisawa, I., Rubio-Huertos, M., and Matsui, C. (1972). *Phytopathology* **62**, 810–811.

Fulton, R. W. (1955). *Plant Dis. Rep.* **39**, 799–800.

Gilbert, W., Gralla, J., Majors, J., and Maxam, A. (1975). *In* "Protein Ligand Interactions" (H. Sund and G. Blauer, eds.), pp. 193–210. de Gruyter, Berlin.

Giles, K. L., and Whitehead, H. C. M. (1975). *Cytobios* **14**, 49–61.

Gillespie, D. (1968). *In* "Nucleic Acids" (L. Grossman and K. Moldave, eds.), Methods in Enzymology, Vol. 12, Part B, pp. 641–668. Academic Press, New York.

Gleba, Y. Y., Khasapov, M. M., Slyusavenko, A. G., Butenko, R. G., and Vinetskii, Y. P. (1974). *Dokl. Akad. Nauk SSSR, Ser. Biol.* **219**, 1478–1481.

Goebel, W., and Schiess, W. (1975). *Mol. Gen. Genet.* **138**, 213–223.

Goeddel, D. V., Kleid, D. G., Bolivar, F., Heyneker, H. L., Yansura, D. G., Crea, R., Hirose, T., Krasewski, A., Itakura, K., and Riggs, A. D. (1979). *Proc. Natl. Acad. Sci. U.S.A.* **76**, 106–110.

Goodman, R. M. (1977). *Virology* **83**, 171–179.

Goodman, R. M., Bird, J., and Thongmeearkom, P. (1977). *Phytopathology* **67**, 37–42.

Gradmann-Rebel, W., and Hemleben, V. (1976). *Z. Naturforsch., Teil II* **31**, 558–564.

Gronenborn, B., and Messing, J. (1978). *Nature (London)* **272**, 375–377.

Grunstein, M., and Hogness, D. (1975). *Proc. Natl. Acad. Sci. U.S.A.* **72**, 3961–3965.

Hamer, D. H., and Thomas, C. A., Jr. (1976). *Proc. Natl. Acad. Sci. U.S.A.* **73**, 1537–1541.

Hanson, R. S., and Chilton, M. D. (1975). *J. Bacteriol.* **124**, 1220–1226.

Harrison, B. D., Finch, J. T., Gibbs, A. J., Hollings, M., Shepherd, R. J., Valenta, V., and Wetter, C. (1971). *Virology* **45**, 356–363.

Harrison, B. D., Barker, H., Bock, K. R., Guthrie, E. J., Meredith, G., and Atkinson, M. (1977). *Nature (London)* **270**, 760–762.

Helinski, D. R. (1977). *In* "Genetic Engineering for Nitrogen Fixation" (A. Hollaender *et al.*, eds.), pp. 19–46. Plenum, New York.

Hershfield, V., Boyer, H. W., Yanofsky, C., Lovett, M. A., and Helinski, D. R. (1974). *Proc. Natl. Acad. Sci. U.S.A.* **71**, 3455–3459.

Hess, D. (1969). *Z. Pflanzenphysiol.* **60**, 348–358.

Hess, D. (1972). *Z. Pflanzenphysiol.* **66**, 155–166.

Hess, D. (1975). *In* "Genetic Manipulations with Plant Material" (L. Ledoux, ed.), pp. 519–537. Plenum, New York.

Hess, D. (1978). *Z. Pflanzenphysiol.* **90**, 119–132.

Hess, D., Gresshoff, P. M., Fielitz, U., and Gleiss, D. (1974). *Z. Pflanzenphysiol.* **74**, 371–376.

Hess, D., Schneider, G., Lorz, H., Blaich, G. (1976). *Z. Pflanzenphysiol.* **77**, 247–254.

Hinnen, A., Hicks, J. B., and Fink, G. R. (1978). *Proc. Natl. Acad. Sci. U.S.A.* **75**, 1929–1933.

Hoffmann-Berling, H., Durwald, H., and Beulke, I. (1963). *Z. Naturforsch., Teil B* **18**, 893–898.

Hoffmann, F. (1973). *Z. Pflanzenphysiol.* **69**, 249–261.

Hoffmann, F., and Hess, D., (1973). *Z. Pflanzenphysiol.* **69**, 81–83.

Hollings, M., and Stone, O. M. (1969). *Rep. Glasshouse Crops Res. Inst., 1968* p. 102.

Hotta, Y., and Stern, H. (1971). *In* "Informative Molecules in Biological Systems" (L. Ledoux, ed.), pp. 176–186. Am. Elsevier, New York.

Howell, S., and Hull, R. (1977). *Plant Physiol.* **59**, Suppl., p. 105. (Abstr.)

Hughes, B. G., White, F. G., and Smith, M. A. (1977). *FEBS Lett.* **79**, 80–84.

Hughes, B. G., White, F. G., and Smith, M. A. (1978a). *Z. Pflanzenphysiol.* **87**, 1–23.

Hughes, B. G., White, F. G., and Smith, M. A. (1978b). *Plant Sci. Lett.* **11**, 199–206.

Hull, R., and Howell, S. H. (1978). *Virology* **86**, 482–493.

Itakura, K., Hirose, T., Crea, R., Riggs, A. D., Heyneker, H. L., Bolivar, F., and Boyer, H. W. (1977). *Science* **198**, 1056–1063.

Johnson, C. B., and Grierson, D. (1974). *Curr. Adv. Plant. Sci.* **9**, 1–12.

Johnson, C. B., Grierson, D., and Smith, H. (1973). *Nature (London), New Biol.* **244**, 105–106.

Jones, K. W., and Murray, K. (1975). *J. Mol. Biol.* **9**, 455–460.

Kado, C. I. (1976). *Annu. Rev. Phytopathol.* **14**, 265–308.

Kado, C. I. (1977). *In* "Beltsville Symposium in Agricultural Research. I. Virology in Agriculture" (J. A. Romberger, gen. ed.), pp. 247–266. Allanheld, Osmun, Montclair, New Jersey.

Kado, C. I. (1979). *In* "Genetic Engineering: Principles and Methods" (J. K. Setlow, ed.), Vol. 1, pp. 223–239. Plenum, New York.

Kado, C. I., and Lurquin, P. F. (1975). *Biochem. Biophys. Res. Commun.* **64**, 175–183.

Kado, C. I., and Lurquin, P. F. (1976). *Physiol. Plant Pathol.* **8**, 73–82.

Kado, C. I., and Lurquin, P. F. (1978). *In* "Microbiology—1978" (D. Schlessinger, ed.), pp. 231–234. Am. Soc. Microbiol., Washington, D.C.

Kellenberger, G., Zichichi, M. L., and Weigle, J. (1961). *J. Mol. Biol.* **3**, 399–408.

Kitajima, E. W., Betti, J. A., and Costa, A. S. (1973). *J. Gen. Virol.* **20**, 117–119.

Kleinhofs, A. (1975). *In* "Genetic Manipulations with Plant Material" (L. Ledoux, ed.), pp. 461–477. Plenum, New York.

Kleinhofs, A. (1977). *In* "Molecular Genetic Modification of Eucaryotes" (I. Rubenstein, R. L. Phillips, C. E. Green, and R. J. Desnick, eds.), pp. 137–147. Academic Press, New York.

Kleinhofs, A., and Behki, R. (1977). *Annu. Rev. Genet.* **11**, 79–101.

Kleinhofs, A., Eden, F. C., Chilton, M. D., and Bendich, A. J. (1975). *Proc. Natl. Acad. Sci. U.S.A.* **72**, 2748–2752.

Kramer, R. A., Cameron, J. R., and Davis, R. W. (1976). *Cell* **8**, 227–232.

Kressman, A., Clarkson, S. G., Pirrotta, V., and Birnstiel, M. L. (1978). *Proc. Natl. Acad. Sci. U.S.A.* **75**, 1176–1180.

Kung, S. D., Gray, J. C., Wildman, S. G., and Carlson, P. S. (1975). *Science* **187**, 353–355.

Lawson, R. H., and Civerolo, E. L. (1976). *Acta Hortic.* **59**, 49–59.

Ledeboer, A. M. (1978). Ph.D. Thesis, Univ. of Leiden, Leiden.

Ledoux, L. (1965). *Nucleic Acid Res. Mol. Biol.* **4**, 231–267.

Ledoux, L., and Charles, P. (1972). *In* "Uptake of Informative Molecules by Living Cells" (L. Ledoux, ed.), pp. 29–46. North-Holland Publ., Amsterdam.

Ledoux, L., Huart, R. (1968). *Nature (London)* **218**, 1256–1259.

Ledoux, L., and Huart, R. (1969). *J. Mol. Biol.* **43**, 243–262.

Ledoux, L., and Huart, R., and Jacobs, M. (1971). *Eur. J. Biochem.* **23**, 96–108.

Ledoux, L., Huart, R., and Jacobs, M. (1972). *In* "The Way Ahead in Plant Breeding" (F. G. Lupton, G. Jenkins, and R. Johnson, eds.), pp. 165–184. Adlard & Son, Dorking, England.

Ledoux, L., Huart, R., and Jacobs, M. (1974). *Nature (London)* **249**, 17–21.

Li, S. L., and Redei, G. P. (1969). *Biochem. Genet.* **3,** 163–170.

Liebke, B., and Hess, D. (1977). *Biochem. Physiol. Pflanz.* **171,** 493–501.

Liebke, B., Blaschek, W., and Hess, D. (1977). *Z. Pflanzenphysiol.* **84,** 265–274.

Lippincott, J. A., and Lippincott, B. B. (1975). *Annu. Rev. Microbiol.* **29,** 377–405.

Lister, R. M. (1959). *Nature (London)* **183,** 1588–1589.

Lung, M. C. Y., and Pirone, T. P. (1972). *Phytopathology* **62,** 1473–1474.

Lurquin, P. F. (1977). *Prog. Nucleic Acid Res. Mol. Biol.* **20,** 161–207.

Lurquin, P. F. (1979). *Nucleic Acids Res.* **6,** 3773–3784.

Lurquin, P. F., and Hotta, Y. (1975). *Plant Sci. Lett.* **5,** 103–112.

Lurquin, P. F., and Kado, C. I. (1977a). *Mol. Gen. Genet.* **154,** 113–121.

Lurquin, P. F., and Kado, C. I. (1977b). *Arch. Int. Physiol. Biochim.* **85,** 999–1000.

Lurquin, P. F., and Kado, C. I. (1979). *Plant. Cell Environ.* **2,** 199–203.

Maitland, N. J., and McDougall, J. K. (1977). *Cell* **11,** 233–241.

Maliga, P., Sz.-Brezhovits, A., Márton, L., and Joo, F. (1975). *Nature (London)* **255,** 401–402.

Márton, L., Wullems, G. J., Molendijk, L., and Schilperoort, R. A. (1979). *Nature (London)* **277,** 129–131.

Maxam, A. M., and Gilbert, W. (1977). *Proc. Natl. Acad. Sci. U.S.A.* **74,** 560–564.

Meagher, R. B., Shepherd, R. J., and Boyer, H. W. (1977). *Virology* **80,** 362–375.

Meeks, J. C., Malmberg, R. L., and Wolk, C. P. (1978). *Planta* **139,** 55–60.

Meiners, S. P., Lawson, R. H., Smith, F. F., and Diaz, A. J. (1975). *In* "Tropical Diseases of Legumes" (J. Bird, ed.), pp. 61–69. Academic Press, New York.

Merril, C. R., Geier, M. R., and Petricciani, J. C. (1971). *Nature (London)* **233,** 398–400.

Mertz, J. E., and Gurdon, J. B. (1977). *Proc. Natl. Acad. Sci. U.S.A.* **74,** 1502–1506.

Messing, J., and Gronenborn, B. (1978). "The Single-Stranded DNA Phages" (D. T. Denhardt, D. Dressler, and D. S. Ray, eds.), pp. 449–453. Cold Spring Harbor Lab., Cold Spring Harbor, New York.

Messing, J., Gronenborn, B., Muller-Hill, B., and Hofschneider, P. H. (1977). *Proc. Natl. Acad. Sci. U.S.A.* **74,** 3642–3646.

Minson, A. C., Wildy, P., Buchan, A., and Darby, G. (1978). *Cell* **13,** 581–587.

Motoyoshi, F., and Hull, R. (1974). *J. Gen. Virol.* **24,** 89–99.

Motoyoshi, F., Bancroft, J. B., Watts, J. W., and Burgess, J. (1973). *J. Gen. Virol.* **20,** 177–193.

Mumford, D. L. (1974). *Phytopathology* **64,** 136–139.

Nagata, T., and Melchers, G. (1978). *Planta* **142,** 235–238.

Nawa, S., Yamada, M.-A., and Ohta, Y. (1975). *Jpn. J. Genet.* **50,** 341–344.

Nilan, R. A. (1964). "The Cytology and Genetics of Barley, 1951–1962." Washington State Univ. Press, Pullman.

Ohyama, K. (1978). *Plant Physiol.* **61,** 515–520.

Ohyama, K., Gamborg, O. L., and Miller, R. A. (1972). *Can. J. Bot.* **50,** 2077–2080.

Ohyama, K., Pelcher, L. E., and Horn, D. (1977). *Plant Physiol.* **60,** 98–101.

Ohyama, K., Pelcher, L. E., and Schaefer, A. (1978). "Proceedings 4th International Congress Plant Tissue and Cell Culture, 1978, Alberta, Canada" (T. A. Thorpe, ed.), pp. 75–84. Int. Assoc. Plant Tissue Culture.

Oleson, A. E., Janski, A. M., and Clark, E. T. (1974). *Biochem. Biophys. Acta* **366,** 89–100.

Osaki, T., and Inouye, T. (1978). *Nippon Shokubutsu Byori Gakkaiho* **44,** 167–178.

Ostro, M. J., Giacomoni, D., and Dray, S. (1977). *Biochem. Biophys. Res. Commun.* **76,** 836–842.

Pandey, K. K. (1975). *Nature (London)* **256,** 310–313.

Pellicer, A., Wigler, M., and Axel, R. (1978). *Cell* **14,** 133–141.

Petes, T. D., and Williamson, D. H. (1975). *Cell* **4,** 249–253.

Pirone, T. P., Pound, G. S., and Shepherd, R. J. (1961). *Phytopathology* **51**, 541–546.

Potrykus, I. (1973). *Z. Pflanzenphysiol.* **70**, 364–366.

Potrykus, I., and Hoffmann, F. (1973). *Z. Pflanzenphysiol.* **69**, 287–289.

Rambach, A., and Hogness, D. S. (1977). *Proc. Natl. Acad. Sci. U.S.A.* **74**, 5041–5045.

Ratzkin, B., and Carbon, J. (1977). *Proc. Natl. Acad. Sci. U.S.A.* **74**, 487–491.

Rebel, W., Hemleben, V., and Seyffert, W. (1973). *Z. Naturforsch., Teil B* **28**, 473–474.

Redei, G. P. (1975). *Annu. Rev. Genet.* **9**, 111–127.

Redei, G. P., Acedo, G., Weingarten, H., and Kier, L. D. (1977). *In* "Cell Genetics in Higher Plants" (D. Dudits, G. L., Farkas, and P. Maliga, eds.), pp. 91–94. Akadémiai Kiadó, Budapest.

Roberts, T. M., Kacich, R., and Ptashne, M. (1979). *Proc. Natl. Acad. Sci. U.S.A.* **76**, 760–764.

Rubio-Huertos, M., Castro, S., Fujisawa, I., and Matsui, C. (1972). *J. Gen. Virol.* **15**, 257–260.

Sager, R., and Kitchum, R. (1975). *Science* **189**, 426–433.

Sanger, F., Nicklen, S., and Cowlson, A. R. (1977). *Proc. Natl. Acad. Sci. U.S.A.* **74**, 5463–5467.

Sanzey, B., Mercereau, O., Ternynck, T., and Kowrilsky, P. (1976). *Proc. Natl. Acad. Sci. U.S.A.* **72**, 3394–3397.

Sarkar, S. (1976). *Virology* **70**, 265–273.

Sarkar, S. (1978). *Abstr., Int. Congr. Plant Pathol., 3rd, Munich* p. 54.

Sarkar, S., Upadhya, M. D., and Melchers, G. (1974). *Mol. Gen. Genet.* **135**, 1–9.

Scandella, D., and Arber, W. (1976). *Virology* **69**, 206–215.

Shepard, J. F., and Totten, R. E. (1975). *Plant Physiol.* **55**, 689–694.

Shepherd, R. J. (1976). *Adv. Virus Res.* **20**, 305–339.

Shepherd, R. J. (1979). *Annu. Rev. Plant Physiol.* **30**, 405–423.

Shepherd, R. J., and Wakeman, R. J. (1971). *Phytopathology* **61**, 188–193.

Shepherd, R. J., Wakeman, R. J., and Romanko, R. R. (1968). *Virology,* **36**, 150–152.

Shepherd, R. J., Bruening, G. E., and Wakeman, R. J. (1970). *Virology* **41**, 339–347.

Salvik, N. S., and Widholm, J. M. (1978). *Plant Physiol.* **62**, 272–275.

Smith, C. A., Vinograd, J. (1972). *J. Mol. Biol.* **69**, 163–178.

Smith, K. M. (1957). "A Textbook of Plant Virus Diseases." Little, Brown, Boston, Massachusetts.

Smith, V. A., and Hindley, J. (1978). *Nature (London)* **276**, 498–500.

Sonoki, S., and Kado, C. I. (1978). *Proc. Natl. Acad. Sci. U.S.A.* **75**, 3796–3800.

Sorsa, V. (1973). *Hereditas* **73**, 312–316.

Soyfer, V. N., Kartel, N. A., Chekalin, N. M., Titov, Y. B., Cieminis, K. K., and Turbin, N. V. (1976). *Mutat. Res.* **36**, 303–310.

Stanfield, S., and Helinski, D. R. (1976). *Cell* **9**, 333–345.

Storey, H. H. (1928). *Ann. Appl. Biol.* **15**, 1–25.

Stroun, M., Anker, P., Charles, P., and Ledoux, L. (1966). *Nature (London)* **212**, 397–398.

Struhl, K., Cameron, J. R., and Davis, R. W. (1976). *Proc. Natl. Acad. Sci. U.S.A.* **73**, 1471–1475.

Sutcliffe, J. G., and Ausubel, F. M. (1978). *In* "Genetic Engineering" (A. M. Chakrabarty, ed.), pp. 83–111. CRC Press, Cleveland, Ohio.

Suzuki, M., and Takebe, I. (1976). *Z. Pflanzenphysiol.* **78**, 421–433.

Suzuki, M., and Takebe, I. (1978). *Z. Pflanzenphysiol.* **89**, 297–311.

Szeto, W. W., Hamer, D. H., Carlson, P. S., and Thomas, C. A. (1977). *Science* **196**, 210–212.

Thomas, M., Cameron, J. R., and Davis, R. W. (1974). *Proc. Natl. Acad. Sci. U.S.A.* **71**, 4579–4583.

Trendelenburg, M. F., and Gurdon, J. B. (1978). *Nature (London)* **276**, 292–294.

Trendelenberg, M. F., Zentgraf, H., Frawke, W. W., and Gurdon, J. B. (1978). *Proc. Natl. Acad. Sci. U.S.A.* **75**, 3791–3795.

Turbin, N. V., Soyfer, V. N., Kartel, N. A., Chekalin, N. M., Dorohov, Y. L., Titov, Y. B., and Cieminis, K. K. (1975). *Mutat. Res.* **27**, 59–68.

Uchimiya, H., and Murashige, T. (1977). *Plant Physiol.* **59**, 301–308.

Villa-Komaroff, L., Efstratiadis, A., Broome, S., Lomedico, P., Tizard, R., Naber, S. P., Chick, W. L., and Gilbert, W. (1978). *Proc. Natl. Acad. Sci. U.S.A.* **75**, 3727–3731.

Volovitch, M., Drugeon, G., and Yot, P. (1978). *Nucleic Acids Res.* **5**, 2913–2925.

Walz, A., Ratzkin, B., and Carbon, J. (1978). *Proc. Natl. Acad. Sci. U.S.A.* **75**, 6172–6176.

Wardell, W. L. (1976). *Plant Physiol.* **57**, 855–861.

Wardell, W. L. (1977). *Plant Physiol.* **60**, 885–891.

Wigler, M., Silverstein, S., Lee, L.-S., Pellicer, A., Cheng, Y.-C., and Axel, R. (1977). *Cell* **11**, 223–232.

Wigler, M., Pellicer, A., Silverstein, S., and Axel, R. (1978). *Cell,* **14**, 725–731.

Wigler, M., Pellicer, A., Silverstein, S., Axel, R., Urlaub, G., and Chasin, L. (1979a). *Proc. Natl. Acad. Sci. U.S.A.* **76**, 1373–1376.

Wigler, M., Sweet, R., Sim, G. K., Wold, B., Pellicer, A., Lacy, E. Maniatis, T., Silverstein, S., and Axel, R. (1979b). *Cell* **16**, 777–785.

Wong, F. Y., and Wildman, S. G. (1972). *Biochim. Biophys. Acta* **259**, 5–12.

Chapter 13

Nitrogen Fixation and Plant Tissue Culture[1]

KENNETH L. GILES[2]

Genetics Department, Iowa State University, Ames, Iowa

INDRA K. VASIL

Botany Department, University of Florida, Gainesville, Florida

I. Introduction

Most of the articles in these volumes describe aspects of tissue culture that have both a well-founded historical background and a discrete form. The role of tissue culture in nitrogen fixation research, however, has had but a short history and diverse aims. The fixation of atmospheric nitrogen by free-living and symbiotic microorganisms is of utmost importance to world agriculture and, indeed, to world economy. There has been a phenomenal increase in nitrogen fixation research since the Arab oil embargo of 1973 and the subsequent rise in energy costs. Most fertilizer nitrogen currently is synthesized industrially from natural gas resources, but it obviously would be more advantageous if nonleguminous plants could be induced to fix atmospheric nitrogen, and if leguminous plants could have their naturally occurring symbiosis improved. The shortage of dietary

[1] Journal Paper No. J-9371 of the Iowa Agriculture and Home Economics Experiment Station, Ames, Iowa. Project No. 2267.

[2] Present address: Department of Life Sciences, Worcester Polytechnic Institute, Worcester, Massachusetts 01609.

81

protein and the high cost of nitrogenous fertilizers has stimulated interest in research in the field so as to gain an understanding of the physiology and genetic control of nitrogen fixation in both symbiotic and free-living organisms (Quispel, 1974).

II. A Historical Perspective

A truly experimental approach to understanding the physiology of nitrogen fixation has always been hampered by the unbelievable complexity of the whole plant system. The roles of translocation and hormonal control of nitrogen fixation still elude description. Raggio *et al.* (1957, 1959) made use of cultures of excised legume roots for nodulation studies. Despite the fact that nodulation did occur on some of these roots under the influence of *Rhizobium,* the system still suffered from many of the complexities of the differentiated state. Ten years elapsed before Velicky and LaRue (1967) were to report the use of soybean root cell cultures in similar experiments. Here they found aspects of cell differentiation and cell lignification, seemingly as a result of infection by *Rhizobium,* but there was no evidence of any intracellular symbiosis with the bacterium. Holsten *et al.* (1971) had more success in initiating apparently normal infection threads in cell cultures of soybeans initiated from explants of the primary roots shortly after germination. Ultrastructurally the method of infection and the final structure of the infection threads were very similar to those observed in normal root nodulation, and 10% of the callus cells were infected by the bacteria. The bacterial cells appeared to penetrate the cell wall and come to lie in the cytoplasm in membrane-bound vesicles derived from the plasmalemma of the cell. Once included in this way, they continued to multiply and divide until they virtually filled the cell cytoplasm. At this point, cell division slowed or ceased, and electron-transparent globules appeared within the cytoplasm of the now modified bacteria. The modified bacteria, called bacteroids when found in the legume nodule, were found to have deposits of poly-β-hydroxybutyric acid within their cytoplasm, a characteristic lipid of bacteroids. The "symbiosis" formed within these callus cells exhibited nitrogen fixation, as measured by acetylene reduction, at rates somewhere around 1% of the normal nitrogenase activity of the nodule. The rate of nitrogenase activity measured in this way was markedly dependent upon the exogenous hormones supplied to the callus in the medium on which it was growing. Although other workers (Child and LaRue, 1974) have confirmed the nitrogenase activity in such forced associations, they have failed to identify the bacteroids as being truly intracellular and have found no conclusive evidence for the penetration of the bacteria into the cytoplasm. Interestingly enough, the presence of leghemoglobin, the pigment responsible for the maintenance of stable oxygen tension within the nodule protecting the nitrogenase enzyme itself, has not been detected in any of these callus systems.

Two important and trend-setting reports were published early in 1975 by Child (1975) and by Scowcroft and Gibson (1975). These reports elegantly demonstrated that effective nitrogen-fixing associations with *Rhizobium* can also be established in tissue culture with nonleguminous plants: bromegrass, tobacco, rapeseed, and wheat. In these systems, bacteria were observed to grow on the surface of the callus tissues or in cracks and intercellular spaces, but no bacteria were ever detected in the intact, viable cells. It was also shown that bacteria in the vicinity of, but not in direct contact with, host tissues, either legume or nonlegume, could also fix nitrogen and show nitrogenase activity. The strains of *Rhizobium* capable of this activity were restricted to slowly growing cowpea varieties. Using these varieties, the workers conclusively showed that any host-produced factors involved in the induction of nitrogenase activity were indeed diffusable and common to both legumes and nonlegumes. Furthermore, these experiments suggest that the species barriers in nitrogen fixation lie at the stages of infection and nodule formation, rather than in the expression of nitrogenase activity itself. That both legumes and nonlegumes can initiate nitrogenase activity in free-living *Rhizobium* indicates that it may be possible to extend rhizobial symbiosis to nonleguminous plants.

Another milestone in the understanding of the biology of nitrogen fixation by rhizobia was the demonstration of nitrogenase activity in pure cultures of *Rhizobium* grown on completely defined medium (Keister, 1975; Kurtz and LaRue, 1975; McComb *et al.*, 1975, Pagan *et al.*, 1975; Tjepkema and Evans, 1975). These findings directly established that all the genes necessary for nitrogenase activity were present in the slowly growing cowpea strains of *Rhizobium*.

These findings taken together have kindled among many a vision of the ability to transfer the genes from *Rhizobium* to other bacteria, ones perhaps not so limited in their host range as is *Rhizobium*. The hope of transferring nitrogen-fixing ability to plants other than legumes at the moment remains a dream. Genetic transfer of nitrogen fixation (*nif*) genes into bacteria that do not naturally fix nitrogen, as in the case from *Klebsiella pneumoniae* to *Escherichia coli* (Cannon *et al.*, 1974), has already been reported. The possibility of transferring the *nif* operon from microorganisms that fix nitrogen directly to cereals, or to other important food plants that do not fix nitrogen, has also been discussed (Hardy and Havelka, 1975;

Shanmugam and Valentine, 1975). These authors speculated that plasmids containing the *nif* operon could be introduced into protoplasts of selected crop plants, followed by the culture of the protoplasts and their ultimate regeneration into whole plants that would carry the newly introduced genetic information in their cells as well as, it is hoped, in subsequent generations through their seeds. At this juncture, it is pertinent to point out that phage-mediated transfer of bacterial genes or operons to higher plant cells has already been reported (Doy *et al.*, 1972).

It is now well documented that strains of *Agrobacterium tumefaciens* induce the crown-gall tumors on many dicotyledonous plants (Lippincott and Lippincott, 1975; Kado, 1976). There is also ample evidence that a large plasmid, called the Ti plasmid, present in tumor-inducing strains of the bacterium determines oncogenicity, and probably at least a part of this plasmid is transferred from a bacterium to target cells during the induction of the crown gall (Gordon *et al.*, 1976; van Larabeke *et al.*, 1977). Transfer of this Ti plasmid between strains of *Agrobacterium* and indeed from *Agrobacterium* to a strain of *Rhizobium trifolii* also has been reported (van Larabeke *et al.*, 1977). In the case in which the Ti plasmid was introduced to *Rhizobium,* the bacterium gained the ability to induce tumors but retained its characteristic of forming nodules on *Trifolium pratense* (Hooykaas *et al.*, 1977). Various large plasmids have also been detected in strains of *Rhizobium* (Nuti *et al.*, 1977). It has been suggested that *nif* genes are plasmid borne in *Rhizobium* (Dunican and Tierney, 1974) and that they control the infection process of legumes. These developments have encouraged discussion of the possibility of transferring *nif* genes to the Ti plasmid of *Agrobacterium* and then using it to infect nonleguminous dicotyledonous plants in order to transfer the *nif* genes into crown-gall tumor tissue. It is sometimes possible to regenerate plants from crown-gall tumor tissues, such as in tobacco (Sacristán and Melchers, 1977). However, a large number of problems must be resolved before such schemes can come to reality (for further discussion see Chapter 12, this volume).

III. Barriers to the Transfer of Nitrogen Fixation

It is essential for the normal functioning of nitrogenase that two proteins be present. One of these is called the MoFe protein. It is a molybdenum protein containing approximately one to two molybdenum atoms and about 20 iron atoms, with an equivalent amount of acid-labile sulfur per molecule and a molecular weight (MW) of approximately 220,000. This protein is an $\alpha^2 \beta^2$ system composed of two subunits. The other protein contains only iron. It is a dimer of identical subunits containing a 4Fe-4S cubic cluster of the type found in bacterial ferredoxins. In every case studied so far, the component proteins have been shown to be quite oxygen sensitive. It is estimated that the half-life of molybdenum–iron protein in oxygenated buffer is of the order of minutes, whereas the half life of the iron protein is of the order of only seconds. In considering problems that may be encountered by nitrogenase enzymes acting in a novel plant host, the assumption is made that the properties of all nitrogenase are similar to those found in the well-studied systems, such as those of *Azotobacter, Clostridium,* or *Klebsiella.* Let us then consider some of the physiological limitations that may be placed upon the activity of nitrogenase when in a new cellular environment.

1. It is clear by casual observation of the structure of the nitrogenase enzyme that a good supply of trace elements (e.g., iron and molybdenum) is required by cells that synthesize the enzyme. It has been suggested that in *Clostridium pasteurianum* the minimal cellular concentration of iron is about 400 mM and that of molybdenum, 20 mM (Orme-Johnson, 1977). These concentrations are much higher than those present in most cells, and it is evident that appropriate transport and storage systems for these ions must be present.

2. That the reduction of one molecule of dinitrogen seems to require a minimum of 12 adenosine triphosphate (ATP) molecules (and this, according to some sources, is a very low estimate of the actual energy expenditure) first suggests that the possession of nitrogenase imposes a considerable energetic burden upon the cell. It is indisputable that this energy consumption is appreciable. In relation to the rest of the cellular economy, however, it may prove to be a somewhat small percentage of effective photosynthate, and of course against this energy expenditure must be balanced the fact that energy is normally expended by the plant on the uptake of nitrate from the soil and its subsequent reduction in the roots. Orme-Johnson (1977) quotes a calculation by Bessel Kok suggesting that a 30% increase of the photosynthate-derived energy requirement of the plant may be sufficient to sustain nitrogen fixation. These calculations are but rough estimates and can only suggest very gross energetic approximations.

3. The oxygen sensitivity of nitrogenase makes it obvious that some form of oxygen protection mechanism must exist before activity can be transferred to aerobic organisms. Apart from the leghemoglobin that is present within the peribacteroid membrane and derived from the legume plasmalemma and that has been shown to constitute a mechanism for the maintenance of low oxygen tension in the contact environment of the bac-

teroid, little is known of the oxygen protection mechanisms in most nitro-gen-fixing organisms. The solution to the problem is obvious for anaerobic nitrogen fixers, but in aerobic nitrogen fixers the question remains. Post-gate (1974) mentions that within *Azotobacter,* in which a high respiration rate had been the explanation for a low oxygen tension, the enzyme was still very oxygen sensitive and the nitrogenase worked best when the oxygen tension was reduced below that of saturation for air in water. Haacker and Veeger (1976) have suggested that one of the components of the *Azotobacter* nitrogenase complex, required for the protection of the activity from oxygen, is one of the *Azotobacter* ferredoxins. This, of course, means that it may be necessary not only to transfer the *nif* genes to a new host, but also to transfer with it the gene for an enzyme or protein capable of binding to the nitrogenase complex conferring protection against oxygen inactivation. Unless some specific protein factor can be uncovered with certainty, this oxygen inactivation of nitrogenase seems likely to be the most severe limitation on the ability to introduce and express the *nif* genes in novel aerobic organisms.

4. It has been shown by Bergerson and Turner (1973) that *in vivo* nitrogenase activity is limited by the ratio of adenosine diphosphate (ADP) and ATP concentrations. It seems not only that a sufficient flux of ATP has to be maintained for the functioning of nitrogenase but also that nitrogenase itself is sensitive to inhibition by adenosine diphosphate (ADP). This raises another problem: Although the relative requirements for energy by the nitrogen-fixing system may not be high, the catalytic capacity of the ATP-generating system has to be relatively much higher than normal to maintain a high ATP/ADP ratio.

Orme-Johnson (1977) also raises the problems of the redox potential environment of the nitrogenase. He points out that although the number of electrons per second required of the nitrogenase system in excess of normal metabolism of the cell may not be large, it may indeed be rate limiting. A thermodynamic requirement imposed by nitrogenase is that a relatively negative redox potential has to be maintained in the cell. This factor requires considerable metabolic adaptation if the enzyme is to function effectively in its new environment. Orme-Johnson also points out that the component ratio of the nitrogenase subunit enzymes has to be controlled to give optimum activity. There is general agreement that the activity of nitrogenase peaks at a high level when the iron protein is present in excess; therefore, some consideration must be given to the transfer of multiple copies of the iron protein to maintain such a ratio and balance within the cell.

The problem then resolves itself to one of not simply transferring genetic information from a prokaryote to a eukaryotic cell, but of transforming the cellular environment of the eukaryotic cell in such a way as to permit the expression of that information. So complex are the changes required that the question arises as to whether such changes can meaningfully be introduced within the organism. The answer to this question is most positively yes, for already there exist plant cells whose environment is adequately transformed for nitrogen fixation to be expressed: the nodule cells of legumes. Whether it will prove possible to transform other plant cells in a similar fashion to accommodate the prokaryotic information and its expression remains to be seen. The legume symbiosis is very specific and complex; the subtle changes required may not easily or rapidly be engineered into nonleguminous plants.

IV. A Review of Work to Date

To date, most of the work done using tissue culture in the field of nitrogen fixation has been aimed at the transfer of fixation ability to nonlegumes rather than at deepening current understanding of the nitrogen fixation processes themselves. Although some of the techniques used have been extremely novel, others have tried to mimic natural events under *in vitro* conditions.

A. ASSOCIATIVE SYMBIOSES

Symbioses of several tropical grasses and cereal plant species with the bacterium *Azospirillum lipoferum* resulting in nitrogen fixation and increased productivity have been reported (Day *et al.*, 1975; von Bulow and Dobereiner, 1975). Some exciting studies also have been reported concerning the establishment and study of the biology of this symbiotic association with tissue cultures of higher plants (Vasil *et al.*, 1977, 1979a,b; Berg *et al.*, 1979, 1980). Associated cultures of *Azospirillum brasilense* and sugar cane callus tissues have been maintained for over 18 months on nutrient media containing very low levels of added nitrogen or completely devoid of nitrogen. The bacteria have been shown to grow on the surface (Figs. 1 and 2) and in the intercellular spaces (Fig. 3) of the callus tissue and do not invade living sugar cane cells. Such associated cultures show appreciable rates of acetylene reduction activity on media markedly deficient in combined nitrogen. The results of these experiments also suggest that bacteria grown in association with sugar cane callus tissue have very high rates of acetylene reduction activity as compared to the control bacterial cultures. A significant proportion of the

FIG. 1.   Scanning electron micrograph showing the growth of *Azospirillum brasilense* on the surface of sugar cane callus tissue. (From Berg *et al.*, 1979.)

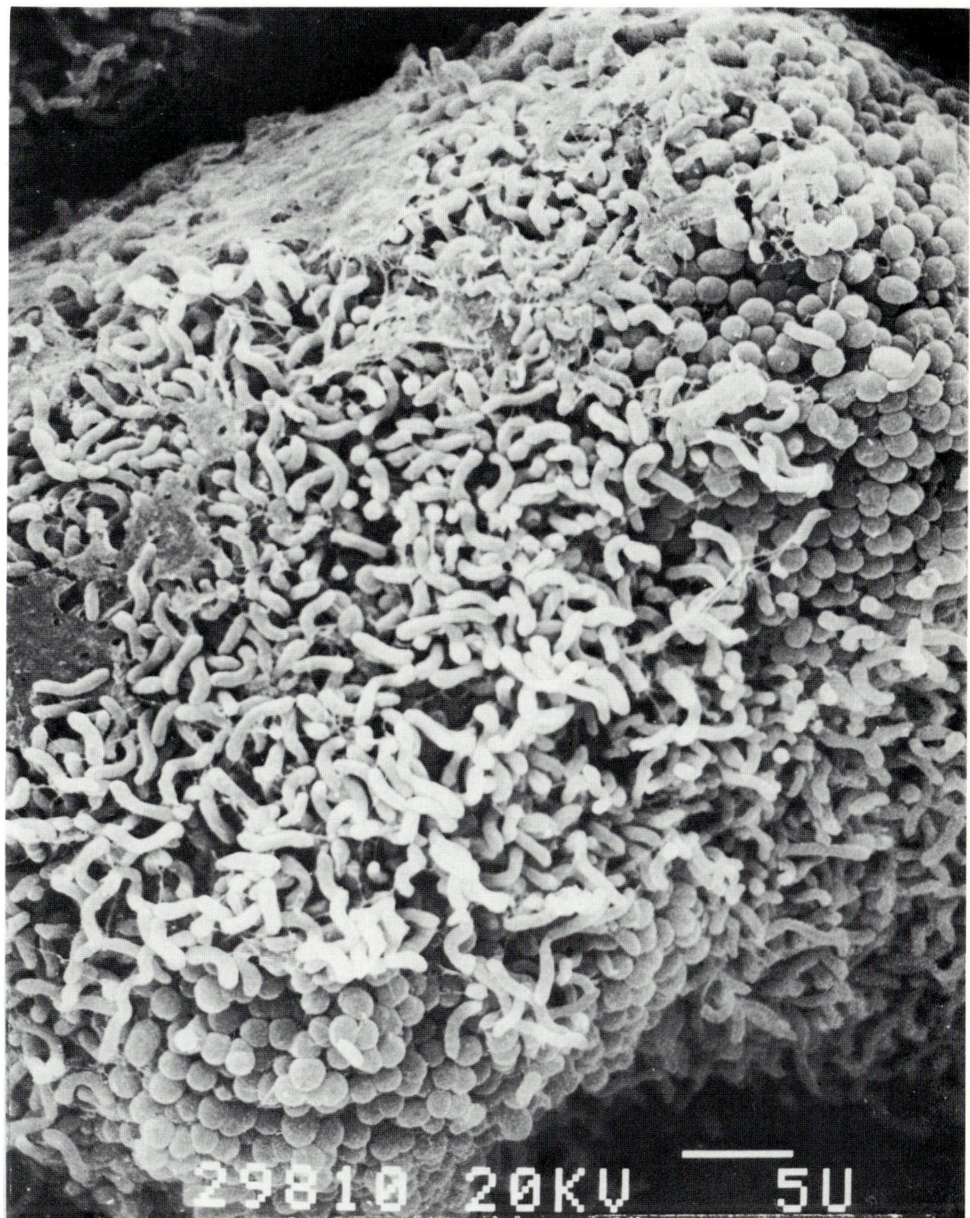

FIG. 2. Scanning electron micrograph showing the wild-type as well as the pleomorphic (spherical) forms of *Azospirillum brasilense* growing on the surface of sugar cane callus tissue *in vitro*. (From Berg *et al.*, 1979.)

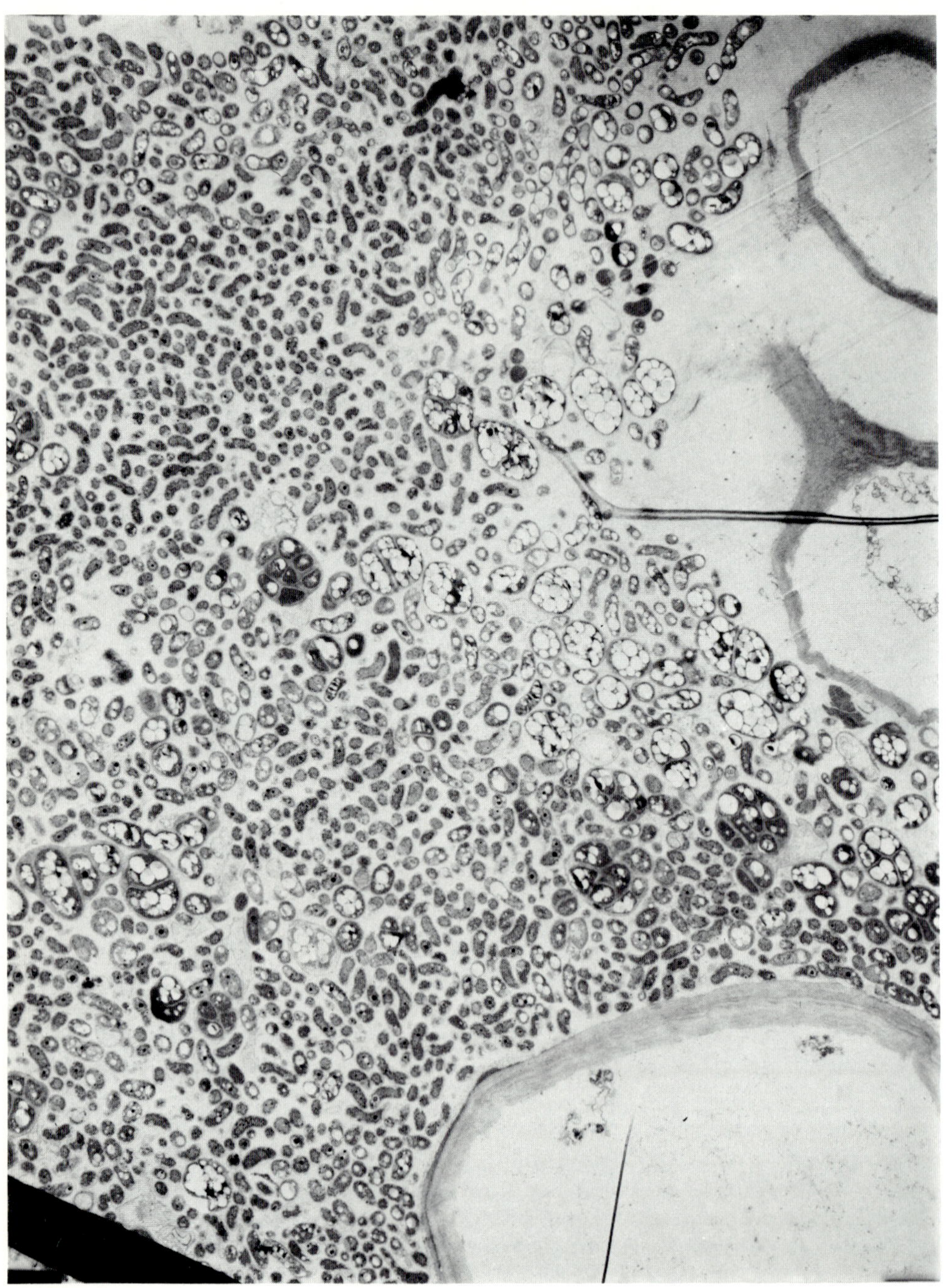

FIG. 3.   Transmission electron micrograph showing the growth of *Azospirillum brasilense* in the intercellular spaces of sugar cane callus tissue *in vitro*. Note the presence of some multicellular, encapsulated pleomorphic forms. (From Berg *et al.*, 1979.)

bacterial population in the associated cultures is pleomorphic and encapsulated, suggesting a possible mechanism for protection against oxygen. At present there is no evidence that these associations are truly symbiotic in nature. Nevertheless, such associative symbioses hold promise because of the much wider host range of *Azospirillum* than of *Rhizobium*. In the field, however, these associations or symbioses have proved erratic and difficult to manipulate, and it remains to be seen whether these will remain tight under somewhat less stringent pressures than those applied *in vitro*.

A rather similar extracellular forced symbiotic association was reported by Carlson and Chaleff (1974) between *Azotobacter vinelandii* and carrot tissue cultures. The bacterium used in this experiment does not normally require a plant host to fix nitrogen, and the precise physiological status of this particular symbiosis is questionable.

B. Uptake of Whole Microorganisms by Higher Plant Protoplasts

Uptake of whole rhizobia by leaf protoplasts was reported by Davey and Cocking (1972), but unfortunately, the experimental conditions described in these experiments were such that the survival of any viable nitrogen-fixing bacteria after uptake was extremely unlikely.

One of the problems associated with the transfer of nitrogen fixation, as described in Section III, is the high energy consumption of the system. In an attempt to circumvent this problem, Burgoon and Bottino (1976) transferred cells of the photosynthetic blue-green alga *Gloeocapsa* into the protoplasts of tobacco and corn. Since that time, other workers (Hughes *et al.*, 1978) have used the cyanelles in transfer experiments. In almost all instances, the included microorganism has become surrounded by an extra membrane, presumed to be of plasmalemmal origin. However, with the included cyanelles no such extra membrane has been found, suggesting either that they are more physiologically suited to such uptake experiments or that their own outer membranes have been damaged and fused during the uptake processes. Unfortunately, there seem to be no good data describing the activity, either photosynthetic or nitrogen fixing, of these included blue-green algae. Indeed, in most uptake experiments, the metabolic activity of both included organelles and microorganisms has not been adequately documented.

One case in which the acetylene reduction activity of intracellular bacteria has been reported is that for *A. vinelandii* in *Zea mays* protoplasts (Giles, 1978). Although corn protoplast culture has proved difficult, a way to get around the problem has been found in this instance, allowing uptake of bacteria into viable protoplasts capable of repeated divisions. Immediately after fertilization, the endosperm of corn exists as a free-nuclear syncytium. If *Azotobacter* cells are injected into this syncytium, bacteria are included in the cells of the developing endosperm at the time of cellularization. This tissue can be cultured as callus and, if bacteria are included, grown on nitrogen-deficient medium. Acetylene reduction activity has been assayed in association with this tissue at about 90 nM ethylene per milligram callus tissue per hour. This rate is variable and decreases to nearly zero 21 days after excision of the developing endosperm from the kernel. When the developing embryo was left associated with the endosperm, acetylene reduction activity continued at a high rate until the browning and withering of the endosperm some 6 weeks after initiation of the culture, and the embryo was supported in its development to a plantlet 15–20 cm tall even on completely nitrogen-deficient medium.

C. PROTOPLAST FUSION STUDIES

It has been possible to fuse plant protoplasts from diverse origin since 1970 (Power and Cocking, 1970), and methods now exist (Kao and Michayluk, 1974) that give rise to reproducibly high percentages of protoplast fusion. Several authors have described the successful fusion of protoplasts of legumes with protoplasts from nonlegumes (Kao *et al.*, 1974; Kartha *et al.*, 1974). Although it has been generally assumed that hybrid cells resulting from such fusions will be able to fix nitrogen as well as possibly maintain other important characteristics, of both economic and nutritional value, there is little or no evidence at present that desirable hybrid plants can indeed be regenerated from the fusion products of such widely diverse protoplasts (Vasil, 1976; Vasil *et al.*, 1979a). Kao *et al.* (1974) have described limited divisions in heterokaryons formed between barley and soybean and between wheat and soybean (but in no case have the cell divisions been sustained over an indefinite period.) Because of the often low uptake percentages recorded for microorganisms uptake into protoplasts (Vasil *et al.*, 1975, 1977; Vasil, 1976), isolated protoplasts from the nitrogen-fixing root nodules of *Lupinus angustifolius* were used by these authors and fused with mesophyll protoplasts isolated from the leaves of *Nicotiana tabacum* with the help of polyethylene glycol. The nodule protoplasts contained modified rhizobia, and the bacteroids contained the nitrogen-fixing form of the bacterium. Although the nodule protoplasts had some contaminating cell wall material still attached to them, sufficient plasmalemma was exposed that fusion between protoplasts could still occur. Fusion products of the nodule and leaf protoplasts remained viable for several days in culture, and ultrastructural examination after 2 or 6 days in culture showed no evidence of cytoplasmic or organelle disintegration. [M. R. Davey (personal communication) has lately shown the mixing of cytoplasms following the fusion of *Petunia* mesophyll protoplasts with root nodule protoplasts of *Vicia* (see Fig. 4).] No cell wall regeneration or visible signs of growth were observed in the fusion products. Although Vasil *et al.* (1975) observed acetylene reduction activity in the root nodule slices during the entire period of incubation in cell wall-degrading enzymes, they were unable to demonstrate such activity in the isolated nodule protoplasts immediately after isolation, whereas Broughton *et al.* (1976) showed that acetylene reduction activity reappeared only after a lag period of 2 or 3 days. The use of nodule protoplasts in fusion experiments with nonlegume protoplasts offers several advantages (Vasil, 1976). It overcomes the difficulty of the infection barrier, a phase of the growth of rhizobia still not clearly understood, and it also allows the continuous protection of the delicate bacteroids and their nitrogenase system within the membranes of the original legume host cell.

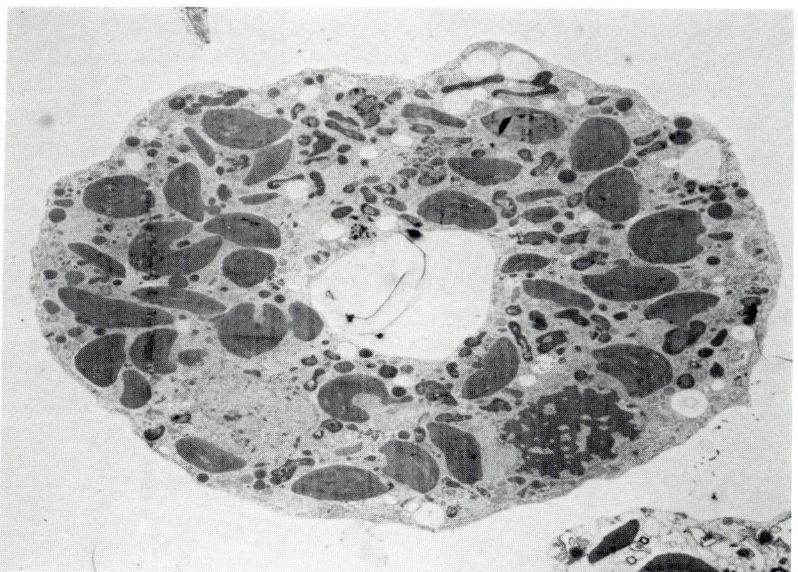

FIG. 4. *Petunia* mesophyll protoplast/*Vicia* root nodule protoplast heterokaryon 24 hours after PEG fusion. *Rhizobium* bacteroids are distributed throughout the heterokaryon following extensive mixing of the two cytoplasms. ×2640. (Photograph courtesy of Dr. M. R. Davey.)

D. THE UPTAKE OF WHOLE MICROORGANISMS INTO LOWER EUKARYOTE PROTOPLASTS

A series of experiments by Giles and Whitehead (1975, 1976, 1977a,b) reports the successful transfer of nitrogen-fixing ability to eukaryotic cells. They introduced the cells of the nitrogen-fixing bacterium *Azotobacter vinelandii,* with the aid of polyethylene glycol, into enzymatically produced protoplasts of the fungus *Rhizopogon.* This fungus forms a mycorrhizal association with the roots of *Pinus radiata.* Several strains of the fungus were regenerated from the treated protoplasts and were shown not only to grow on nitrogen-deficient medium but also to reduce acetylene in the acetylene reduction assay for nitrogenase, and they also gave positive results in the nitrogen-15 assay for nitrogen fixation. It was at first thought that the bacterial genetic information responsible for these activities had been housed in modified mitochondria, but subsequent fluorescent antibody experiments showed that L forms of the bacteria were actually present in the modified strains of the fungus and that it was these that were responsible for its ability to fix atmospheric nitrogen. All the strains isolated showed a regular peak of acetylene reduction activity associated with their subculture to fresh medium. The period of acetylene reduction activity lasted between 6 and 8 days from the period of transfer and could be repeated for over a period of 18 months. Initially, before the mechanism of control of the bacterial genetic information was understood, the authors called this an example of transgenosis, a term coined by Doy *et al.* (1973) for the reported transfer of bacterial genetic information to higher plant cells.

The most important aspect of this work was that, because *Rhizopogon* forms a mycorrhizal association with the roots of *Pinus radiata,* when the modified nitrogen-fixing strains were reassociated with the roots of seedlings of this species, a nitrogen-fixing symbiosis was set up between the fungus and the higher plant. This gave rise to increased nitrogen levels in the nitrogen-deprived seedlings and to more vigorous growth and dry-matter accumulation over the 6-month period of the experiments. Although there were some deaths among seedlings associated with a modified form of the fungus, the relevance of these deaths was greatly exaggerated both by the press and by certain factions of the scientific population. They were interpreted as implying that genetically engineered organisms would inevitably cause death and disease in the environment. Later controlled experiments showed that under aseptic conditions *Rhizopogon* itself, without any genetic modification, could cause the death of very young sterile seedlings, and it is now thought that the death of the seedlings in the treated experiments resulted from premature reassociation with seedlings younger than 3 months old. To date, this remains the most successful example of transfer of nitrogen-fixing ability to species other than legumes. It should be pointed out that most crops of agronomic value reassociate their roots with endomycorrhizal fungi that are extremely difficult to manipulate in culture and, hence, are presently beyond the scope of this technique.

It is perhaps disappointing that so much work has been directed toward the use of tissue culture techniques to study the transfer of nitrogen-fixing ability, whereas so little use seems to have been made of the technique in the study of legume nitrogen fixation. Raggio *et al.* (1959) believed that tissue culture could be used to simplify the complexities of the whole plant system, and such an idea is still realistic today. Very little is known of the physiology and genetics of the initial infection process at the legume root hair, and because there have been reports, although not always repeatable, of successful infection thread formation in tissue culture, this process would bear investigation again at the tissue culture level in the hope of obtaining a simplified system in which infection thread formation could be studied. The current trend toward the acceptance of the role of lectins in recognition, infection, and nodulation could easily be investigated in such a simplified system. The growth and culture of whole excised roots on agar might allow for the isolation of diffusable substances escaping from the roots, which might have some physiological or morphological effect upon vegetative rhizobial cells. A study of the formation of "swarmers" would be particularly amenable to such a system of study.

Some of the early experiments (Child, 1975; Scowcroft and Gibson, 1975) indicate that the genetic information for the expression of the nitrogenase genes is encoded in the rhizobia in the slow-growing cowpea strains and that the expression of the genetic information in these strains is dependent upon a diffusable factor, or factors, produced by both legume and nonlegume tissues. That nitrogenase activity can be experimentally induced in rhizobia associated with nonlegume tissues raises the possibility that a functional nitrogen-fixing association could be developed between rhizobia and nonlegume species, provided that recognition barriers could be overcome. It would seem that further work with separate systems and the culture of nitrogen-fixing *Rhizobium* in an appropriate nutrient medium could provide valuable information about the factors responsible for the induction of nitrogenase in the bacteria.

V. Future Developments

This is the age of recombinant DNA technology, and it is inevitable that there be great interest in using this technique to transfer and investigate the genetic material responsible for nitrogen fixation into higher plant cells. It would seem, however, that some workers run into a fundamental error in attempts to transfer this information to higher plants. It has long been maintained that suspension cell cultures of higher plants can be treated, and thought of, much in the same way as cultures of microorganisms. This, in fact, is untrue. Because of the necessity of inserting molecular DNA into individual cells, and of selecting out those cells that have taken up, transcribed, translated, and expressed that DNA faithfully, one is forced to use cultures at the single-cell level of plant organization. The evidence that development in eukaryotic, multicellular organisms is at least in part, if not wholly, epigenetic suggests that the expression of genes at one level of development does not necessarily reflect their expression at a later stage of development. The whole concept of development is, after all, that different genes are switched on and off at different levels of development. That a single cell capable of expressing inserted genetic information for nitrogen fixation can be selected does not necessarily mean that the genetic information will be expressed in the same way once the cell is grown into callus and differentiated into a shoot or a whole plant. Isoenzyme patterns have been shown to change during the regeneration of plants from 5-methyltryptophan-resistant tobacco and potato cultures (Widholm, 1978; Carlson and Widholm, 1978).

There are those who believe that the techniques and concepts of genetic engineering in *Escherichia coli* cells suggest the possibility of developing a molecular cloning system for higher organisms. They suggest that it should be possible to use the replicative properties of a DNA virus, such as cauliflower mosaic virus, to construct a molecular cloning vehicle for cloning in the host organism of the virus. Synthetic A-T inserted into the *Eco*RI endonuclease site of simian virus 40 (SV40) DNA has been replicated in monkey culture cells. It has been suggested that cauliflower mosaic virus (CaMV) can be adapted as a molecular cloning vehicle for higher plant cells, especially those in the Cruciferae. There is a very big difference, however, between introducing genetic information into a cell and allowing it to replicate, and getting it to be expressed correctly and functionally. Very few workers seem to have addressed themselves to this problem. It does indeed seem clear that, in many cases, higher eukaryotic DNA contained in recombinant vehicles is not correctly expressed in *E. coli*. Fragments of histone genes, mouse mitochondria DNA, adenovirus DNA, and *Drosophila melanogaster* DNA seem to be artifactually expressed or not expressed at all when contained in recombinant vehicles in *E. coli* (Kedes *et al.*, 1975; Chang *et al.*, 1975; Tiollais *et al.*, 1977; Meagher *et al.*, 1977). Fareed and Upcroft (1977) have reported the use of defective SV40 replicons as efficient vectors for the propagation and cloning of prokaryotic DNA segments in mammalian cells. The present lack of understanding concerning the controls of expression of prokaryotic and eukaryotic DNA in foreign cytoplasms and the constraints placed upon our ability to select those cell lines in which expression is occurring, constraints placed upon the system by the epigenetic control of multicellular development, severely limit our ability in this field.

On a more positive note, however, it should not be overlooked that small circular DNA molecules, not unlike bacterial plasmids, have been reported in eukaryote cells, such as yeast (Billheimer and Avers, 1969; Petes and Williamson, 1975), *Neurospora* (Agsteribbe *et al.*, 1972), and tobacco (Wong and Wildman, 1972). It is therefore possible that these molecules may be used in the future to carry and clone information within eukaryotic cells themselves. Of course, these systems would also suffer from the problems of uptake into the plant cell or protoplast and, possibly also, the epigenetic constraints placed upon the tissue by differentiation. It is not untimely, however, considering the implications of recombinant DNA technology, to consider some of these problems in the light of their possible future application in the area of nitrogen fixation and tissue culture.

REFERENCES

Agsteribbe, E., Kroos, A. M., and van Bruggen, E. F. J. (1972). *Biochem. Biophys. Acta* **269**, 299–303.

Berg, R. H., Vasil, V., and Vasil, I. K. (1979). *Protoplasma* **101**, 143–163.

Berg, R. H., Tyler, M., Novick, N., Vasil, V., and Vasil, I. K. (1980). *Environ. Appl. Microbiol.* **39**, 642–649.

Bergerson, F. J., and Turner, G. L. (1973). *Biochem. J.* **131**, 61.

Billheimer, F. E., and Avers, C. J. (1969). *Proc. Natl. Acad. Sci. U.S.A.* **64**, 739–746.

Broughton, W. J., Wooi, K. C., and Holi, C. H. (1976). *Nature (London)* **262**, 208.

Burgoon, A. C., and Bottino, P. J. (1976). *J. Hered.* **67**, 223–226.

Cannon, F. C., Dixon, R. A., Postgate, J. R., and Primrose, S. B. (1974). *J. Gen. Microbiol.* **80**, 227–239.

Carlson, J. E., and Widholm, J. M. (1978). *Abstr., Int. Congr. Plant Tissue Cell Cult., 4th, Calgary* No. 1610, p. 140.

Carlson, P. S., and Chaleff, R. S. (1974). *Nature (London)* **252**, 393–394.

Chang, A. C. Y., Lansman, R. A., Clayton, D. A., and Cohen, S. N. (1975). *Cell* **6**, 231.

Child, J. J. (1975). *Nature (London)* **253**, 350–351.

Child, J. J., and LaRue, T. A. (1974). *Plant Physiol.* **53**, 88–90.

Davey, M. R., and Cocking, E. C. (1972). *Nature (London)* **239**, 455–456.

Day, J. M., Neves, M. C. P., and Dobereiner, J. (1975). *Soil Biol. Biochem.* **7**, 107–112.

Doy, C. H., Gresshoff, P. M., and Rolfe, B. G. (1972). *Proc. Aust. Biochem. Soc.* **5**, 3.

Doy, C. H., Gresshoff, P. M., and Rolfe, B. G. (1973). *Proc. Natl. Acad. Sci. U.S.A.* **70**, 723–726.

Dunican, L. K., and Tierney, A. B. (1974). *Mol. Gen. Genet.* **126**, 187–190.

Fareed, G. C., and Upcroft, P. (1977). *In* "Genetic Manipulation as it Affects the Cancer Problem" (J. Schultz and Z. Brada, eds.), Miami Winter Symposia, No. 14, p. 49. Academic Press, New York.

Giles, K. L. (1978). *In* "Frontiers of Plant Tissue Culture" (T. A. Thorpe, ed.), pp. 67–73. Univ. of Calgary Press, Calgary, Alberta.

Giles, K. L., and Whitehead, H. C. M. (1975). *Cytobios* **14,** 49–61.

Giles, K. L., and Whitehead, H. C. M. (1976). *Science* **193,** 1125.

Giles, K. L., and Whitehead, H. C. M (1977a). *Plant Soil* **48,** 143–152.

Giles, K. L., and Whitehead, H. C. M. (1977b). *Plant Sci. Lett.* **10,** 367–372.

Gordon, M. P., Farrand, S. K., Sciaky, D., Montoya, A., Chilton, M. D., Menlo, D., and Nester, E. W. (1976). *In* "Molecular Biology of Plants" (I. Rubinstein, ed.). Academic Press, New York.

Haacker, H., and Veeger, C. (1976). *Eur. J. Biochem.* **63,** 499–507.

Hardy, R. W. F., and Havelka, U. D.(1975). *Science* **188,** 633–634.

Holsten, R. D., Burns, R. C., Hardy, R. W. F., and Hebert, R. R. (1971). *Nature (London)* **232,** 173–176.

Hooykaas, P. J., Klapwijk, P. M., Nuti, M. P., Schilperoort, R. A., and Rorsch, A. (1977). *J. Gen. Microbiol.* **98,** 477–484.

Hughes, B. G., White, F. G., Vernon, L. P., and Smith, M. A. (1978). *Abstr., Annu. Meet. Am. Soc. Plant Physiol., Blacksburg, Va.* No. 300.

Kado, C. I. (1976). *Annu. Rev. Phytopathol.* **14,** 265–308.

Kao, K. N., and Michayluk, M. R. (1974). *Planta* **115,** 355–367.

Kao, K. N., Constabel, F., Michayluk, M. R., and Gamborg, O. L. (1974). *Planta* **120,** 215–227.

Kartha, K. K., Gamborg, O. L., Constabel, F., and Kao, K. N. (1974). *Can. J. Bot.* **52,** 2435–2436.

Kedes, L. H., Chang, A. C. Y., Houseman, D., and Cohen, S. N. (1975). *Nature (London)* **255,** 533.

Keister, D. L. (1975). *J. Bacteriol.* **123,** 1265–1268.

Kurtz, W. G. W., and LaRue, T. A. (1975). *Nature (London)* **256,** 407–409.

Lippincott, B. B., and Lippincott, J. A. (1975). *Plant Physiol.* **56,** 213–215.

McComb, J. A., Elliot, J., and Dilworth, M. J. (1975). *Nature (London)* **256,** 409–410.

Meagher, R. B., Shepherd, R. J., and Boyer, H. W. (1977). *Virology* **80,** 362–375.

Nuti, M. P., Ledeboer, A. M., Lepidi, A. A., and Schilperoort, R. A. (1977). *J. Gen. Microbiol.* **100,** 241–248.

Orme-Johnson, W. H. (1977). *In* "Genetic Engineering for Nitrogen Fixation" (A. Hollaender *et al.,* ed.), pp. 317–332. Plenum, New York.

Pagan, J. D., Child, J. J., Scowcroft, W. R., and Gibson, A. H. (1975). *Nature (London)* **256,** 406–407.

Petes, T. D., and Williamson, D. H. (1975). *Cell* **4,** 249–253.

Postgate, J., (1974). *In* "The Biology of Nitrogen Fixation," Frontiers of Biology, Vol. 33. North-Holland Publ., Amsterdam.

Power, J. B., and Cocking, E. C. (1970). *Nature (London)* **225,** 1016–1018.

Quispel, A. (1974). *In* "The Biology of Nitrogen Fixation" (Quispel, ed.), Frontiers of Biology, Vol. 33. North-Holland Publ., Amsterdam.

Raggio, N., Raggio, M., and Torrey, J. G. (1957). *Am. J. Bot.* **44,** 325–334.

Raggio, N., Raggio, M., and Burris, R. H. (1959). *Biochim. Biophys. Acta* **32,** 274–275.

Sacristán, M. D., and Melchers, G. (1977). *Mol. Gen. Genet.* **152,** 111–118.

Scowcroft, W. R., and Gibson, A. H. (1975). *Nature (London)* **253,** 351–352.

Shanmugam, K. T., and Valentine, R. L. (1975). *Proc. Natl. Acad. Sci. U.S.A.* **72**, 136–139.

Tiollais, P., Perricaudet, M., Pettersson, U., and Philipson, L. (1977). *Gene* **1**, 49.

Tjepkema, J., and Evans, H. J. (1975). *Biochem. Biophys. Res. Commun.* **65**, 625–628.

van Larabeke, N., Genetello, C., Harralsteens, J. P., De Picker, A., Zaenen, I., Messens, E., van Montagu, N., and Schell, J. (1977). *Mol. Gen. Genet.* **152**, 119–124.

Vasil, I. K. (1976). *Adv. Agon.* **28**, 119–160.

Vasil, I. K., Vasil, V., Sutton, W. D., and Giles, K. L. (1975). *Proc. Int. Symp. Yeasts Other Protoplasts, Univ. Nottingham* p. 81.

Vasil, I. K., Vasil, V., and Hubbell, D. H. (1977). *In* "Genetic Engineering for Nitrogen Fixation" (A. Hollaender *et al.*, eds.), pp. 197–211. Plenum, New York.

Vasil, I. K., Vasil, V., White, D. W. R., and Berg, R. H. (1979a). *In* "Plant Growth Regulation and World Agriculture" (T. K. Scott, ed.). Plenum, New York.

Vasil, V., Vasil, I. K., Zuberer, D. A., and Hubbell, D. H. (1979b). *Z. Pflanzenphysiol.* **95**, 141–147.

Velicky, I., and LaRue, T. A. (1967). *Naturwissenschaften* **54**, 96–97.

von Bulow, J. F. W., and Dobereiner, J. (1975). *Proc. Natl. Acad. Sci. U.S.A.* **72**, 2389–2393.

Widholm, J. M. (1978). *Abstr. Int. Congr. Plant Tissue Cell Cult., 4th, Calgary* No. 1609, p. 139.

Wong, F. Y., and Wildman, S. G. (1972). *Biochim. Biophys. Acta* **259**, 5.

Chapter 14

Preservation of Germplasm

LYNDSEY A. WITHERS

Friedrich Miescher Institut, Basel, Switzerland

I. Introduction

World germplasm resources are subject to a range of threats deriving from natural selection, the destruction of habitats, and the often merely expedient agricultural practices of mankind. A farsighted approach is required to store for later exploitation the vast pool of genetic variability presently available. It is not possible to define, except in the broadest terms, the requirements of future plant breeding programs, and we should therefore, to quote Hooker (1977), adopt a policy in which "Germplasm diversity affords some protection against the unknown." In recognition of this need, active programs are under way for the conservation of seed and vegetative organs (Frankel, 1970; Frankel and Hawkes, 1975). Through meristem propagation, plant tissue culture aids such programs by offering a means of eradicating pathogens and of maintaining and cloning valuable genome stocks (Morel, 1975; Murashige, 1978; Walkey, 1978; Wang, 1977; Westcott *et al.*, 1977).

101

Meristem culture is just one of a number of tissue culture methods for *in vitro* propagation of a specific genome. However, although the opportunity for the introduction of genetic variability through sexual reproduction is eliminated, maintenance of somatic tissues in active growth can result in progressive genetic changes involving changes in ploidy, chromosome and gene mutations, and resultant alterations in the phenotype of the culture (D'Amato, 1978; see also Withers, 1980). Maintenance even under the strictest subculturing routines cannot guarantee long-term genetic uniformity. Therefore, where meristem culture is either unavailable or inappropriate, there is a serious risk of losing valuable *in vitro* genomes. The limitation of growth by manipulation of the culture conditions has been attempted with some success. A discussion of this approach to tissue culture genome conservation is given in Section V. Whereas such investigations should be encouraged, it is undeniably desirable that a method such as ultra-low-temperature storage be developed to entirely suspend metabolic activity, thereby effectively suspending time. For many workers the attraction of such a storage method is not so much the prevention of genetic change but the virtual elimination of the weightly demands of culture maintenance. The proportion of a laboratory's resources in terms of finance and personnel and equipment time thus released for experimentation must offer enormous scope for furthering the study of plant tissue culture and its applications. The risk of losing valuable culture stocks through contamination, equipment failure, and human error also would be greatly reduced.

From the above it can be seen that ultra-low-temperature storage— *freeze preservation*—of plant tissue cultures has a valuable role to play both in the conservation of stocks for pure and applied *in vitro* studies and in serving the wider concern of world germplasm conservation by providing a simple, stable, and economic means of storing and distributing unique genomes.

II. Freeze Preservation—The Background

Freeze preservation at ultra-low temperatures (normally at or near −196°C, the temperature of liquid nitrogen, or LN, is a valuable aid to genome storage in microbiology, medicine, and animal husbandry (Ashwood-Smith and Farrant, 1980; Meryman, 1966). The freeze preservation of plant material is a younger and less widely exploited field, but specimens covering virtually the entire phylogenetic spectra of higher and lower plants can be committed to freeze storage (Gresshoff, 1977; Morris, 1980; Takeuchi, 1978; Withers, 1980). The successful freeze preservation of cultured plant cells was first reported by Quatrano (1968). Since that time considerable progress has been made. Several reviews have been published in recent years indicating, with practical details, the range of specimens that can be so preserved (Bajaj and Reinert, 1977; Sakai and Sugawara, 1978; Withers, 1978c, 1980; Withers and Street, 1977b). This information will not exhaustively be reiterated here, but an overview of the level of expertise in the freeze preservation of higher plant tissue cultures will be presented, indicating the areas in which difficulties are encountered and in which further study is required.

Many organisms are able to tolerate chilling and exposure to temperatures that would induce freezing of the extracellular milieu, but tolerance of intracellular freezing in nature is highly exceptional, particularly in plants (Burke *et al.*, 1976; Levitt, 1978). In wishing to store plant material at ultra-low temperatures, therefore, we are imposing upon it abnormal stresses. Very few examples of successful freeze preservation have involved simply a transfer to and from a low temperature. It is normally necessary to carry out specific operations prior and subsequent to freezing and thawing at appropriate rates. A number of pretreatments and aspects of postthaw handling, relating to subculturing routines and manipulation of the tissue's environment, are described in Section III,B; more general points are considered here.

A. CRYOPROTECTANTS

One of the most common pretreatments applied to specimens shortly before freezing is *chemical cryoprotection*. A large number of miscellaneous compounds have cryoprotectant properties but the most commonly used are dimethylsulfoxide (DMSO) and glycerol (Meryman, 1966). Findings in a number of laboratories support the use of mixtures of cryoprotectants. Of course, where cryoprotectants are made up in the culture medium, the constituents of that medium are all contributing to the net effect, if only as osmotica. Finkle and Ulrich (1978, 1979) have successfully used mixtures of glucose and DMSO and of polyethylene glycol, glucose, and DMSO for the preservation of cultures of *Saccharum* sp. (see also Sugawara and Sakai, 1974). Another generally successfully employed mixture includes DMSO and glycerol (see Withers and Street, 1977b). Mixtures incorporating the novel cryoprotectant proline have been adopted in this laboratory for the preservation of suspension culture cells of a wide range of species, including *Zea mays, Rosa* (Paul's Scarlet), *Acer pseudoplatanus,* and *Daucus carota* (Withers and King, 1979a,b, and unpublished observations). Mixtures of 5% DMSO, 5% glycerol, and 10% proline, or of 10% glycerol and 10% proline, as well as proline alone at 10%

are particularly effective. There are also indications that other amino acids and derivatives, such as asparate, lysine · HCl, glycine betaine, threonine, hydroxyproline, and γ-aminobutyric acid, may have useful cryoprotectant properties (Heber *et al.*, 1971; Withers and King, unpublished observations). The alga *Euglena gracilis* has successfully been preserved after cryoprotection with methanol (Morris, 1980) and there is scope for its application to higher plant cells. There is little evidence to suggest that polymeric cryoprotectants may be effective in the preservation of viability in plant cells (Nag and Street, 1975a; Morris, 1980), although they may be effective in preservation for structural and analytical studies (e.g., Skaer *et al.*, 1978).

There is no single explanation for the efficacy of cryoprotectants. Some are effective both in cellular and noncellular systems and some are more effective in one type of freezing program than in another (Withers, 1980). Bearing in mind the two major causes of cryoinjury, *ice damage* and *solution effects* (see Section II,B), the cryoprotective compounds may operate through a reduction in ice crystal size and growth rate, depression of the freezing point of the cytoplasm, colligative action protecting against high intra- and extracellular solute levels, and stabilization of membrane components and other macromolecules exposed to ice and concentrated solutes. The membranes of the cell are particularly susceptible to cryodamage, suffering rupture, qualitative and quantitative loss of their molecular components, loss of enzyme activities, and altered permeability properties (Cella *et al.*, 1978; Heber, 1968; Heber *et al.*, 1971; Lyons, 1973; Mazur, 1977; Meryman, *et al.*, 1977; Palta *et al.*, 1977; Wiest and Steponkus, 1978; Yoshida and Sakai, 1974). Further detailed discussion of cryodamage, the range of cryoprotectant compounds, and their modes of action can be found in Bajaj and Reinert (1977), Finkel and Ulrich (1979), Mazur (1969), Meryman (1966), Nash (1966), and Withers (1980).

Long- and medium-term exposure to, for example, DMSO may be damaging (Dougall and Wetherell, 1974; Morris, 1980; Nag and Street, 1975a). Glycerol may be heavily plasmolyzing, because of reduced uptake at the temperature of application (McGann, 1978), as well as possibly predisposing cells to deplasmolysis injury during and after thawing (see Withers 1980).

In the context of cryoprotectant toxicity, the use of proline is particularly relevant. Many similarities can be drawn between low-temperature stress, osmotic stress, and the cellular response to such natural environmental situations as exposure to thermal extremes, drought, and high external salt concentrations (Levitt, 1978; Towill and Mazur, 1976; Wiest and Steponkus, 1978). Under such conditions certain plants, including a number of important cereals, respond by the cellular accumulation of proline or glycine betaine (Blum and Ebercon, 1976; Chu *et al.*, 1978; Stewart and Lee, 1974; see also Pollard and Wyn-Jones, 1979; Withers and King, 1979a). As an exogenously applied cryoprotectant, therefore, proline may be acting primarily against the solution concentration stresses of slow freezing rather than specifically against ice damage (see Section II,B). In cryoprotectant mixtures, proline may protect against potentially damaging effects of other components of the mixture (Withers and King, 1979b).

The total cryoprotectant concentration normally falls between 0.5 and 2 *M*. The optimum level and mode of application of cryoprotectants should be determined empirically for the system under examination. A suggested procedure, suitable for the preservation of suspension culture cells, is described in Section III,A. Removal of cryoprotectants need not always be carried out. Separate consideration of this point is given in Section III,C.

B. FREEZING AND THAWING RATES

The precise pattern of events occurring during freezing and thawing will depend upon the rate of change of temperature. During relatively slow freezing, as the temperature is lowered, the extracellular medium will reach its freezing point and will either supercool or spontaneously form ice. Seeding or other physical means will initiate ice formation and prevent excessive supercooling. The cytoplasm is unlikely to freeze at the same temperature as the extracellular medium and therefore a vapor pressure deficit is established, causing the flow of water out of the cytoplasm. The consequent concentration of intracellular solutes further depresses the intracellular freezing point. This process can continue to temperatures well below the initial freezing point of the cytoplasm as long as the rate of cooling does not exceed that consistent with dehydration, and as long as water transport out of the cell can be maintained. Eventually a temperature is reached at which solidification of the cytoplasm occurs and any remaining free water forms intracellular ice. Some intracellular ice can be tolerated in slowly frozen cells and it is likely that the total amount of ice, its location, and the size of ice crystals are factors that determine whether damage will be revealed upon thawing (see Farrant *et al.*, 1977; Mazur *et al.*, 1972; Withers, 1980). Excessively slow freezing reduces the amount of intracellular ice but may lead to extreme concentration of intracellular solutes and injury by "solution effects" (Meryman *et al.*, 1977).

Very rapid freezing will prevent dehydration by the formation of ice *in situ* from intracellular water. At an adequately rapid rate, especially if cryoprotective compounds have been employed, the ice crystal size may be small enough to prevent physical damage, but the presence of this ice will predispose the cell to damage during thawing (Mazur, 1977). Slow thawing, particularly at temperatures above about −50°C, leads to ice recrystallization and the formation of progressively larger ice masses, which may rupture cell boundaries, the vacuolar membrane, and the cytoplasmic organelles (Sakai and Otsuka, 1967). Therefore, it is necessary to pass through the recrystallization temperature zone as rapidly as possible. The slowly frozen cell, in theory, may be thawed at either a rapid or a slow rate. However, in the case of most plant cells, the empirical finding is that rapid thawing is beneficial (see Withers, 1980), suggesting that at a slow rate of freezing, consistent with the prevention of solution effect damage, sufficient free water remains within the cell to freeze and be available for recrystallization during slow thawing (but see Dougall and Wetherell, 1974; Sugawara and Sakai, 1974).

The compromise between the damaging effects of ice and of concentrated solutes is the basis for the two-factor hypothesis of freezing damage (Mazur *et al.*, 1972), which describes the response of many biological systems to freezing and thawing. It can be predicted from this hypothesis that two fundamental approaches to cryopreservation can be made in designing freeze-preservation protocols. This is indeed the case, the protocols employed for the preservation of microbial systems, animal cells, tissues and organisms, and botanical specimens all being based on either a slow or a rapid freezing rationale (Mazur *et al.*, 1972; Meryman, 1966). The actual definition of slow or rapid freezing may vary widely from system to system, but for the present purposes, "slow" will normally mean a rate between 0.1 and 10°C per minute, and "rapid" a rate from ca. 40°C per minute upwards.

A number of purpose-built freezing units, such as the Planer R 201 and 202 series (Planer Products, Windmill Road, Sunbury-on-Thames, England), can be obtained (Withers, 1980). Henshaw (1975) describes a simple, laboratory-constructed, controlled-freezing apparatus in which the specimen is frozen at a predetermined rate by contact with an aluminium block cooled by partial immersion in LN and warmed by a heating coil. An appropriate choice of machine will enable from several to many hundred specimens to be frozen at a single controlled rate, at a series of rates, or through holding periods at intermediate temperature stages from room temperature down to about −180°C. Reproducible rates of freezing can be achieved using improvised units consisting of, for example, (1) a Dewar flask inside a deep freeze at −70°C, giving a freezing rate of about 0.1°C per minute (Sala *et al.*, 1979), or (2) an inner glass tube to which the specimen is attached, enclosed by an outer glass tube, the two being held-together by a cork and separated from a bath of coolant liquid by insulating material; this provides ultra-slow freezing at rates as low as 0.2°C per hour (Finkle *et al.*, 1974).

As an alternative to gradual slow freezing, the specimen can be transferred through a series of coolant baths at progressively lower temperatures before storage in LN (Sugawara and Sakai, 1974). Where only one or two intermediate temperatures are used we are, in fact, carrying out stepwise or prefreezing rather than slow freezing (Mazur, 1977; Sakai and Otsuka, 1972). In this laboratory a combination of slow and stepwise freezing (1°C per minute to −30°C or −35°C, holding for, e.g., 30 minutes) has been employed using a commercial freezing unit and an improvised unit with equal success. The latter consists of an alcohol bath cooled by an electrically powered cooling coil (Withers and King, 1979a, 1980).

Vapor-phase cooling, in which the specimen is suspended in the vapor over LN, is the principle upon which the Biological Freezing Plug modification of the Union Carbide LR-33-10 operates (Union Carbide U.K. Ltd., Aycliffe Industrial Estate, Darlington Co., Durham, England). Successful preservation using this apparatus has been reported by Nag and Street (1973) and, using an improvised unit in which the specimen is suspended at a known distance above the LN in a Dewar flask, by Seibert and Wetherbee (1977) and Grout *et al.* (1978). Relatively rapid or slow rates of freezing may be achieved.

Rapid freezing is the least technically demanding of all methods. The specimen suspended in liquid medium in a vial may simply be dropped into LN or dipped into LN as more LN is poured into the open end of the vial (Seibert and Wetherbee, 1977). Direct exposure of the specimen by propelling it into LN has been used by Grout and Henshaw (1978, 1979) to maximize the rate of freezing of meristems. Such methods result in freezing at rates in the region of 1000°C per minute.

Long-term storage must be carried out at a suitably low temperature using a LN-cooled refrigerator in which the specimens are either immersed in the liquid or suspended in the LN vapor. Deep freezers running at −20 to −30°C are entirely unsuitable (Bajaj, 1976; Weatherhead *et al.*, 1979). Storage at about −78°C may be adequate for the short term but there is evidence of progressive deterioration at such a temperature (Nag and Street, 1973). Stability in storage is essential to justify entrusting valuable genome stocks to freeze preservation. At ultra-low temperatures, the only serious threat to very long-term storage is a risk of accumulated mutations resulting from background radiation. Investigations into this phenomenon suggest that, for decades at least, there will be no major problem (Whittingham *et al.*, 1977), but it is important that this source of variability be minimized. Dimethyl sulfoxide has radioprotectant properties and storage in glass vials may reduce the exposure (Whittingham *et al.*, 1977), although this must be set against the risk of explosion upon warming of glass containers because of liquid nitrogen penetration.

Rapid thawing is most easily achieved by agitating the specimen vial in warm water (+ 30 to + 40°C; rate of ca. 120°C per minute between − 50 and − 10°C; Nag and Street, 1975a,b). Slower rates may be produced by exposing the specimen vial to blown warm air (Seibert and Wetherbee, 1977), air at room temperature, or LN vapor followed by air at room temperature (Withers 1979). These latter methods increase the total thawing time to up to 20 minutes.

III. The Freeze Preservation of Suspension Cultures

In many laboratories, the maintenance of suspension cultures represents one of the major demands upon time and equipment yet is essential for continued experimental work. It is fortunate, therefore, that much of the developmental work in freeze preservation has involved suspension cultures. The stages of freeze preservation can conveniently be divided into pretreatment, freezing and thawing, and postthaw treatment. Strictly speaking, effective pretreatments may extend well back into the pregrowth period of the culture, and supportive postthaw treatments extend forward beyond the initial postthaw period. However, for the present purposes, attention is given to the basic practical manipulations that are necessary to commit a suspension culture to storage and subsequently recover it. Treatments refining the method and enhancing survival are described in Section III,C.

A. The Basic Method

The most appropriate freezing method is slow freezing. There are no reports of successful preservation of suspension cultures by freezing at a rate greater than about 10°C per minute (Latta, 1971; Quatrano, 1968; Withers, 1980). Cells from a growing culture are chilled in a suitable container in an ice bath. An equal volume of cryoprotectant (single or mixture) prepared at double strength in liquid culture medium and sterilized by an appropriate method (Withers, 1980) is also chilled. Over a period of about 1 hour, the cryoprotectant solution is added to the cell suspension in several aliquots and the two are mixed thoroughly. The cryoprotected cell suspension is then dispensed into the storage containers, for example, a 1-ml volume of cell suspension in a 2-ml capacity polypropylene screw-top ampule (Nag and Street, 1975a; Withers and Street, 1977a). It is important that the density of cells in the ampule be adequate to reinoculate a culture after thawing, taking into account the minimum cell density (Sala *et al.*, 1979; Stuart and Street, 1969), the loss of viability upon freezing and thawing, and the dilution of the cells into fresh medium. The ampules are labeled, placed into a suitable rack, and transferred to the freezing unit. Freezing at a rate of 1 or 2°C per minute to a temperature between −40 and −100°C is generally satisfactory. When slow cooling is complete, the specimen is quenched in LN and transferred to storage.

After storage, the specimen is thawed by agitation of the ampule in a water bath. As thawing proceeds the contents of the ampule should be well mixed to prevent local heating and the ampule removed to an ice bath as soon as only a small ice plug remains frozen (Withers and Street, 1977a). At this stage a viability test should be carried out (see Section III,B). The frozen and thawed cells are then returned to culture by inoculation into a larger volume of medium (e.g., 1 ml into 10 ml) with or without prior washing (see Section III,C). Alternatively, regrowth may be initiated by layering the cell suspension over semisolid medium (Withers and King, 1979a).

In the days following thawing it is desirable to monitor recovery by repeated viability testing (Withers and King, 1979a) or measurement of a parameter of growth (Dougall and Wetherell, 1974; Sala *et al.*, 1979; Withers and Street, 1977a). As soon as growth has been reestablished, normal subculturing routines can be adopted.

B. The Response of Cells to Freezing and Thawing

A precise characterization of frozen, thawed, and recovered cells is essential to the development and exploitation of ultra-low-temperature storage. It is of the utmost importance to be able to determine: (1) how many cells are surviving; (2) how the survivors relate to the total population; (3) the nature of cryoinjury at the cellular level; (4) the pattern of recovery growth in quantitative terms and again at the cellular level; and, finally, (5) whether the special features of the culture indicating its storage as a valuable genome (e.g., morphogenic potential or a high production of a specific compound) are retained. Below are described methods for carrying out such characterization and observed responses of cultures that have been recovered after freeze preservation.

1. *Viability Tests*

Several viability tests can be employed to rapidly ascertain the initial viability of a culture, assess cryoprotectant toxicity, and determine the percentage of cells surviving. A wide range of tests is reviewed critically in the literature (see Gressel *et al.*, 1978; Palta *et al.*, 1978; Withers, 1980). It is clear that such tests should be used with caution because no one test comprehensively determines the physiological state of the cell or its capacity to recover normal growth processes.

Flourescein diacetate staining, developed for use with plant tissue cultures by Widholm (1972), gives a rapid visual assessment of percentage viability. Cells are treated with a dilute (0.01%) solution of the stain and then observed under ultraviolet and white light illumination. "Viable" cells fluoresce brightly and "dead" cells do not (Fig. 1a). An approximate percentage survival can be calculated bearing in mind that intermediate levels of fluorescence can be observed and that the central cells in aggregates cannot clearly be seen. The test is complementary to those, such as Evan's blue staining, which indicate viability by the exclusion of a dye by undamaged cells. Cells are again treated with a dilute (0.025%) solution of the stain and observed in the light microscope. "Viable" cells remain unstained, whereas "dead" cells accumulate the stain to visible levels (Fig. 1b). Withers (1978a) has used this stain to mark lethally dam-

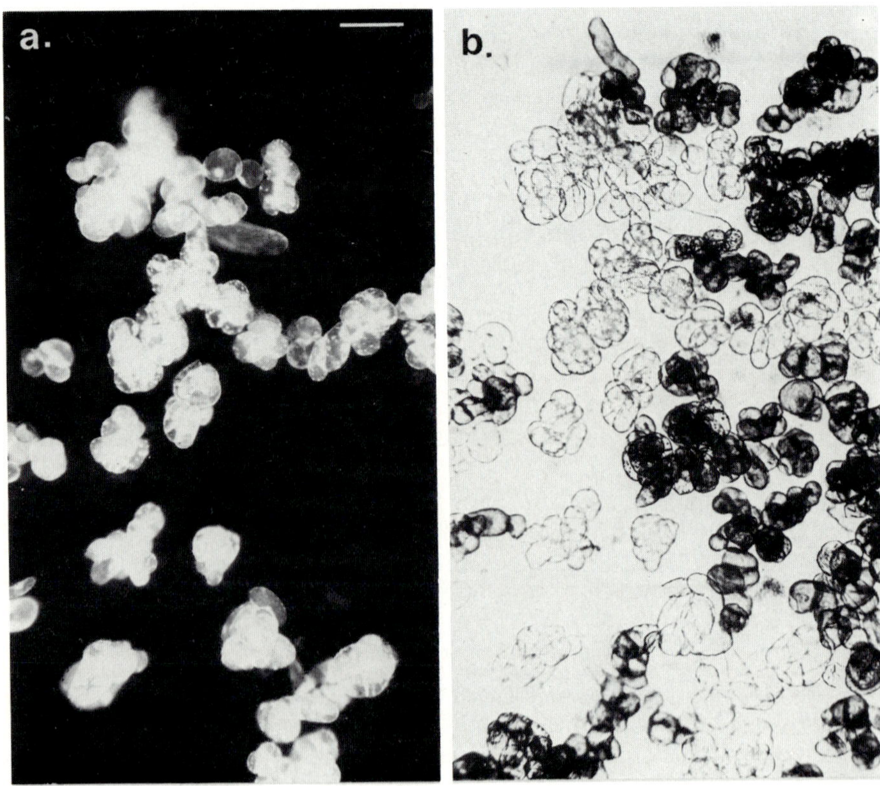

Fig. 1.   A mixture of live and dead cells stained with fluorescein diacetate and Evan's blue and viewed under (a) ultraviolet illumination or (b) white light (tungsten) illumination (bar = 50 μm). (From Withers, 1978a.)

aged cells as an aid to histological examination of the surviving population (see Fig. 5 and Section III,B,3).

A more time consuming but arguably less subjective test involves the estimation of respiratory efficiency by reduction of 2,3,5-triphenyltetrazolium chloride (TTC) to the red dye formazan (Towill and Mazur, 1974). Formazan can be extracted and measured spectrophotometrically. Caution in the use of this test is indicated by the fact that a high-positive response may signal cell damage (Gressel *et al.*, 1978; Withers, 1980), correlation with other viability tests or with recovery potential may be poor (Bajaj and Reinert, 1977; Sala *et al.*, 1979; Withers, 1978a, 1980), and certain changes in response occurring in the immediate postthaw period cannot be explained by alterations in the viable cell population (Fig. 2; see also Cella *et al.*, 1978; Sala *et al.*, 1979).

Viability tests reveal that after freezing and thawing 100% viability is never recovered. Among the highest viability values recorded for suspension cultures are 75% (*Zea mays;* Withers and King, 1979a), 60–70% (*Daucus carota;* Nag and Street, 1973, 1975a,b; Withers and Davey, 1978; Withers and Street, 1977a), 60–65% (*Oryza sativa;* Sala *et al.*, 1979), 50% *Glycine max;* Bajaj, 1976), and 40–50% (*Acer pseudoplatanus;* Withers, 1978a; Withers and Street, 1977a). Many reports indicate that survival values of 10–20% are not uncommon (Bajaj and Reinert, 1977; Withers and Street, 1977b), suggesting vast scope for improvement. Recovery growth is unlikely to ensue from very low-percentage survivals and it is indeed undesirable that it should because the risk of selection for a resistant genotype would be contrary to the principle of genome preservation.

2. Recovery Growth—Quantitative Aspects

A lag phase is a common feature of batch culture growth and it is therefore not surprising that frozen and thawed cells do not embark upon re-

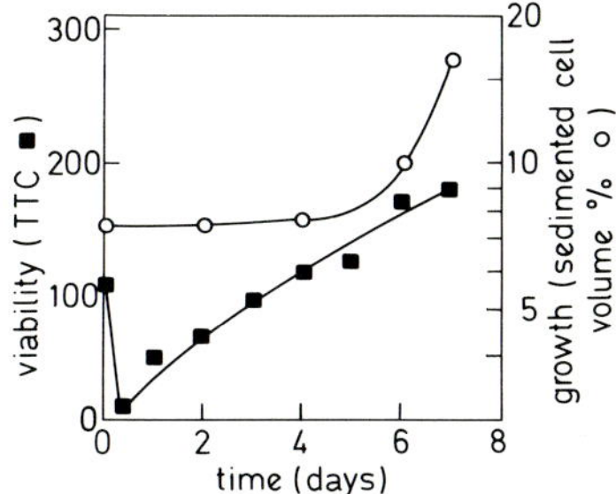

FIG. 2. Viability (■) and cell growth (○) of freeze-preserved cells of *Oryza sativa* during the first days in culture after thawing. Growth is expressed as the volume in milliliters of cells per 100 ml of culture. The TTC response is expressed as A_{485} units per gram fresh weight of cells. The initial viability (ca. 60% of unfrozen control) drops to ca. 20% by 7 hours and returns to the initial value before growth is detected. (From Sala *et al.*, 1979.)

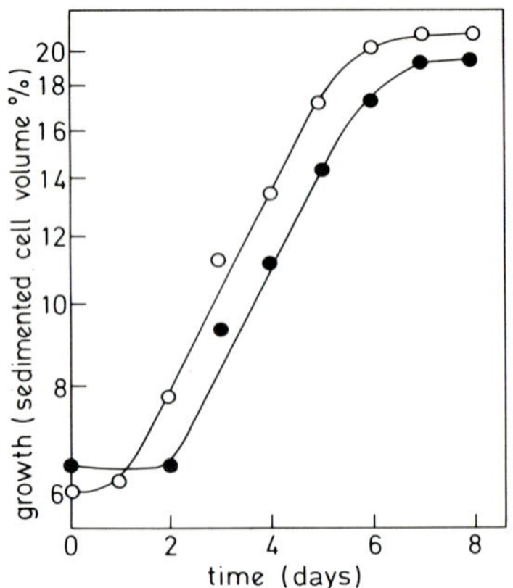

FIG. 3.   Recovery growth of frozen and thawed cells of *Oryza sativa* (●) in comparison with an unfrozen control culture (○). Estimation of growth as in Fig. 2. (From Sala *et al.*, 1979.)

growth immediately. Lag periods of from several days to several months are reported (Bajaj, 1976; Withers and Street, 1977a), the briefest—2 days—occurring in recovered suspensions of *Oryza sativa* (Fig. 3; Sala *et al.*, 1979). The reduction in the viable population may mean that perception of growth is not possible until a substantial increase in, for example, cell number has occurred. Regrowth therefore may be initiated at an earlier stage than is immediately apparent (see Fig. 4b). Sala *et al.* (1979) describe a rate of recovery growth equivalent to unfrozen control cells (Fig. 3), whereas in recovered suspensions of *Daucus carota*, Withers and Street (1977a) observed a reduced rate of growth (Fig. 4). In these two reports, sedimented cell volume and cell number were used to estimate the rate of recovery growth, but any means of measuring biomass appropriate to the system under examination would suffice. Where recovery is initiated in or over semisolid medium (Nag and Street, 1975a; Withers and King, 1979a), a sample may be taken for percentage viability estimation (see Figs. 13 and 14) or the plating efficiency may be estimated (Nag and Street, 1975a).

3. *Cytology and Morphogenic Potential*

Cytology and morphogenic potential are linked together because they both relate to the possibility of selection or modification of the cell's genome as a result of freeze preservation. Little detailed information can be offered to categorically state that freeze preservation does not alter or select from within a population. Wherever the morphogenic potential of a culture has been assessed after freezing and thawing, it has been found to be retained (Bajaj, 1976; Dougall and Wetherell, 1974; Nag and Street, 1973, 1975b; Withers, 1979). Similarly karyotypic studies indicate no chromosome alterations at the level of detection (Nag and Street, 1973, 1975a). However, suspension cultures frequently contain several related genotypes and there is some evidence that freeze preservation may select for cells that have a different cell size by virtue of a lower DNA content. Therefore, the lower ploidy in a mixed population may be preserved preferentially (Shillito, 1978).

A more subtle distinction is between stages of the cell cycle. In an exponentially growing culture all stages of the cell cycle are likely to be present at all times. However, rapid subculturing or pretreatment, such as nitrate starvation, may induce a degree of synchrony in nuclear phase, cell division, or both (Gould and Street, 1975; Withers, 1978a; Withers and Street, 1977a). Withers (1978a) has found that there may be a difference in response to freezing and thawing between the various stages of the cell cycle, cells in early G_1 phase having an increased survival potential (Fig. 5). This increase can only partially be explained by changes in cell size and may be related to an enhanced nuclear stability and consequent tolerance of freeze–thaw stresses. Studies of synchronized animal cell cultures indicate that the response described above is unique to plant tissues cultures (see Withers, 1980). Although a preferential survival by

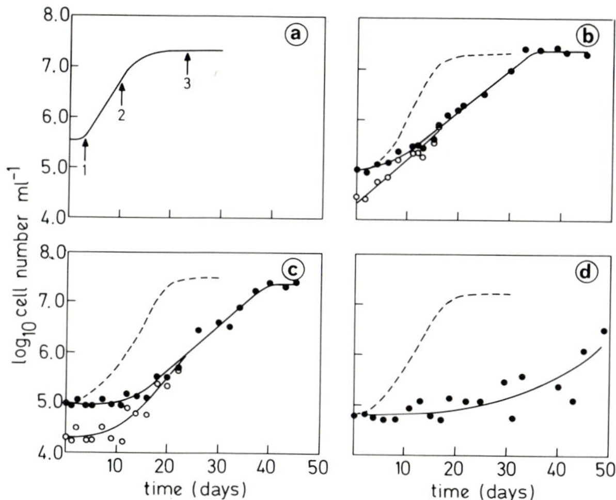

FIG. 4. Recovery growth of frozen and thawed cells of *Daucus carota*. (a) Growth curve of parent culture indicating times at which the cells were harvested for freezing. (b–d) Patterns of recovery growth of cells harvested at times 1, 2, and 3, respectively, in (a). Dotted line indicates unfrozen control cells; ●, total cell density; ○, corrected cell density taking into account percentage viability. (From Withers and Street, 1977a.)

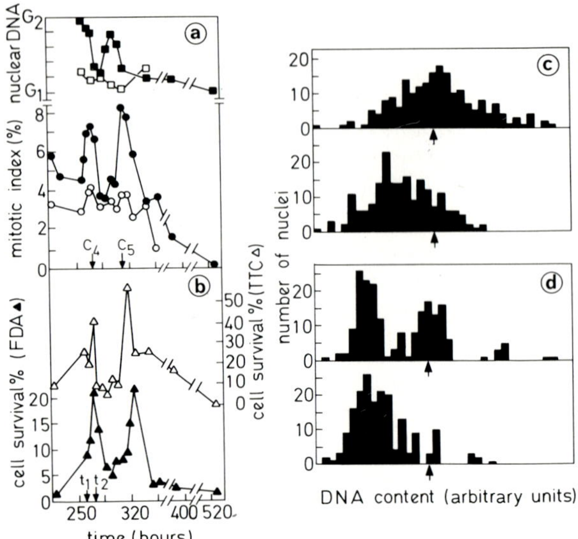

FIG. 5.   Cytological and survival data for cells harvested from a synchronous culture of
Acer pseudoplatanus. (a) Relative nuclear DNA levels and mitotic index values recorded
during the period of growth including the fourth and fifth cytokineses (c_4 and c_5). The total
population values (solid symbols) show clear oscillations, unlike the surviving cells (open
symbols). (b) Cell survival as estimated by FDA (▲) and TTC (△) tests. Samples from time
t_1 and t_2 were stained with Evan's blue and Feulgen reagent to reveal the DNA levels of the
surviving cells. (c,d) Nuclear DNA profiles of cells harvested at times t_1 and t_2, respectively,
in (b). In each case, the upper histogram relates to the total population and the lower histogram to the surviving population. Arrows indicate metaphase (G_2) DNA values. (From Withers 1978a.)

one stage of the animal cell cycle may be observed, the most likely is late
G_1 or S phase, and the oscillations in survival, where present, are of a much
lower order. The survival values recorded for a synchronized plant cell
culture do not greatly exceed those obtained in exponential growth and
therefore there is little to be gained by undertaking the complex manipulations necessary to induce synchrony.

4. *Ultrastructural Examination of Frozen, Thawed, and Recovered Cells*

The likely course of events during rapid or slow freezing and thawing as
described in Section II,B is confirmed empirically by carrying out electron microscopical examination of freeze-preserved suspension culture
cells. Cells have been examined before, during, and after preservation in
order to determine the nature and extent of ultrastructural modification
induced by the transitions to and from the frozen state (Withers, 1978b;
Withers and Davey, 1978; Withers and Street, 1977b). The scheme outlined in Fig. 6 illustrates the points in the freeze–thaw cycle at which sam-

ples were taken, the modified specimens which were used for examination of material in the frozen state, and the type of ultrastructural image which was obtained by fixation in the unfrozen state, freeze etching, and freeze substitution.

Conventional freeze-etch preparation using high levels of cryoprotectants and ultra-rapid freezing produces a specimen with apparently well-preserved ultrastructure (compare Fig. 7a,b). However, the specimen cannot be recovered after thawing. Similarly, cells frozen rapidly with levels of cryoprotectant that would be consistent with survival under a slow freeze regime, or frozen either rapidly or slowly without chemical cryoprotection, suffer lethal damage. In the former case, gross damage

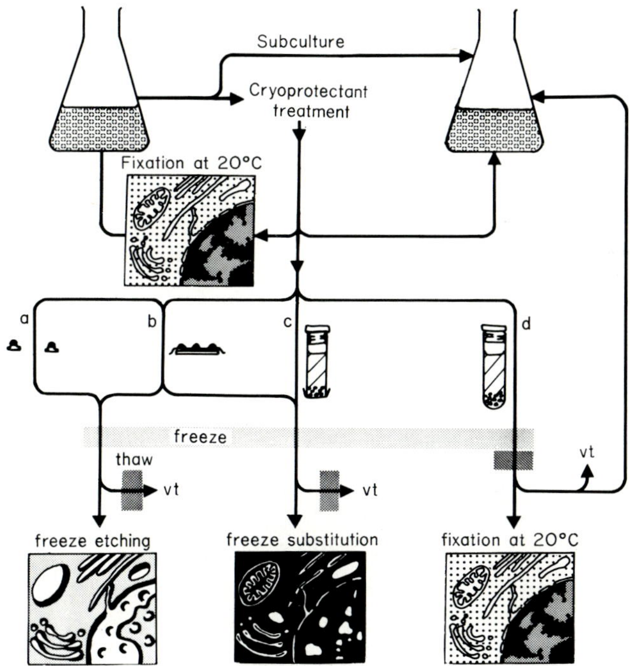

FIG. 6. Scheme adopted to examine cells before and after freeze preservation and in the frozen state. Conventional fixation with glutaraldehyde and osmium followed by embedding and thin sectioning was employed to examine unfrozen or thawed cells. Freeze etching and freeze substitution were employed to examine cells at ca. −100 and −78°C, respectively. For the former technique cells were frozen either (a) on gold stubs or (b) in droplets on foil. For the latter technique they were frozen in droplets or (c) ampules modified by the removal of the base and its replacement with *Nescofilm* (Withers and Davey, 1978). The fixative for freeze substitution was osmium tetroxide in acetone. Cells for fixation after thawing were frozen in unmodified ampoules (d). Viability tests (vt) were used to check the survival values of cells processed in parallel to microscopic preparations.

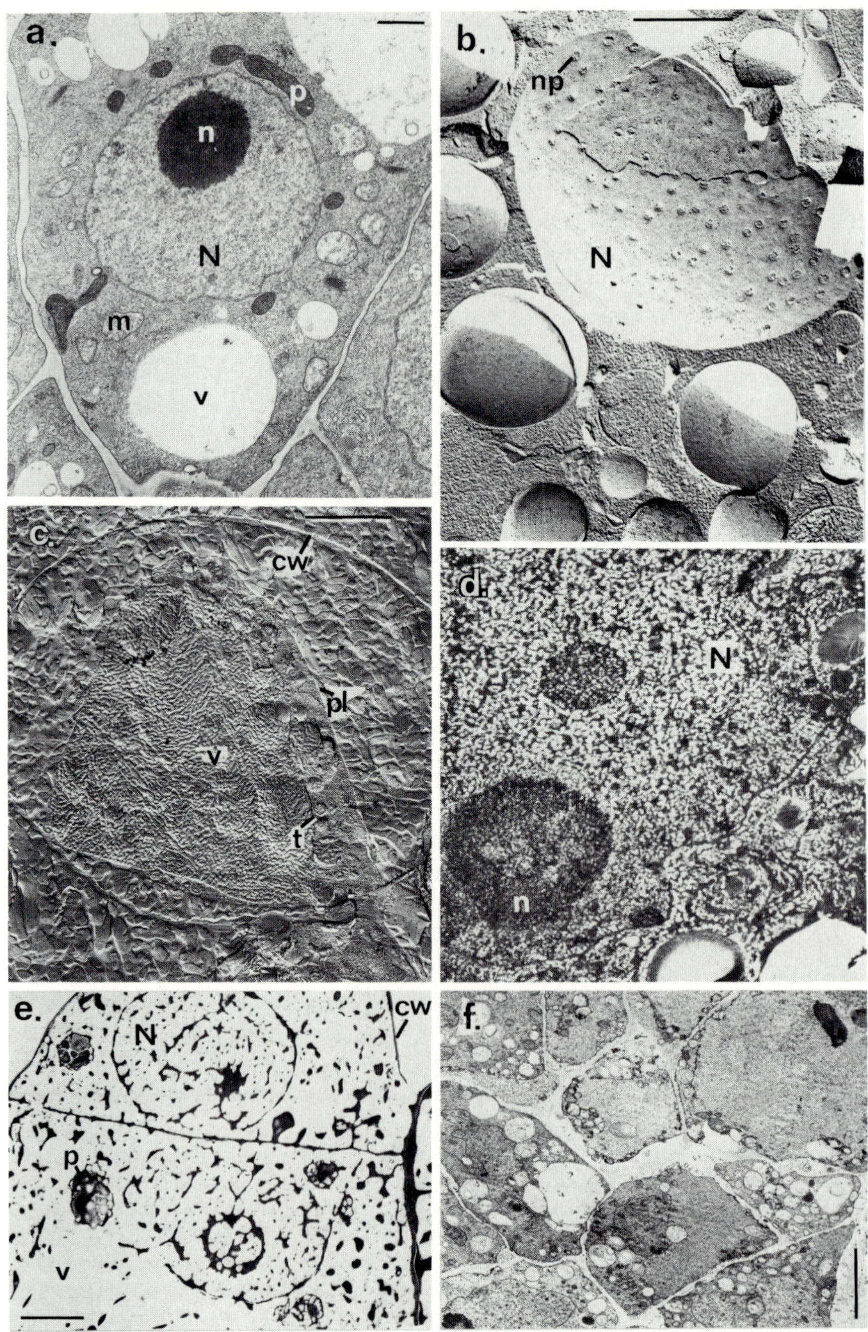

Fig. 7. (a) Unfrozen cells of *Daucus carota* (bar = 1 μm). (b) Freeze-etch replica of a cell of *D. carota* frozen rapidly after treatment with 25% glycerol (bar = 1 μm). (c) Freeze-etch replica of cell of *Acer pseudoplatanus* frozen rapidly after treatment with 5% DMSO

may not be apparent in the frozen state, because the ice formed is normally confined by cell and organelle boundaries (Fig. 7c,d). However, in the latter case ice damage is extensive in the frozen cell (Fig. 7e). The appearance of the cells in Fig. 7c,d represents a predisposition for damage during thawing, when, as stated earlier (Section II,B), ice recrystallization occurs resulting in massive cytoplasmic disorganization (Fig. 7f).

In complete contrast to the examples quoted above, cells that are frozen slowly after appropriate cryoprotectant treatment can be recovered. Their appearance is modified in both the frozen and the thawed state. Cellular collapse and an increase in cytoplasmic density manifested as increased electron density in freeze-substituted preparations (Fig. 8a) and modified freeze-fracture characteristics (Fig. 8b) indicate that protective dehydration has occurred. Ice crystals can be seen in the cytoplasm but they are large and generally are confined to the cytoplasmic matrix or the nucleoplasm (Fig. 8a,c). The plasmalemma membrane often is furrowed, presumably in response to the changed volume of the cell, and has an assymetrical appearance in freeze-substituted preparations (Fig. 8d,e). The tonoplast bears large protrusions into the vacuole and there is a redistribution of the membrane granules (Fig. 8b) that may be indicative of changes in pH and concentrations of cryoprotectants or other solutes during cooling (see Withers and Davey, 1978). Further, it may represent phase transitions in the membrane as solidification took place (Verkleij et al., 1972). The small cytoplasmic organelles and endoplasmic reticulum appear relatively unchanged in the frozen state despite a negative contrast appearance in freeze-substituted preparations (Fig. 8c,e).

Upon thawing the majority of the slowly frozen cells are able to resume a volume equivalent to or approaching that of the unfrozen cell, although some may remain collapsed and dense initially (Fig. 9a). Some cells, particularly highly vacuolated ones and those lying centrally in aggregates, show severe damage, including rupture of the tonoplast and plasmalemma. Less seriously damaged cells exhibit characteristic modifications, the most common of which is dilation of the endoplasmic reticulum, nuclear envelope, and small cytoplasmic organelles (Fig. 9a–c). Disturbances in solute and solvent transport relating to organelle volume regulation are indicated (Trump et al., 1971). This type of damage is common to many injury situations induced by physical trauma, toxins, or pathogens

---

and 10% glycerol. The plasmalemma, tonoplast, and cell wall are intact (bar = 10 μm). (d) Freeze-substituted preparation showing a region of a cell frozen as in (c). Ice cavities can be seen in the nucleus and cytoplasm (bar = 1 μm). (e) Freeze-substituted preparation of cells of *D. carota* frozen rapidly without cryoprotectant treatment (bar = 2 μm). (f) Cells frozen as in (c) and (d) fixed after thawing (bar = 5 μm). Abbreviations: cw, cell wall; m, mitochondrion; n, nucleolus; N, nucleus; np, nuclear pore; p, plastid; pl, plasmalemma; t, tonoplast; v, vacuole/vesicle. (From Withers, 1978b; Withers and Davey, 1978.)

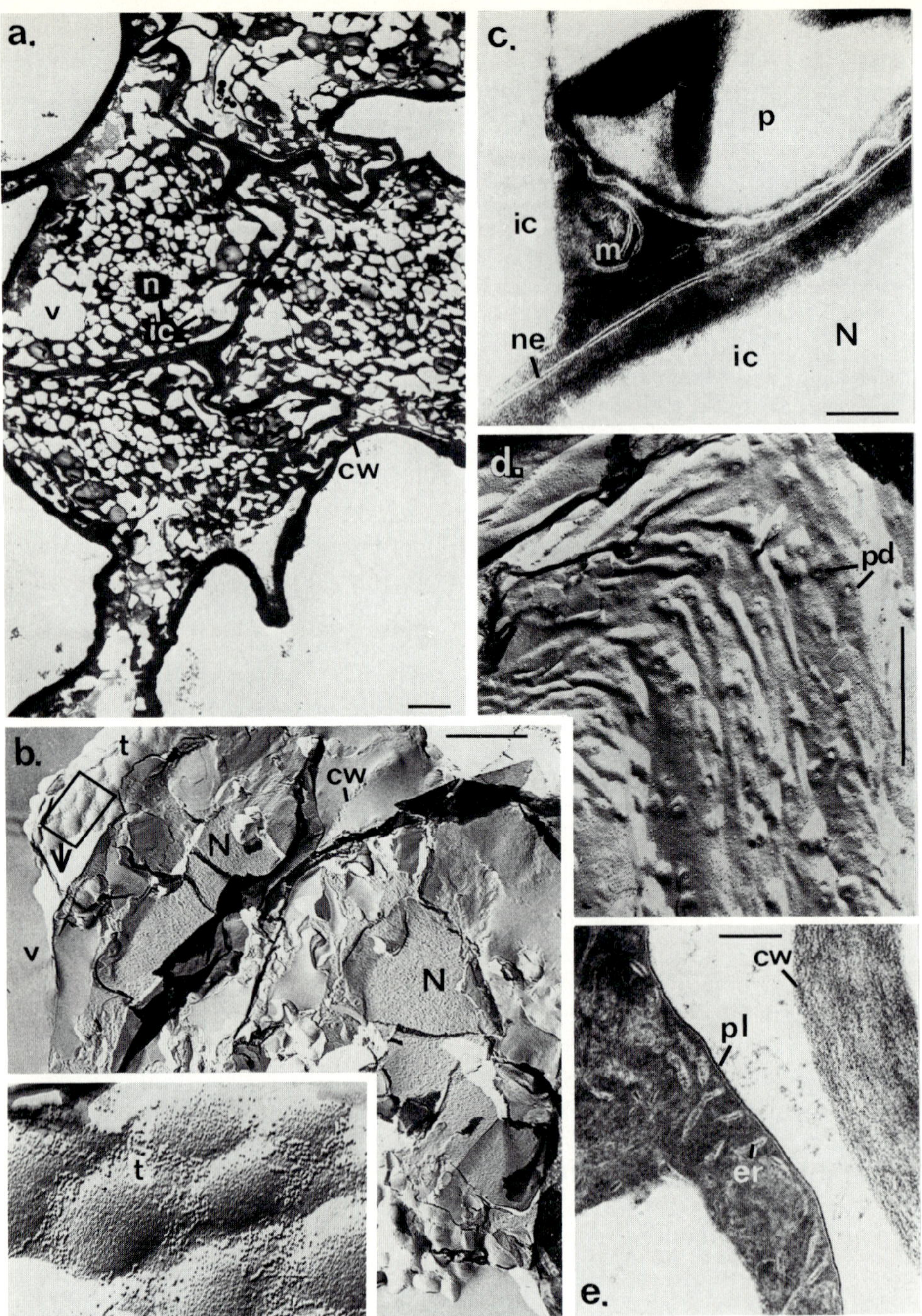

Fig. 8. (a) Cells of *Acer pseudoplatanus* frozen slowly after treatment with 5% DMSO and 10% glycerol and then fixed by freeze substitution (bar = 5 μm). (b) Freeze-etch replica of cells of *Daucus carota* frozen as in (a). Inset shows detail of tonoplast (boxed area in b) (bar = 1 μm). (c) Detail of a cell prepared as in (a) (bar = 200 nm). (d) Freeze-etch replica of cell frozen as in (a) showing furrowed plasmalemma (bar = 1 μm). (e) Detail of a cell

(see Withers, 1978b). Other common modifications in the frozen–thawed cells include polysome dissociation and an increase in cytoplasmic microfilaments (Fig. 9d). Both cortical and mitotic spindle microtubules appear undamaged (Fig. 9e–g).

Observations of suspension cultured cells of *Zea mays* (Withers, 1980; Withers and King, unpublished observations) indicate that the pattern of injury observed in the cells of *Acer pseudoplatanus* and *Daucus carota* is not an inevitable consequence of freezing and thawing. The protoplasts in the cells of *Z. mays* are highly plasmolyzed upon thawing and only resume their full volume some 24–48 hours later. By this time some cells have suffered lethal damage but the remainder are virtually undamaged. It is likely that the reduced level of injury in these cells results from the cryoprotectant treatment (including proline)and the induction of recovery growth over semisolid medium (see Section III,C; Withers and King, 1979a,b). The main ultrastructural modification in the cells is plasmodesmatal breakage caused by the severity of the plasmolysis and a slight vacuolar deposit corresponding to browning of the culture (Fig. 10a).

The cells of *D. carota* described earlier embark upon recovery growth by a much more complex sequence of events. It is evident that many cells undergo cytological deterioration in the days following thawing (Withers, 1980; Withers and Street, 1977b). Senescence, typified by an increased osmiophilia, secretion of excessive wall material, and accumulation of lipidlike material in the cytoplasm, is the fate of the majority of the cells (Fig. 10b,c). A minority are able to undergo ultrastructural recovery preceeding the onset of cell division. A temporary increase is observed in the spherosome complement of the cytoplasm, followed by their erosion as cytoplasmic membranes are regenerated (Fig. 10d,e). Many cells accumulate vast quantities of new membrane (Fig. 10f), which is dispersed only after several cell generations. This diverted cellular activity, the small proportion of cells contributing to the growing population, and a potentially toxic effect of the senescing cells may all account for the slower rate of growth observed in the recovery growth passage (Fig. 4; Withers and Street, 1977a).

It is clear from the observations described above that certain cells have an increased probability of surviving freezing and thawing. In general terms these cells are relatively small, are highly cytoplasmic, contain proplastids rather than amyloplasts and little endoplasmic reticulum, and lie

---

prepared as in (a) showing a section through the plasmalemma and cell wall (bar = 200 nm). cw, cell wall; ic, intracellular ice cavity; m, mitochondrion; n, nucleolus; N, nucleus; ne, nuclear envelope; p, plastid; pd, plasmodesma; pl, plasmalemma; t, tonoplast; v, vacuole/vesicle. (From Withers and Davey, 1978.)

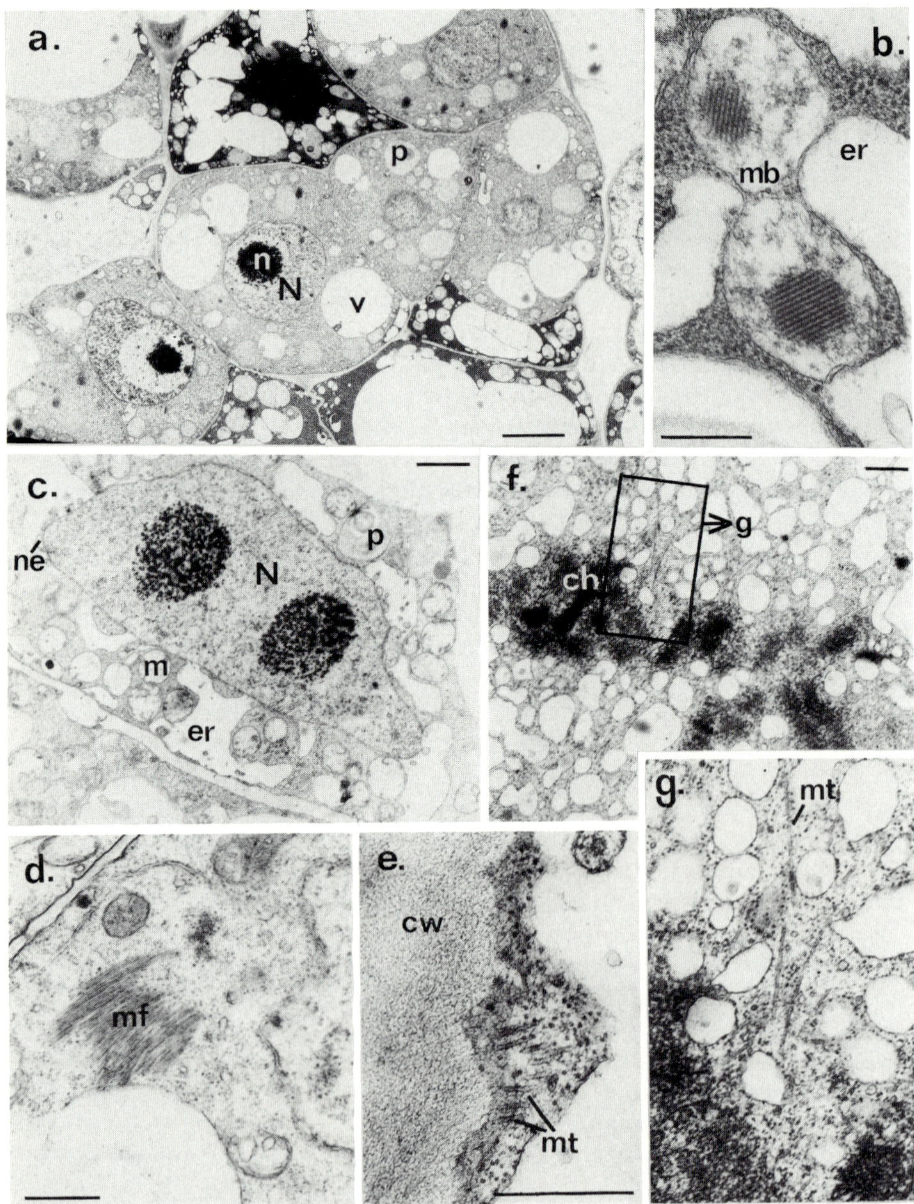

FIG. 9. (a) Cells of *Daucus carota* frozen as in Fig. 8 and then thawed rapidly (bar = 2.5 μm). (b,c) Regions of cells prepared as in (a), showing organellar dilation (bars = 500 nm). (d) Region of a cell prepared as in (a), showing bundles of microfilaments (mf) (bar = 500 nm). (e–g) Regions of cells prepared as in (a) showing cortical and spindle microtubules (mt). (g) Enlargement of boxed area in (f) (bars = 500 nm). ch, chromosome; cw, cell wall; er, endoplasmic reticulum; m, mitochondrion; mb, microbody; n, nucleolus; N, nucleus; ne, nuclear envelope; p, plastid; v, vacuole/vesicle. (From Withers, 1978b.)

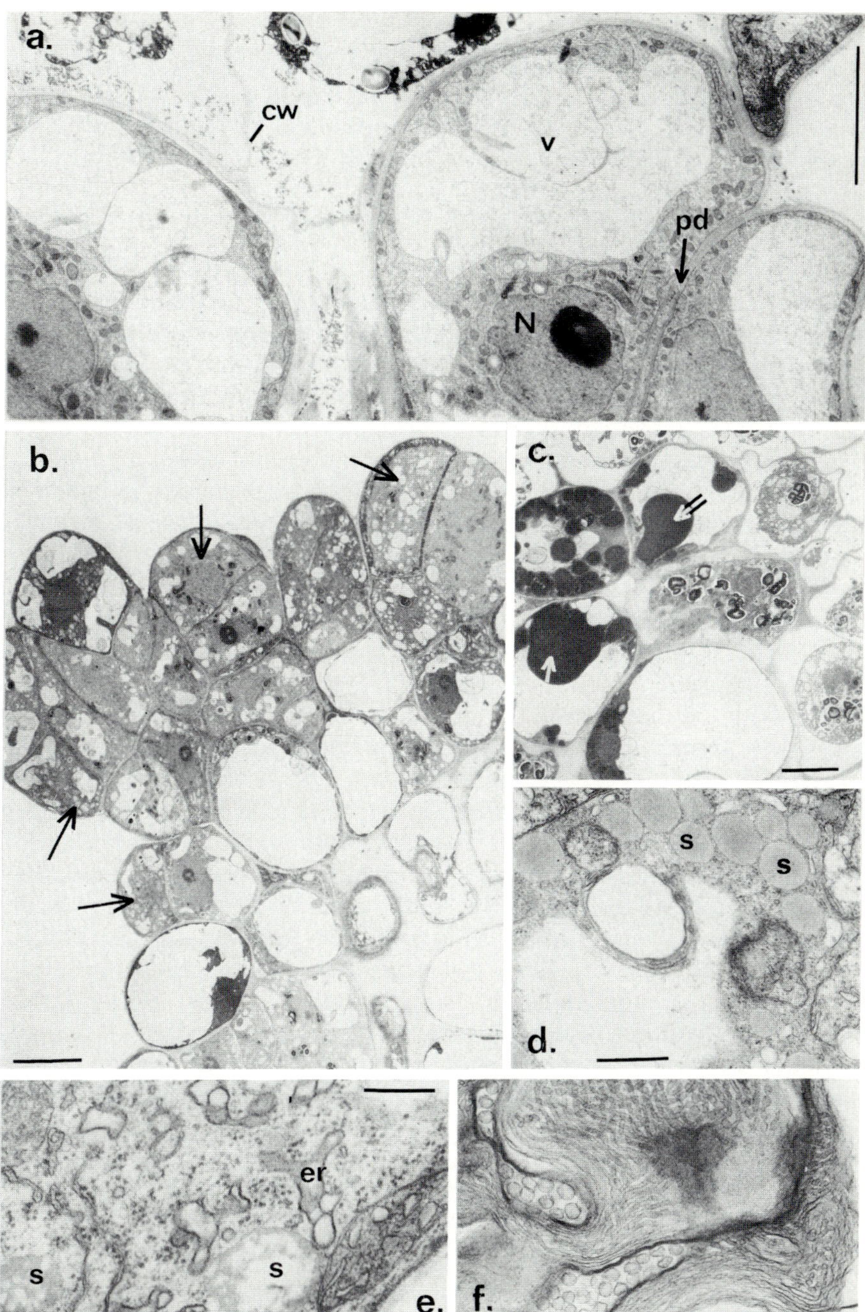

FIG. 10. (a) Cells of *Zea mays* returned to culture over semisolid medium after treatment with 5% DMSO, 5% glycerol, and 10% proline; stepwise freezing; and then rapid thawing. Cells were fixed 2 days after thawing. Note vacuolar granularity and plasmodesmatal breakage between adjacent cells (bar = 5μm). (b–f) Cells of *Daucus carota* fixed 10 days after

on the surface of cell aggregates rather than centrally. Such a definition enables a prediction to be made of the likelihood of a particular specimen's tolerance of freezing and thawing and also indicates potential means of improving the survival potential of that specimen.

C. The Improvement of Survival Potential

The suspension cultures that can be preserved exactly as described in Section III,A are very few in number and even they may need to be chosen with care to ensure survival. A variation in response during the batch culture growth cycle is widely reported (Bajaj and Reinert, 1977; Sakai and Sugawara, 1978; Withers, 1978c, 1979a, 1980; Withers and Street, 1977b). This is evidenced both as a variation in immediate postthaw survival (Fig. 11) and extent of recovery growth in a given period of time (Fig. 12a), and as a variation in the lag period before growth ensues, as indicated in Fig. 4 (Withers and Street, 1977a,b). Generally, cells in late lag phase or exponential growth are most likely to recover growth rapidly. These correspond to the most highly cytoplasmic stages in the growth cycle; are most responsive to viability tests, such as TTC reduction in the unfrozen state (Sala *et al.*, 1979); and are most suitable for the rapid initiation of growth upon passaging.

Adoption of rapid subculturing routines may maintain cells in exponential growth, thereby providing a more suitable subject for freeze preservation (Finkle *et al.*, 1975; Withers and Street, 1977a). However, the risk of synchronization, discussed in Section III,A, should be considered (see Fig. 11). Supplementation of pregrowth culture medium with such additives as mannitol, sugars, or proline is an effective method for increasing freeze tolerance (Fig. 11 and 13; see Latta, 1971; Withers and King, 1979a,b; Withers and Street, 1977a; see also Bannier and Steponkus, 1972; Sakai and Yoshida, 1968). Although cell size reductions may result from the medium supplementation (e.g., Fig. 11), this is unlikely to explain totally the improvement in survival potential. Growth of plants and plant cells in hypertonic media may lead to uptake of the medium supplement, to the intracellular accumulation of other osmotically active compounds, and to an alteration in the membrane fatty acid composition in a

---

returning to liquid culture. Freezing and thawing were under conditions detailed in Figs. 8a and 9a. (b) Some cells (arrows) are recovering, others are senescing (bar = 10 μm). (c) Some cells show large osmiophilic accumulations (arrows) (bar = 5 μm). (d) Regions of a cell showing many spherosomes (bar = 500 nm). (e) Region of a cell showing erosion of spherosomes and regeneration of endoplasmic reticulum (bar = 200 nm). (f) Accumulation of regenerated membrane in the cytoplasm of a recovering cell (bar = 200 nm). cw, cell wall; er, endoplasmic reticulum; N, nucleus; pd, plasmodesma; s, spherosome; v, vacuole/vesicle. (From Withers and Street, 1977b; Withers, unpublished observations.)

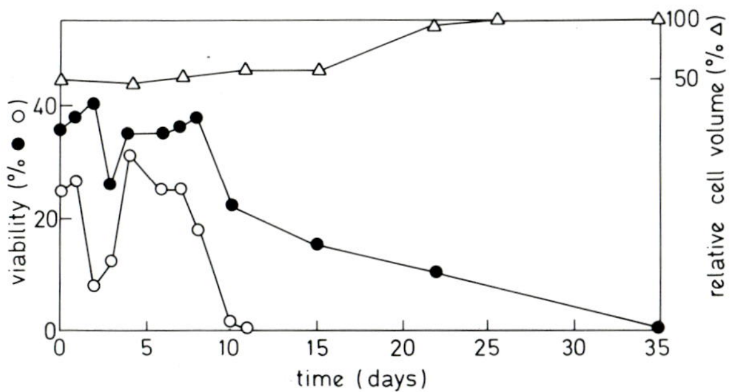

FIG. 11. Cells of *Acer pseudoplatanus* were grown on a 7-day subculturing routine in a basic medium and that medium containing 3.3% mannitol. In the latter medium the mean volume of the cells showed marked reduction compared with cells grown in the basic medium, even in an extended 35-day growth passage. During this passage the survival after freezing and thawing was monitored. Pregrowth in the supplemented medium (●) resulted in generally higher survival values compared with the basic medium (○), but in both media the earlier, exponential phase of growth yielded cells with a higher freeze tolerance. The reduced survival at ca. day 3 may result from partial synchronization. (From Withers and Street, 1977a.)

manner conducive to enhanced survival of freeze–thaw damage (see Morris, 1980). The effect may in fact mimic natural cold hardening. It is noteworthy that cells of *Zea mays* pregrown in medium supplemented with 10% proline are able to survive freezing and thawing without further cryoprotection and recover growth more rapidly than cells cryoprotected with proline for about 1 hour before freezing (Fig. 13). Nag and Street (1975a) reported an enhanced survival capacity in cells of *Atropa belladonna* that had been pregrown for one passage in the presence of 5% DMSO.

Little attention has been given to other means of modifying the pregrowth environment. Cold hardening, which is successful in increasing the freeze tolerance of callus and meristems (see Section IV,B), has not

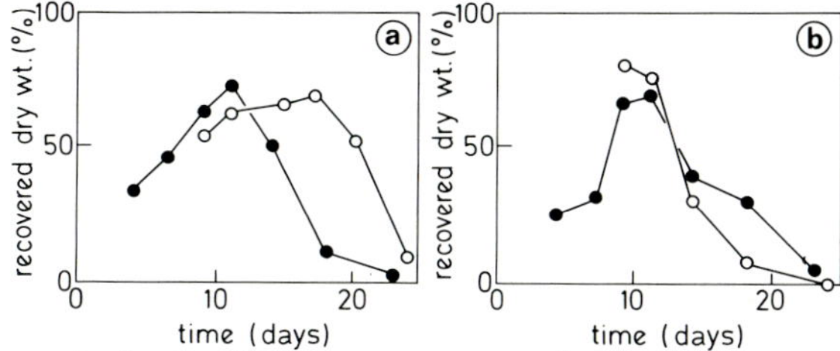

FIG. 12. Regrowth of cells of *Daucus carota* sampled during two passages [starting at day 0 (●) and day 7 (○), respectively] of growth in medium (a) containing 2,4-D and (b) lacking 2,4-D. Dry weight after 30 days of regrowth is expressed as a percentage of unfrozen control cells. Subculturing at day 7 prolonged exponential growth in (a) but accellerated embryogenesis in (b). (From Withers and Street, 1977b.)

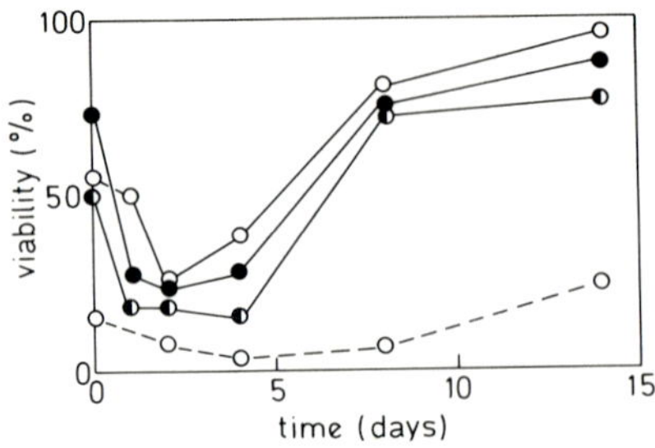

FIG. 13.   Recovery growth over semisolid medium of cells of *Zea mays* pregrown in basic medium and cryoprotected with 10% proline (O–O); pregrown in medium supplemented with 10% proline and frozen after no further cryoprotection (O—O) or cryoprotection with 10% proline (●); or pregrown in medium containing 5% proline and cryoprotected with 10% proline (◑). (From Withers and King, 1979a.)

been applied to suspension cultures, although the efficacy of the prolonged period of cryoprotectant application at a low temperature (ca. 20 hours on ice) in the case of cells of *Capsicum annuum* studied by Withers and Street (1977a) may involve a mechanism related to cold hardening. Modification of the pregrowth medium by alteration of the hormones supplied to the cells may be a fruitful line of approach in future work. Suspension cultured cells of *Daucus carota* subjected to withdrawal of 2,4-dichlorophenoxyacetic acid (2,4-D) are induced to undergo embryogenesis. In the early stages of induction there is a peak in regrowth potential (Fig. 12b; see also Withers and Street, 1977b), although as soon as visible embryos appear the potential declines markedly. Two other reports suggest that, through alteration of the cellular membrane composition, the presence or absence of this growth hormone might influence freeze tolerance (Sugawara and Sakai, 1978; Warren and Fowler, 1979). Further scope for modification might lie in the application of growth inhibitors, such as abscisic acid (ABA), which can increase intracellular proline levels and perhaps thereby enhance cold tolerance (Rajagopal and Anderson, 1978). This compound has been employed in pretreatment prior to storage by desiccation (Nitzsche, 1978; see Section V) and in the cold storage maintenance medium for meristem cultures (Henshaw *et al.*, 1978; see Section V).

The quantitative data relating to the postthaw period, the ultrastructural observations of frozen and thawed cells, and the frequent failure of regrowth (Bajaj, 1976; Sala *et al.*, 1979; Withers and King, 1979a; Withers and Street, 1977a) suggest a high degree of vulnerability in the postthaw period. There are indications that, although inadequacies in viability estimation may be giving deceptively high values in some cases, in others survival will only persist if the cells are given appropriate postthaw treatment.

It is evident that reinoculation into liquid medium, with or without postthaw washing, is consistent with recovery for a number of species (*Acer pseudoplatanus,* Sugawara and Sakai, 1974; *Atropa belladonna,* Nag and Street, 1975a,b; *Datura stramonium,* Bajaj, 1976; *Daucus carota,* Withers and Street, 1977a; *Glycine max,* Bajaj, 1976; *Nicotiana tabacum,* Bajaj, 1976; *Oryza sativa,* Sala *et al.,* 1979). However, suggestions that rapid dilution of the suspending medium may lead to deplasmolysis injury (Gresshoff, personal communication; Shillito, 1978; Towill and Mazur, 1976; Wiest and Steponkus, 1978) and that such dilution or washing may greatly reduce the proportion of viable cells, even precluding recovery (Withers and King, 1979a), challenge the validity of washing as an essential postthaw treatment. Withers and King (1979a) have examined the response of frozen and thawed suspension cultured cells of *Z. mays* to a range of postthaw treatments. Cells cryoprotected with DMSO and glycerol recovered growth only when layered over semisolid medium, whereas those cryoprotected with proline were more tolerant, recovering in liquid medium, in droplet culture, and, at a very rapid rate, over semisolid medium (Fig. 14). This general pattern has been confirmed with suspension cultures of *A. pseudoplatanus* and *D. carota* (Withers and King, unpublished observations), indicating that the basic method, so widely used, may in fact lead to unecessary loss of viability.

A further consideration in the handling of cells during the immediate postthaw period is the possibility of leakage of vital solutes. It is well known that low-temperature treatment can lead to the loss of water, ions, sugars and amino acids (Lyons, 1973; Palta *et al.,* 1977; Siminovitch *et al.,* 1964). Washing of a freshly thawed preparation would render these solutes unavailable for uptake during recovery. Cells of *Z. mays* are able to recover growth more rapidly and at a higher percentage efficiency if regrown in the presence of the suspending medium (Withers and King, 1979a). Removal of lost solutes has been suggested as a component of

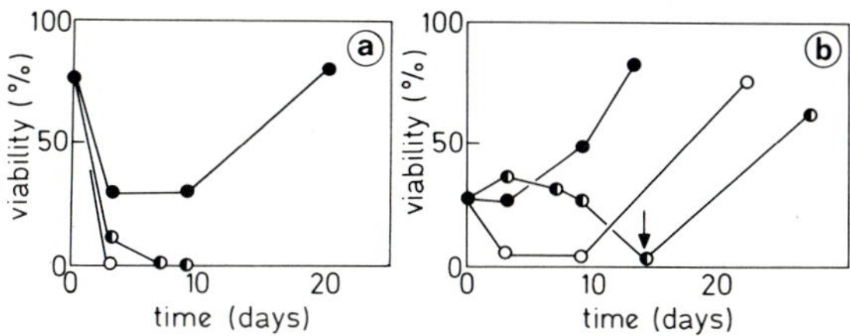

FIG. 14. Recovery growth estimated by percentage viability values during regrowth of cells of *Zea mays* cryoprotected with (a) DMSO and glycerol or (b) proline. Regrowth was initiated over semisolid medium (●), by dilution in liquid medium (○), or in droplet culture (◑). In the latter case the addition of fresh medium (arrow) was necessary to prevent senescence in proline-treated cells. (From Withers and King, 1979a.)

washing injury in freeze-preserved plantlets of *D. carota* (Section IV,B; Withers, 1979). Future experimentation might usefully give attention to nursing freshly thawed cells by application of osmotica and ions conducive to the alleviation of cryoinjury (Palta and Li, 1978; Shillito, 1978; Sugawara and Sakai, 1975; Wiest and Steponkus, 1978) and of conditioning factors or nutrients, such as amino acids (Withers and Street, 1977a).

IV. The Wider Application of Freeze Preservation within Plant Tissue Culture

The observation that cell and aggregate size can influence the capacity of a specimen to survive freezing and thawing implies that special procedures may be required to preserve successfully larger and organized structures. The structural differences at a cellular level between suspension cultures and protoplasts or pollen similarly suggest a requirement for modified treatment.

A. CALLUS CULTURES

Callus cultures present difficulties through their size and consequent impediment to protective dehydration and cryoprotectant penetration, and through their culture habit on semisolid medium, which is inappropriate for the direct application of cryoprotectants. Only limited success is reported in the preservation of callus. Tumanov *et al.* (1968) have successfully exposed callus of *Prunus cerasus* to temperatures as low as $-30°C$, and survival of callus masses of *Populus euramericana* preserved at the temperature of LN after stepwise freezing through -30, -70, and $-120°C$ has been reported by Sakai and Sugawara (1973). In both cases cold hardening preceeded exposure to lower temperatures and cryoprotectant treatment was ommitted. Small callus pieces of *Zea mays* cryoprotected with DMSO have successfully been preserved by Withers (unpublished observations) using the dry method described in Section III, B. Finkle and Ulrich (1978) make a preliminary report of effective preservation at $-196°C$ of callus of *Saccharum* sp. However, further experimentation is necessary before any generally applicable methods can be offered for the preservation of callus.

B. EMBRYOS, PLANTLETS, AND MERISTEMS

Embryos, plantlets, and meristems embody the difficulties described for callus plus a requirement for the maintenance of symplastic continuity within the tissue mass. Two approaches have been taken to develop freeze-preservation methods for such material: (1) rapid freezing and (2) dry freezing.

Rapid freezing avoids the problems of plasmolysis and solution effects inherent in slow freezing and is particularly suitable for meristems in which the cells are highly cytoplasmic, and therefore less prone to ice damage. Seibert and Wetherbee (1977) describe the successful preservation of shoot apices of *Dianthus caryophyllus* by treatment with 5% DMSO, cooling at a rate of at least 50°C, per minute, and warming rapidly (rate, ca. 1450°C per minute). Apices with two to three leaf pairs survived optimally but only the meristem proper yielded new growth. Recovery rates of about 40% could be increased to over 90% by an appropriate cold-hardening treatment. Grout and colleagues (Grout *et al.*, 1978; Grout and Henshaw, 1978, 1980) have successfully used freezing at a relatively rapid rate to preserve seedling meristems of *Lycopersicon esculentum* (15%, DMSO; freezing rate ca. 35°C per minute; recovery up to 45%) and isolated shoot tips of *Solanum goniocalyx* (10% DMSO; freezing rate 1000°C per minute; 20–60% recovery; Fig. 15). In the former example, organized outgrowth of the original meristem was promoted by recovery growth on medium supplemented with gibberellic acid at 3 mg/liter, and in the latter case recovery was enhanced by regrowth under reduced lighting. In all of the examples of rapid freezing, postthaw washing preceeded recovery on semisolid medium, indicating the absence of deplasmolysis injury.

In contrast with the examples noted above, Kartha (personal communication) has successfully preserved meristems of *Pisum sativum* and *Cicer arietinum* cryoprotected with 5% DMSO by slow freezing at a rate of 0.6°C per minute to −40°C before immersion in LN and rapid thawing.

Conventional slow freezing of suspension cultures of *Daucus carota* that have been induced to undergo embryogenesis by 2,4-D withdrawal reveals a serious loss of recovery potential as embryogenesis progresses (Fig. 12). The use of rapid freezing will not overcome this problem but by the use of a modified slow-freezing method survival of early stage embryos and survival of the meristems of more advanced embryos and plantlets can be achieved (Withers, 1979).

The simple method, dry freezing, involves removing the specimen from the cryoprotectant solution, blotting it dry, and enclosing it in an aluminium foil envelope before freezing. The slow freezing can be terminated at −40°C or at a lower temperature, such as −100°C. In the former case rapid thawing is essential but in the latter, slow thawing (see Section II,B) can greatly enhance survival values (Fig. 16). Slow thawing is beneficial because it avoids physical damage to these very brittle specimens. The interrelationship between degree of dehydration of the specimen and its response to either rapid or slow thawing is further demonstrated by desiccation of the cryoprotected specimen before freezing (Fig. 17). Postthaw washing will entirely prevent recovery in dry frozen preparations. Recovery of organized growth is promoted by regrowth on medium supplemented with activated charcoal. Withers (1978c) has used dry freezing to

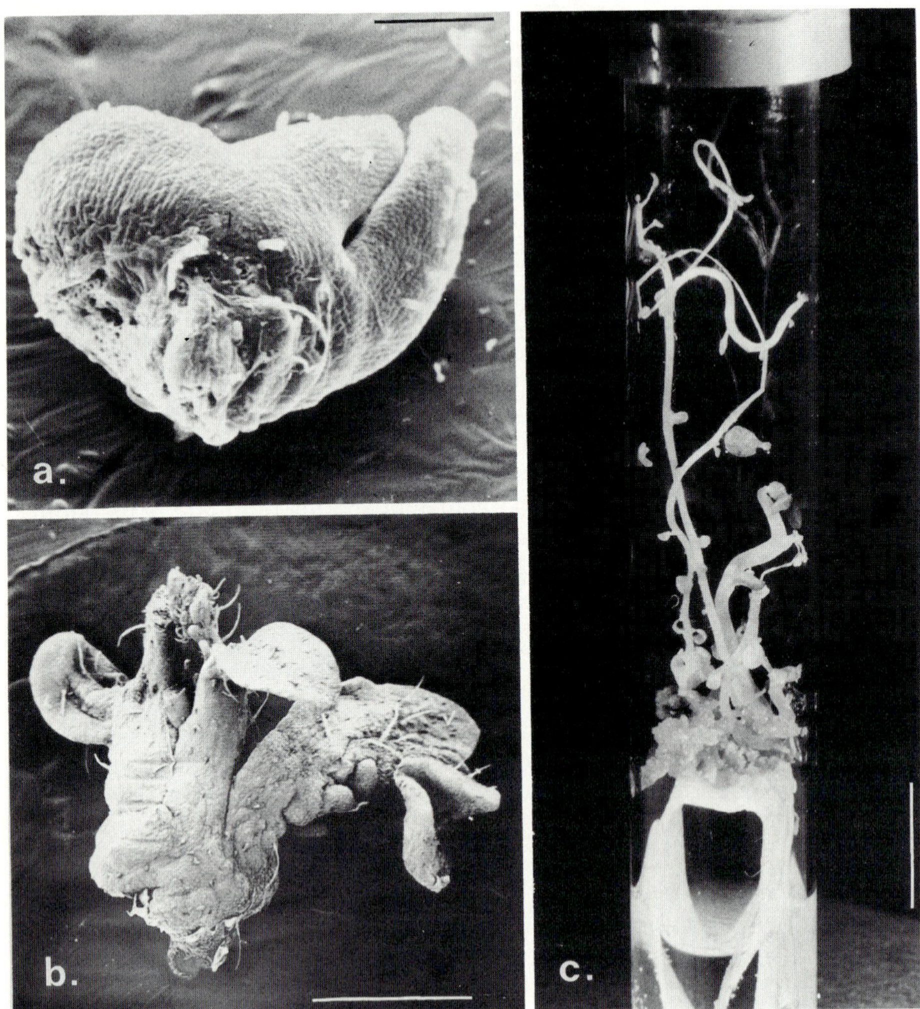

FIG. 15.   (a,b) Scanning electron micrographs of isolated shoot apices of *Solanum gonio-calyx* taken (a) immediately after rapid freezing and thawing (bar = 250 μm), and (b) after 5 weeks of regrowth producing a small plantlet (bar = 5 mm). (c) Established plantlet after 8 months of recovery growth (bar = 10 mm). (Figures kindly supplied by B. W. W. Grout, G. G. Henshaw, and J. A. Stamp.)

preserve zygotic embryos of *Zea mays* and dried or partially hydrated seeds. (For other examples of seed preservation, see references in Withers 1978c, 1980.)

Dry freezing is limited in that the axial region of advanced plantlets fails to survive. It is highly likely that preservation of such material will never

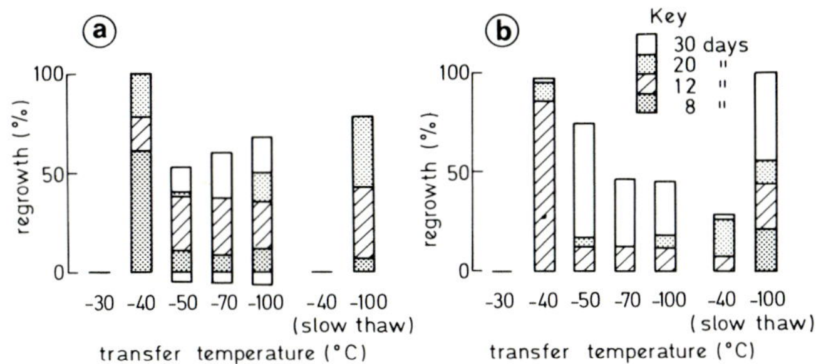

FIG. 16. Time course of regeneration of meristems (a) or callus (b) in clonal plantlets of *Daucus carota* dry frozen at a rate of 1°C per minute to the temperatures indicated and then transferred to LN. Thawing was at a rapid rate, except where indicated. Regrowth was initiated (a) on medium lacking 2,4-D or (b) on medium containing 2,4-D. Callus formation was rare in (a); the extent is indicated by values below baseline. (From Withers, 1979.)

be achieved because the morphological characteristics of the cells in the various regions of the structure are so different. This is a major impediment to the preservation of organized structures in all branches of cryobiology.

C. POLLEN AND POLLEN EMBRYOS

Pollen are very small, dense structures with a low water content and, predictably, are most successfully preserved by rapid freezing. No cryoprotectant treatment is necessary; the pollen may simply be placed in a vial and frozen by direct immersion in liquid nitrogen or in the vapor phase (Nath and Anderson, 1975; Weatherhead *et al.*, 1979). There is some evidence (King, 1965) that pollen may, uniquely in the present context, be preserved by freeze drying.

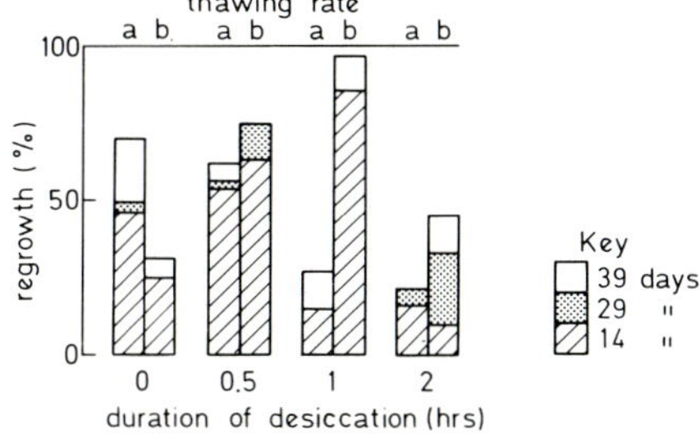

FIG. 17. Time course of regeneration of meristems in clonal plantlets of *Daucus carota* dry frozen at a rate of 2°C per minute to 100°C, transferred to LN, and then thawed relatively rapidly (a) or slowly (b). Before freezing the meristems were subjected to partial desiccation for the periods of time indicated. (From Withers, 1979.)

Pollen embryos isolated from anthers of *Atropa belladonna* and *Nicotiana tabacum* that have been precultured for 3–5 weeks under inductive conditions have successfully been preserved by slow freezing (Bajaj, 1977). After treatment with DMSO (5–7%), freezing at a rate of 1–3°C per minute, storage in LN, and thawing rapidly the embryos were washed and returned to culture. Survival values declined with the progressive development of the embryos: 31% for globular embryos, 9% for early heart-shaped embryos, and 2% for late heart-shaped embryos.

As yet there are no reports of the successful freeze preservation of anthers. There may indeed now be little need for the development of a preservation technique for intact anthers because pollen for fertilization can be stored successfully after isolation, and the development of *in vitro* techniques for the culture of immature pollen/microspores from fresh or precultured anthers (Keller and Stringam, 1978) means that isolated pollen may usefully be stored for the future induction of haploid plants or cultures.

D. PROTOPLASTS

The freeze preservation of isolated protoplasts may have a very limited role to play in genome preservation or as an ancilliary technique for plant tissue culture. However, where cultured tissues or whole plant material are unresponsive to available techniques, single isolated protoplasts may provide a suitable storage form.

Withers has preserved successfully protoplasts of *Daucus carota* and *Zea mays* isolated from suspension cultures (Withers, unpublished observations; Withers and Street, 1977b). When compared with the parent tissue, the protoplasts are tolerant of a wider range of conditions during freezing and thawing and may even be preserved with no chemical cryoprotection beyond the osmotic stabilizer present in their isolation medium. No other reports have been made of genome preservation through protoplast freeze storage, but studies using protoplasts to investigate the effects of freezing and thawing on the structure and function of the plasmalemma (Siminovitch *et al.*, 1978; Wiest and Steponkus, 1978) have highlighted some important factors in the freeze–thaw cycle and will certainly in the future aid our understanding of the response of plant cells to freeze preservation.

V. Alternative Preservation Techniques

Freeze preservation may be an excellent storage facility for some suspension cultures and relatively small callus pieces or organized structures, but it will be evident from the foregoing that there are specific limitations to its use. Consequently, when a worker wishes to store material that is unsuitable for freeze preservation or when a laboratory cannot offer freeze preservation facilities, an alternative method must be sought.

The value of shoot meristem tip culture has been clearly demonstrated for the preservation of foundation stocks of, for example, *Solanum* spp. (Henshaw *et al.*, 1978; Wang, 1977; Westcott *et al.*, 1977). This method avoids the pathogen vector control and strict sanitation requirements inherent in greenhouse or field growth but may actually become a practical encumbrance when many varieties have to be stored for future use. Maintenance of meristem tip cultures or plantlets derived from them at low temperatures can provide a solution. Plantlets of *Fragaria* sp. (Mullin and Schlegel, 1976) have been stored for up to 6 years at 4°C in the dark, with only infrequent attention (every 3 months) to check for medium depletion. Annual transfer is required for shoot tip cultures of *Solanum* spp. maintained at low temperatures (6–8°C) and/or in the presence of growth inhibitors (Henshaw *et al.*, 1978), or for meristem plantlets of several herbage grass species maintained at 2–4°C under low light intensity and an 8-hour photoperiod (Dale, 1978).

Cold storage has been attempted for callus cultures with limited success. Withers (1978c) reports the maintenance of callus of *Capsicum annuum* at 4°C for 25 days extending to 70 days after pretreatment with ethanolamine, a compound known to enhance chill tolerance in whole plants (Ilker *et al.*, 1976). Bannier and Steponkus (1972, 1976) have stored callus cultures of *Chrysanthemum morifolium* at −3.5°C for 28 days and also exposed them to temperatures as low as 16.1°C after acclimation at 4–5°C. Successful cold storage of suspension cultures has not been reported. A drawback to storage at temperatures such as those quoted above is the possibility of selection for resistant genotypes (Dix and Street, 1976) and for this reason cold storage cannot be recommended unless practical or biological reasons preclude the adoption of freeze preservation.

Isolated reports of other methods for suppressing growth have been made. A reduction in the atmospheric pressure or partial pressure of oxygen (Bridgen *et al.*, 1978), the use of mineral oil overlay (Caplin, 1959), and partial desiccation (Jones, 1974; Nitzsche, 1978) have all been reported to give significant reductions in subculturing requirements. As long as the drawbacks to these methods and to cold storage are born in mind, the pursuit of variety in storage methodology can only aid the overall development of a comprehensive genome conservation program.

VI. Conclusions and Prospects

For a number of years, studies of the freeze preservation of plant tissue cultures concentrated upon the storage of model systems. In this way much valuable information has been gained concerning the mechanisms of injury and its avoidance, but difficulties have been encountered in extrapolating from the basic studies to include other plant species. Lately, however, attention has been directed toward the preservation of horticulturally and agriculturally valuable species, such as *Dianthus caryophyllus, Lycopersicon esculentum, Solanum* spp., and *Zea mays*. In each case the studies have resulted in innovative methodology. This is a very encouraging trend and may signal optimism for the future.

Even bearing in mind the progress indicated above, we are still at a stage where no simple recipe can be offered for the successful preservation of a chosen specimen. Broad guidelines can be drawn, indicating the most suitable stage of growth, the likelihood of damage being incurred by cells of a particular morphological character, and some supplementary treatments aimed at improving either the methodology or the response of the specimen to freeze preservation. There is still, however, considerable scope for improving both our technical expertise and our understanding the response of plant cells to the stresses of freezing and thawing.

There are several important areas in which emphasis should be placed in future work and where valuable relevant information may be sought for application to cultured plant cells: (1) Continued study of the nature of chilling injury, its relationship to osmotic stress, and the promotion of healing in cold-damaged tissues will aid the development of appropriate supportive postthaw treatments. (2) Further exploration of the scope for "natural cryoprotection" using such compounds as proline and also for the induction of endogenous compounds that increase freeze tolerance may extend the range of species amenable to freeze preservation and increase the general efficiency of preservation. (3) The development of simple laboratory methods for the routine preservation of tissue cultures will, in parallel with the organized cataloguing of material in storage, ensure the greatest benefit to plant tissue culture workers and the most comprehensive storage program. (4) Continued awareness of progress in all aspects of cryopreservation, and in the very active fields of animal cryobiology and cryosurgery, will enable technical and physiological advances relevant to plant material to be incorporated into plant genome preservation technology. (5) Finally, it is essential that interest be maintained in applying freeze preservation to the wealth of valuable genomes that can be introduced into tissue culture and may even be generated by tissue culture somatic cell genetics.

VII. Addendum

Several pertinent reports have recently come to the attention of the author: Kartha *et al.* [Kartha, K. K., Leung, N. L., and Gamborg, O. L.

(1979). *Plant Sci. Lett.* **15,** 7–15] have detailed the successful preservation method appropriate to meristems of *Pisum sativum* L. (see Section IV,B). It should be noted that 48 hours of preculture in the presence of 5% DMSO preceeded controlled freezing (survival, 60%). Cold conditioning at 4°C prior to freezing without preculture in the presence of DMSO reduced the survival rate (to 28%), and ultra-rapid freezing was lethal (see Seibert and Wetherbee, 1977). Sakai *et al.* [Sakai, A., Yamakawa, T. Sakado, D., Harada, T., and Yakuwa, T. (1978). *Teion Kagaku, Seibutsu-Hen* **36,** 31–38] report the successful preservation (60–80%) of excised runner apices of *Fragaria* × *ananassa* Dutch, cv. "Hokowase." After prefreezing to -20 or -30°C in the presence of 16% DMSO, the apices were frozen rapidly (10^5°C per minute) in LN, and then thawed rapidly by immersion in culture medium at $+40$°C. K. K. Kartha, N. L. Leung, and K. Pahl [(1980). *J. Am. Soc. Hortic. Sci.*, in press] have preserved successfully *in vitro* propagated meristems of this species, by a slow freezing method. In a study of the cryopreservation of excised buds of *Manihot utilissima* Pohl (cv M-4), Bajaj achieved some success [21% recovery consisting of 13% plantlet production and 8% callusing and rooting: Bajaj, Y.P.S. (1977). *Crop Improv.* **4,** 198–204]. Cryoprotection with glycerol (10%) plus sucrose (5%) was followed by rapid freezing and thawing.

Observations on the processes of intracellular and extracellular freezing, described by D. Siminovitch and G. W. Scarth [(1938). *Can. J. Res.* **16,** 467–481], should have been cited in Section II,B. A further publication by Siminovitch, relevant to Section IV,D, describes the survival of black locust tree protoplasts that had been subjected to freezing to -196°C and strong osmotic dehydration [Siminovitch, D. (1979). *Plant Physiol.* **63,** 722–725].

ACKNOWLEDGMENTS

The author was introduced to this field of study by the late Herbert E. Street and gratefully acknowledges his encouragement, guidance, and kindness. Linda Cowling, Nic Everett, Pat King, Angela Monckton, and Sue Pearcey are thanked for helpful discussion and assistance with the preparation of the manuscript.

This article was written during the tenure of a Friedrich Miescher Institut Fellowship in the Botanical Laboratories, University of Leicester, LE1 7RH, U.K., under an arrangement made with the late Professor H. E. Street and continued by Professor H. Smith.

REFERENCES

Ashwood-Smith, M. J., and Farrant, J., eds. (1980). "Low Temperature Preservation in Medicine and Biology." Pitman, London.
Bajaj, Y. P. S. (1976). *Physiol. Plant.* **37,** 263–268.
Bajaj, Y. P. S. (1977). *Curr. Sci.* **46,** 305–307.
Bajaj, Y. P. S., and Reinert, J. (1977). *In* "Applied and Fundamental Aspects of Plant Cell,

Tissue and Organ Culture'' (J. Reinert and Y. P. S. Bajaj, eds.), pp. 757–777. Springer-Verlag, Berlin and New York.

Bannier, L. J., and Steponkus, P. L. (1972). *Hortscience* **7**, 194.

Bannier, L. J., and Steponkus, P. L. (1976). *J. Am. Soc. Hortic. Sci.* **101**, 409–412.

Blum, A., and Ebercon, A. (1976). *Crop Sci.* **16**, 428–430.

Bridgen, M. P., Sharp, W., and Staby, G. L. (1978). *Abstr., Int. Assoc. Plant Tissue Cult., Calgary* p. 106.

Burke, M. J., Gusta, L. V., Quamme, H. A., Weiser, C. J., and Li, P. H. (1976). *Annu. Rev. Plant Physiol.* **27**, 507–528.

Caplin, S. M. (1959). *Am. J. Bot.* **46**, 324–329.

Cella, R., Sala, F., Nielsen, E., Rollo, F., and Parisi, B. (1978). *Abstr. Fed. Eur. Soc. Plant Physiol., Edinburgh* pp. 127–128.

Chu, T. M., Jusaitis, M., Aspinall, D., and Paleg, L. G. (1978). *Physiol. Plant.* **43**, 254–260.

Dale, P. J. (1978). *Abstr. Int. Assoc. Plant Tissue Cult., Calgary,* p. 106.

D'Amato, F. (1978). *In* "Frontiers of Plant Tissue Culture, 1978" (T. A. Thorpe, ed.), pp. 288–296. Int. Assoc. Plant Tissue Cult., Calgary, Alberta.

Dix, P. J., and Street, H. E. (1976). *Ann. Bot.* *(London)* **40** 903–910.

Dougall, D. K., and Wetherell, D. F. (1974). *Cryobiology* **11**, 410–415.

Farrant, J., Walter, C. A., Lee, H., and McGann, L. E. (1977). *Cryobiology* **14**, 273–286.

Finkle, B. J., and Ulrich, J. M. (1978). *Abstr. Int. Assoc. Plant Tissue Cult., Calgary* p. 103.

Finkle, B. J., and Ulrich, J. M. (1979). *Plant Physiol.* **63**, 598–604.

Finkle, B. J., Sa, E., Pereira, B., and Brown, M. S. (1974). *Plant Physiol.* **53**, 705–708.

Finkle, B. J., Sugawara, Y., and Sakai, A. (1975). *Plant Physiol.* **56**, Suppl., 80.

Frankel, O. H. (1970). *Biol. Conserv.* **2**, 162–169.

Frankel, O. H., and Hawkes, J. G., eds. (1975). "Crop Genetic Resources for Today and Tomorrow." Cambridge Univ. Press, London and New York.

Gould, A. R., and Street, H. E. (1975). *J. Cell Sci.* **17**, 337–348.

Gressel, J., Zilkah, S., and Ezra, G. (1978). *In* "Frontiers of Plant Tissue Culture, 1978" (T. A. Thorpe, ed.), pp. 44–53. Int. Assoc. Plant Tissue Cult., Calgary, Alberta.

Gresshoff, P. M. (1977). *Plant Sci. Lett.* **9**, 23–25.

Grout, B. W. W., and Henshaw, G. G. (1978). *Ann. Bot.* *(London)* **42**, 1227–1229.

Grout, B. W. W., and Henshaw, G. G. (1980). *Ann. Bot.* *(London)*, in press.

Grout, B. W. W., Westcott, R. J., and Henshaw, G. G. (1978). *Cryobiology* **15**, 478–483.

Heber, U. (1968). *Cryobiology* **5**, 188–201.

Heber, U., Tyankova, L., and Santarius, K. (1971). *Biochim. Biophys. Acta* **241**, 578–592.

Henshaw, G. G. (1975). *In* "Crop Genetic Resources for Today and Tomorrow" (O. H. Frankel and J. G. Hawkes, eds.), pp. 349–357. Cambridge Univ. Press, London and New York.

Henshaw, G. G., Westcott, R. J., and Roca, W. M. (1978). *Abstr., Int. Assoc. Plant Tissue Cult., Calgary* p. 105.

Hooker, A. L. (1977). *Crop Sci.* **17**, 689–694.

Ilker, R., Waring, A. J., Lyons, J. M., and Breidenbach, R. W. (1976). *Protoplasma* **90**, 229–252.

Jones, L. H. (1974). *Plant Sci. Lett.* **2**, 221–224.

Keller, W. A., and Stringam, G. R. (1978). *In* "Frontiers of Plant Tissue Culture, 1978" (T. A. Thorpe, ed.), pp. 113–122. Int. Assoc. Plant Tissue Cult., Calgary, Alberta.

King, J. R. (1965). *Bull Torrey Bot. Club* **92**, 270–287.

Latta, R. (1971). *Can. J. Bot.* **49**, 1253–1254.

Levitt, J. (1978). *In* "Plant Cold Hardiness and Freezing Stress" (P. H. Li and A. Sakai, eds.), pp. 3–15. Academic Press, New York.

Lyons, J. M. (1973). *Annu. Rev. Plant Physiol.* **24**, 445–466.

McGann, L. E. (1978). *Cryobiology* **15**, 382–390.

Mazur, P. (1969). *Annu. Rev. Plant Physiol.* **20**, 419–448.

Mazur, P. (1977). *Cryobiology* **14**, 251–272.

Mazur, P., Leibo, S. P., and Chu, E. H. Y. (1972). *Exp. Cell Res.* **71**, 345–355.

Meryman, H. T. (1966). *In* "Cryobiology" (H. T. Meryman, ed.), pp. 2–106. Academic Press, New York.

Meryman, H. T., Williams, R. J., and Douglas, M. St. J. (1977). *Cryobiology* **14**, 287–302.

Morel, G. (1975). *In* "Crop Genetic Resources for Today and Tomorrow" (O. H. Frankel and J. G. Hawkes, eds.), pp. 327–332. Cambridge Univ. Press, London and New York.

Morris, G. J. (1980). *In* "Low Temperature Preservation in Medicine and Biology" (M. J. Ashwood-Smith and J. Farrant, eds.), pp. 253–284. Pitman, London.

Mullin, R. H., and Schlegel, D. E. (1976). *Hortscience* **11**, 100–101.

Murashige, T. (1978). *In* "Frontiers of Plant Tissue Culture, 1978" (T. A. Thorpe, ed.), pp. 15–26. Int. Assoc. Plant Tissue Cult., Calgary, Alberta.

Nag, K. K., and Street, H. E. (1973). *Nature (London)* **245**, 270–272.

Nag, K. K., and Street, H. E. (1975a). *Physiol. Plant.* **34**, 254–260.

Nag, K. K., and Street, H. E. (1975b). *Physiol. Plant.* **34**, 261–265.

Nash, T. (1966). *In* "Cryobiology" (H. T. Merymann, ed.), pp. 179–211. Academic Press, New York.

Nath, T., and Anderson, J. O. (1975). *Cryobiology* **12**, 81–88.

Nitzsche, W. (1978). *Z. Pflanzenphysiol.* **87**, 469–472.

Palta, J. P., and Li, P. H. (1978). *In* "Plant Cold Hardiness and Freezing Stress" (P. H. Li and A. Sakai, eds.), pp. 93–115. Academic Press, New York.

Palta, J. P., Levitt, J., and Stadelman, E. J. (1977). *Plant Physiol.* **60**, 393–397.

Palta, J. P., Levitt, J., and Stadelmann, E. J. (1978). *Cryobiology* **15**, 249–255.

Pollard, A., and Wyn-Jones, R. G. (1979). *Planta* **144**, 291–298.

Quatrano, R. S. (1968). *Plant Physiol.* **43**, 2057–2061.

Rajagopal, V., and Anderson, A. S. (1978). *Planta* **143**, 85–88.

Sakai, A., and Otsuka, K. (1967). *Plant Physiol.* **42**, 1680–1694.

Sakai, A., and Otsuka, K. (1972). *Plant Cell Physiol.* **13**, 1129–1133.

Sakai, A., and Sugawara, Y. (1973). *Plant Cell Physiol.* **14**, 1201–1204.

Sakai, A., and Sagawara, Y. (1978). *In* "Plant Cold Hardiness and Freezing Stress" (P. H. Li and A. Sakai, eds.), pp. 345–359. Academic Press, New York.

Sakai, A., and Yoshida, S. (1968). *Cryobiology* **5**, 160–174.

Sala, F., Cella, R., and Rollo, F. (1979). *Physiol. Plant.* **45**, 170–176.

Seibert, M., and Wetherbee, P. J. (1977). *Plant Physiol.* **59**, 1043–1046.

Shillito, R. D. (1978). Ph.D. Thesis, Univ. of Leicester, Leicester.

Siminovitch, D., Therrien, H., Gfeller, F., and Rheaume, B. (1964). *Can. J. Bot.* **42**, 637–649.

Siminovitch, D., Singh, J., and de la Roche, I. A. (1978). *Cryobiology* **15**, 205–213.

Skaer, H. le B., Franks, F., and Echlin, P. (1978). *Cryobiology* **15**, 589–602.

Stewart, G. R., and Lee, J. A. (1974). *Planta* **120**, 279–289.

Stuart, R., and Street, H. E. (1969). *J. Exp. Bot.* **20**, 556–571.

Sugawara, Y., and Sakai, A. (1974). *Plant Physiol.* **54**, 722–724.

Sugawara, Y., and Sakai, A. (1975). *Teion Kagaku, Seibutsu-Hen* **33**, 21–28.

Sugawara, Y., and Sakai, A. (1978). *In* "Plant Cold Hardiness and Freezing Stress" (P. H. Li and A. Sakai, eds.), pp. 197–210. Academic Press, New York.

Takeuchi, M. (1978). *Abstr. Int. Assoc. Plant Tissue Cult., Calgary* p. 1105.

Towill, L. E., and Mazur, P. (1974). *Can. J. Bot.* **53**, 1097–1102.

Towill, L. E., and Mazur, P. (1976). *Plant Physiol.* **57**, 290–296.

Trump, B. F., Croker, B. P., and Mergner, W. J. (1971). *In* "Cell Membranes, Biological and Pathological Aspects" (G. W. Richter and D. G. Scarpelli, eds.), pp. 84–129. Williams & Wilkins, Baltimore, Maryland.

Tumanov, I. I., Butenko, R. G., and Ogolevets, I. V. (1968). *Fiziol. Rast.* **15,** 749–756.

Verkleij, A. J., Ververgaert, P. H. J., Van Deenen, L. L. M., and Elbers, P. F. (1972). *Biochim. Biophys. Acta* **288,** 326–332.

Walkey, D. G. A. (1978). *In* "Frontiers of Plant Tissue Culture, 1978" (T. A. Thorpe, ed.), pp. 245–254. Int. Assoc. Plant Tissue Cult., Calgary, Alberta.

Wang, P. J. (1977). *In* "Plant Tissue Culture and its Bio-technological Application" (W. Barz, E. Reinhard, and M. H. Zenk, eds.), pp. 386–341. Springer-Verlag, Berlin and New York.

Warren, G. S., and Fowler, M. W. (1979). *Planta* **144,** 451–454.

Weatherhead, M. A., Grout, B. W. W., and Henshaw, G. G. (1979). *Potato Res.* **21,** 97–100.

Westcott, R. J., Henshaw, G. G., and Roca, W. M. (1977). *Plant Sci. Lett.* **9,** 309–315.

Whittingham, D. G., Lyon, M. F., and Glenister, P. H. (1977). *Genet. Res.* **29,** 171–181.

Widholm, J. M. (1972). *Stain Technol.* **47,** 189–194.

Wiest, S. C., and Steponkus, P. L. (1978). *Plant Physiol.* **62,** 699–705.

Withers, L. A. (1978a). *Cryobiology* **15,** 87–92.

Withers, L. A. (1978b). *Protoplasma* **94,** 235–247.

Withers, L. A. (1978c). *In* "Frontiers of Plant Tissue Culture, 1978" (T. A. Thorpe, ed.), pp. 297–306. Int. Assoc. Plant Tissue Cult., Calgary, Alberta.

Withers, L. A. (1979). *Plant Physiol.* **63,** 460–467.

Withers, L. A. (1980). *Adv. Biochem. Eng.* **18,** in press.

Withers, L. A., and Davey, M. R. (1978). *Protoplasm* **94,** 207–219.

Withers, L. A., and King, P. J. (1979a). *Plant Physiol.* **64,** 675–678.

Withers, L. A., and King, P. J. (1979b). *Experientia* **35,** 984.

Withers, L. A., and King, P. J. (1980). *Cryoletters,* in press.

Withers, L. A., and Street, H. E. (1977a). *Physiol. Plant.* **39,** 171–178.

Withers, L. A., and Street, H. E. (1977b). *In* "Plant Tissue Culture and its Bio-technological Application" (W. Barz, E. Reinhard, and M. H. Zenk, eds.), pp. 226–244. Springer-Verlag, Berlin and New York.

Yoshida, S., and Sakai, A. (1974). *Plant Physiol.* **53,** 509–511.

NOTE ADDED IN PROOF. Several pertinent reports have recently come to the attention of the author: Kartha *et al.* [Kartha, K. K., Leung, N. L., and Gamborg, O. L. (1979). *Plant Sci. Lett.* **15,** 7–15] detail the successful preservation method appropriate to meristems of *Pisum sativum* L. (see Section IV,B). It should be noted that 48 hours of preculture in the presence of 5% DMSO preceded controlled freezing (survival 60%). Cold conditioning at 4°C prior to freezing without preculture in the presence of DMSO reduced the survival rate (to 28%), and ultra-rapid freezing was lethal (cf. Seibert and Wetherbee, 1977). Sakai *et al.* [Sakai, A., Yamakawa, T., Sakado, D., Harada, T., and Yakuwa, T. (1978). *Low Temp. Sci. Ser. B* **36,** 31–38] report the successful preservation (60–80%) of excised runner apices of *Fragaria* × *ananassa* Dutch, cv. 'Hokowase.' After prefreezing to −20 or −30°C in the presence of 16% DMSO, the apices were frozen rapidly (10^5°C min^{-1}) in LN, and then thawed rapidly by immersion in culture medium at +40°C. In a study of the cryopreservation of excised buds of *Manihot utilissima* Pohl (cv M-4), Bajaj achieved some success [21% recovery consisting of 13% plantlet production and 8% callusing and rooting; Bajaj, Y. P. S. (1977). *Crop Improv.* **4,** 198–204]. Cryoprotection with glycerol (10%) plus sucrose (5%) was followed by rapid freezing and thawing. A further publication by Siminovitch, relevant to Section IV,D, describes the survival of black locust tree protoplasts which had been subjected to freezing to −196°C and strong osmotic dehydration [Siminovitch, D. (1979). *Plant Physiol.* **63,** 722–725].

Chapter 15

Intraovarian and *in Vitro* Pollination

M. ZENKTELER

Institute of Biology, Adam Mickiewicz University, Poznan, Poland

I. Introduction

Sexual reproduction constitutes a very important mechanism for the introduction of variability in a population. For a long time the precise role of pollen grains of reproduction remained obscure, but even in ancient times plant breeders were aware of the fact that pollen was indispensable for seed set. Possibilities of producing new combinations between the crossed plants are limited because of barriers existing during the process of pollination, fertilization, and embryogenesis. Efforts to obtain inter- and intraspecific, intergeneric, or interfamiliar hybrids face such formidable barriers that in most cases they cannot be overcome by the available methods. In some species, whether self- or cross pollinated, obstacles present on the stigma or style make it impossible for the pollen tube to grow inside the ovary and later through the micropyle of the ovule. As far as the initial germination of the pollen is concerned, most stigmas are not too specific. The main problem lies in the subsequent growth of the pollen tube in the style.

The length the pollen tube has to travel to reach the embryo sac is, in most plants, far too great for the necessary tube wall material to be supplied from the comparatively small store of carbohydrates in the pollen grain itself (Johri and Vasil, 1961). It has often been debated whether the

137

cells of the stylar canal, or the transmitting tissue, play a nutritive or chemotropic role in directing the growth of the pollen tube (Vasil, 1974). Pollen tubes on their way to the embryo sac grow over or between transmitting tissue. From a comparison of the composition of expended exudate adhering to the pollen tubes, of pollen tube cytoplasm, and of pollen tube walls, the following sequence is suggested: (1) Utilization of exudate by growing pollen tubes involves the uptake of polysaccharide fragments from the stylar canal into pollen tube cytoplasm; (2) this is followed by extensive metabolism to mono- and oligosaccharides; and (3) these metabolized fragments are then reassembled into the tube wall (see reviews in Vasil, 1973, 1974; Mascarenhas, 1975).

As compared to extensive studies on growth of pollen tube in the pistil in various plants, the mechanism of the entry of the tube into the embryo sac is still not well understood. There are only some electron microscopic studies describing the growth of the pollen tube through the stigma and style, its penetration of the embryo sac, and the discharge of the sperm cell. According to Jensen and Fisher (1968), in cotton, the best studied plant so far, one of the two synergids begins to degenerate approximately 6–8 hours after pollination but 6–8 hours before the pollen tube penetrates the embryo sac. Degeneration of the synergid prior to fertilization probably does not occur in all plants; however, it cannot be excluded as detailed embryological studies are still lacking.

The pathway of the pollen tube, which starts at the stigma and finishes by bursting inside the embryo sac, is only one important, but not indispensable, factor responsible for the formation of the zygote. Much more complicated events occur during and just after syngamy, but our knowledge about this process is very meager and we have still not been able to overcome barriers that hinder the fusion of both gametes when the plants are pollinated with foreign pollen grains. Even in those cases when fusion takes place, the zygote may or may not divide.

In plant breeders, haploids and intergeneric hybrids are of great importance. Therefore, techniques have been worked out to obtain desirable results. The method of *in vitro* culture offers an opportunity to analyze in detail the processes of pollination, fertilization, and embryogenesis in completely controlled conditions. Beside this area of investigation there is a second one of a particular importance: to achieve hybrids among plants that cannot be crossed by conventional methods *in vivo*. When the barriers hindering the process of pollen tube growth lie in the stigma and style, then one of the remedies suggested is to cut away a part of the style, or the whole style, and put pollen onto the surface of the ovary. Another method for stimulating the growth of the pollen tube is the application of certain chemical substances (auxins, plant hormones, vitamins) to the stigma or the direct introduction of pollen into the ovary. This chapter reviews some methods already applied for obtaining embryos and plants by intraovarian pollination and through pollination and fertilization in ovules in test tubes.

II. Intraovarian Pollination

Strasburger (1886) was the first to have the idea of applying the technique of intraovarian pollination. Later on, Dahlrgen (1926) was able to obtain fertilization in ovules of *Codonopsis ovata* by cutting away the style and putting the pollen on the upper part of the ovary. Bosio (1940) also used this technique in *Helleborus* and *Paeonia* and observed fertilization in ovules. At the University of Delhi, Kanta (1960) and Kanta *et al.* (1962) successfully conducted intraovarian pollination with *Papaver rhoeas*, *Papaver somniferum*, *Eschscholzia californica*, *Argemone mexicana*, and *Argemone ochroleuca*. Before the experiments were begun the times of flower opening and dehiscence of the anthers were determined. After this the anthers were removed from selected flower buds, which were bagged in cellophane in order to prevent pollination. Meanwhile, the best medium for pollen grain germination was worked out. A pollen suspension was prepared in sterile, double-distilled water with 100 mg/liter of boric acid. The surface of ovaries used for the experiments was sterilized with ethanol and the following sets of experiments were conducted: (1) Flowers were allowed to become pollinated as in nature; (2) the ovary wall was pricked with a syringe but without introducing pollen; (3) pollen suspensions in 100 and 200 mg/liter of boric acid solution were injected into the ovary; or (4) pollen grains were introduced into the ovary through a slit made on its wall or through an opening made from above after the stigma had been cut away. When the suspension was injected into the ovaries the pollen grains germinated and some of the pollen tubes entered the micropyle. Fertilization occurred and consequently normal seeds developed. The largest number of seeds was obtained when the pollen grains were introduced directly into the ovary the stigma was removed. In all the above experiments it was found that by-passing the stigma and style and directly injecting pollen into the ovary could also bring about fertilization and the formation of seeds. Theoretically, this technique may be useful when the barriers to fertilization lie in the stigma or style. Unfortunately, in the literature there are no available data showing whether this technique is useful for overcoming the barriers of infertility in plants except the information gathered from the experiments carried out by Niemirowicz-Syczytt and Kubrick (1979), who were trying to obtain hybrids in a cross between *Cucurbita maxima* × *Cucumis melo* (Fig. 1A). Pollen

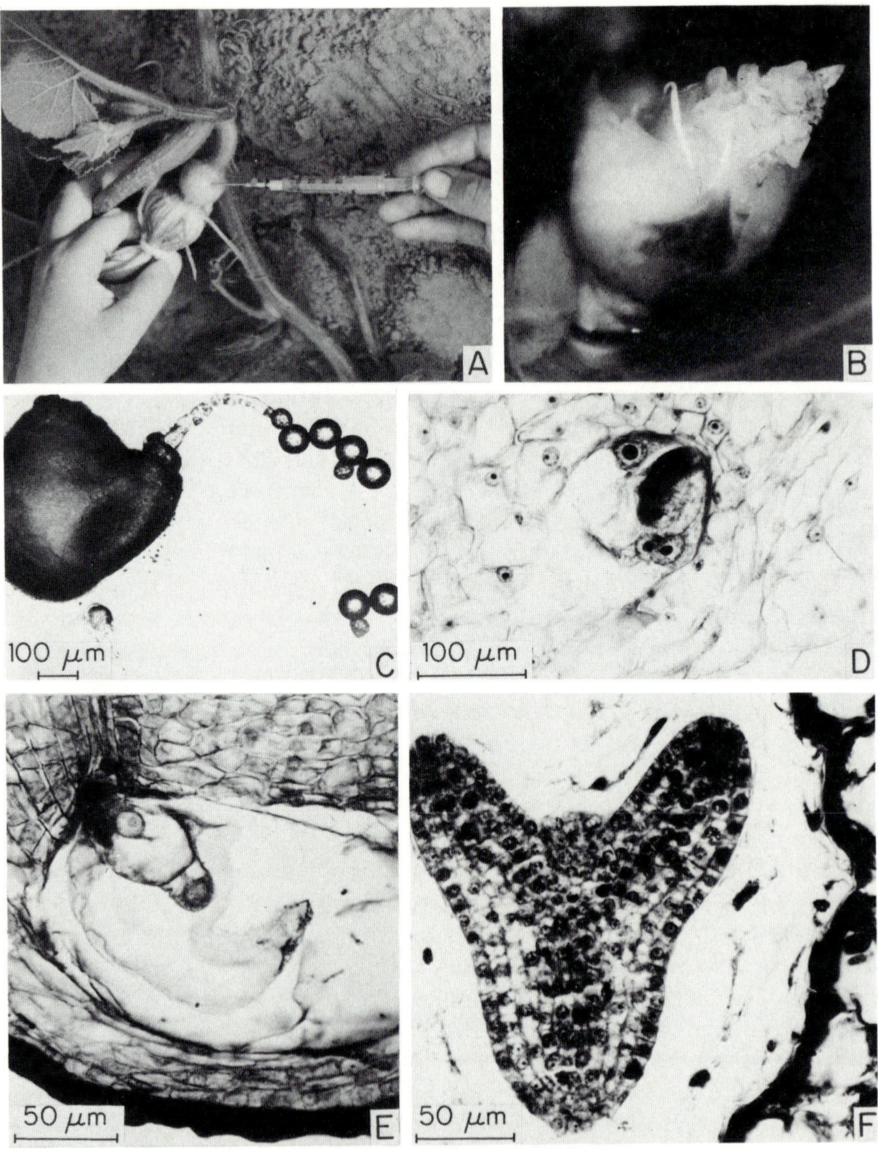

FIG. 1. (A) Intraovarian pollination of *Cucurbita maxima* with pollen grains of *Cucumis melo*. (B) Developing ovules situated on the placenta of *Dianthus caryophyllus*, 8 days after ttp (test tube pollination) (×5.4). (C) Pollen tube of *Melandrium rubrum* entering the micropyle of the ovule of *Melandrium album*. (D) Pollen tube of *Silene schafta* in the micropylar end of the embryo sac of *M. album* 24 hours after ttp. (E) Two-celled proembryo of *M. album* 24 hours after self-pollination *in vitro*. (F) Embryo with cotyledons (*M. album*) 9 days after self-pollination *in vitro*.

grains were introduced inside the ovary but unfortunately no hybrid embryos were obtained.

III. *In Vitro* Pollination

A. SELF-POLLINATION

A more sophisticated technique is the inoculation *in vitro* of ovules or ovules together with the placenta. In both cases transferred ovules are later directly pollinated. The efficacy of this technique depends to some extent on the composition of media that promote, in the case of fertilization, the development of embryos and endosperm. The preliminary steps in the process are the same as in intraovarian pollination. Flower buds are emasculated at the bud stage before the dehiscence of the anthers and the pollen are collected under aseptic conditions. Usually anthers are dissected from buds just before opening and kept in sterile conditions until they burst. The ovaries are sterilized and later on the ovules are excised and transferred to a medium. It was found with several species that it was much more suitable to inoculate ovules still attached to the placenta (Fig. 1B). This prevents the injury of the ovules and also ensures a better supply of the nutritive substances from the gynoecial tissue. When the placenta is well developed and covered with a large number of ovules, as in *Nicotiana tabacum* and *Melandrium album,* it can be cut into small pieces and then placed onto the medium. Little is known about the effects of the medium on the process of fertilization *in vitro* and the development of embryos. The roles of different components present in the medium, such as minerals, vitamins, cytokinins, and auxins, or the concentration of sucrose also are not well known. The basic media used are those of White (1963), Nitsch (1951, 1969), modified White's medium (Rangaswamy, 1961), and Murashige and Skoog (1962).

There is also a lack of data about the precise effects of physical factors, such as light, humidity, and temperature, on test tube fertilization. Zenkteler (1969), working with *M. album,* found that light did not effect pollen tube growth on ovules. Fertilization took place and seeds developed normally in dark as well as in light. Similar observations were made when ovules of *Melandrium album* and *Melandrium rubrum* were pollinated with pollen of other species.

The normal growth of pollen tubes when pollen stick to the naked ovules is often dependent on the presence of water on the surface of the ovules. When ovaries of *Solanum tuberosum* or *Lycopersicum esculen-*

tum were inoculated onto an agar medium, without regared to agar concentration, a film of water soon appeared on the surface of ovules or placentas. Efforts to dry the material by filter paper were unsuccessful, as water continuously accumulated on the surface of the ovules. In such cases pollen grains usually do not germinate, or even when they start germinating they later burst and, therefore, the chances of fertilization are very poor to none (Zenkteler, unpublished observations).

Usually pollen grains are put directly on the surface of ovules, but, as reported by Balatkova and Tupy (1972), who cultured placentas of *N. tabacum*, better results can be obtained when a certain distance between pollen grains and ovules is maintained. Some modification of the method of direct pollination of ovules has been tried by Niimi (1976), who analyzed the effect of placental and stylar pollination on *in vitro* seed setting of *Petunia hybrida*. Pollen grains were put directly onto the placenta or the styles were excised, cut to about 7–10 mm in length after pollen grains were applied to the stigma, and then were placed on or around the ovules. These experiments showed that, *in vitro*, stylar pollination was much more successful than placental pollination and the rate of seed setting was much higher. It is possible that the technique of stylar pollination, because of a normal growth of pollen tube and a normal division of generative cells, is superior to placental pollination. However, these experiments can be carried out only styles that support normal pollen tube growth.

In general ovules, together with the placentas, are cultured *in vitro* because only the presence of the placenta ensures the development of seeds. It is not known whether ovules, when cultured and pollinated, undergo fertilization and are capable of further growth. Kameya *et al*. (1966) are the only ones who have succeeded to some extent in culturing single ovules. They inoculated ovules of *Brassica oleracea* on a medium of gelatin film, and pretreated them with 1% $CaCl_2$ solution before transfer. As soon as the ovules were put on the gelatin film, pollen grains from flowers that had just bloomed were applied. The slides with pollinated ovules were put into a glass dish and covered with a lid having a sheet of filter paper moistened with water. The ovules that were penetrated by pollen tubes were later transplanted into test tubes containing Nitsch's medium. Two out of the total 75 ovules pollinated grew into seeds and germinated later. The authors supposed that the pretreatment of ovules with 1% $CaCl_2$ solution played an important role in these experiments. It seems that in the case of culturing single ovules the composition of the medium and other environmental factors play important roles in their further development.

Ovules that are fertilized in test tubes must later be transplanted and cultured in fresh medium, which again constitutes a great difficulty. In some cases, as in *Melandrium,* embryos and endosperm developed in ovules still attached to the placenta (Fig. 1C–F), and seeds produce seedlings *in situ*. When ovules pollinated *in vitro* were excised 2 days after test tube pollination and transferred to Murashige and Skoog's medium containing various concentrations of 2,4-dichlorophenoxyacetic acid (2,4-D) and kinetin, ovules of *Nicotiana* developed further and during the next 25 days embryos started to germinate and later produced seedlings. In contrast, in ovules of *Melandrium* during the same period only globular embryos developed, and these later degenerated (Zenkteler, unpublished observations).

It is still difficult to culture ovules soon after fertilization *in vivo* or even at the stage of several celled proembryos. Only a few successful attempts have been made on *Zephyranthes* by Kapoor (1959), *P. somniferum* by Maheshwari and Lal (1961), *Gynandropsis gynandra* and *Impatiens balsamina* by Chopra and Sabharwal (1963), *Abelmoschus esculenthus* by Bajaj (1964), *Trifolium repens* by Nakajima *et al.* (1969), *P. hybrida* by Nakajima and Wakizuka (1977), and *Gossypium hirsutum* by Joshi (1962). Misiura and Zenkteler (unpublished observations) were able to obtain fully developed embryos from ovules of *Atropa belladonna* inoculated at the stage of zygote or several celled proembryo. They concluded that the zygote or a very young globular proembryo present in the ovules was able to develop further if the osmotic value of the medium and other nutritive components were favorable.

In vitro self-pollination of placentae with ovules has been performed successfully with several species (Table I). As the table shows, the process of sexual reproduction *in vitro* occurred only in those species where ovaries contained a large number of ovules. This is probably one of the main conditions for achieving any success in this technique. When ovaries with only one ovule are isolated, the ovule is usually damaged, probably because of the presence of a poor placenta and physiological disturbances occurring in tissues surrounding the female gametophyte, which, therefore, is not able to undergo fertilization. The processes of fertilization and subsequent embryogenesis are so complicated that, at present, we are not able to ensure the optimal growth of embryos *in vitro*. The results obtained up to now indicate that the method of pollination and fertilization *in vitro* produces seeds with fully developed embryos only in certain species. Different methods of cultivation and a larger series of experiments might help to obtain more positive results in a wider range of species.

TABLE I

FERTILIZATION AND THE FORMATION OF SEEDS AFTER SELF-POLLINATION OF
PLACENTAS WITH ATTACHED OVULES

Species	Viable seeds	Nongerminable seeds	Basic medium[a]	References
Agrostemma githago	+		N	Zubkova and Sladky (1975)
Antirrhinum majus		+	N	Zubkova and Sladky (1975)
Argemone mexicana	+		N	Kanta and Maheshwari (1963)
Clivia miniata		+	N	Zubkova and Sladky (1975)
Cucubalus baccifer		+	N	Zenkteler (1969)
Dianthus caryophyllus	+		W	Zenkteler (1965)
Dicranostigma franchetianum	+		RS	Rangaswamy and Shivanna (1969)
Eschscholzia californica	+		N	Kanta and Maheshwari (1963)
Galanthus nivalis		+	N	Zubkova and Sladky (1975)
Melandrium album	+		WN	Zenkteler (1967); Zubkova and Sladky (1975)
Melandrium rubrum	+		W	Zenkteler (1967)
Narcissus pseudonarcissus		+	N	Zubkova and Sladky (1975)
Nicotiana alata	+		LS	Zenkteler (1980)
Nicotiana rustica	+		N	Kanta and Maheshwari (1963)
Nicotiana tabacum	+		N	Kanta and Maheshwari (1963); Dulieu (1963)
Paeonia officinalis		+	N	Zubkova and Sladky (1975)
Papaver somniferum	+		W	Kanta *et al.* (1962)
Petunia axillaris	+		RS	Rangaswamy and Shivanna (1967)
Petunia hybrida	+		Ni	Niimi (1970)
Tulipa gesneriana		+	N	Zubkova and Sladky (1975)
Zea mays	+		MS	Dhaliwal and King (1978)

[a] LS, Linsmaier and Skoog (1965); MS, Murashige and Skoog (modified 1962; in Green and Phillips, 1975); N, Nitsch (1951); Ni, Niimi (1970); RS, Rangaswamy and Shivanna (1967); W, White (1963; modified in Rangaswamy, 1961).

B. CROSS POLLINATION

Another field in which the technique of *in vitro* pollination and fertilization may be especially useful is in raising interspecific and intergeneric hybrids that cannot be obtained in nature by conventional methods. At present it is possible to produce somatic hybrid plants by protoplast fusion but largely in those cases where the species involved can also be crossed sexually *in vivo*. However, the method of sexual *in vitro* hybridization yields hybrid embryos when normal *in vivo* hybridization fails. This promising method allows the culture of both ovules and pollen in an artificial medium under closely controlled conditions. Although many embryological investigations on intergeneric hybridization have been made *in vivo,* there is still very little information on or understanding of intergeneric fertilization in test tubes.

During the course experiments carried out *in vivo,* it often has been found impossible to bring about fertilization following cross pollination by the application of standard pollinating methods. Before experiments on cross pollination of ovules in test tubes are begun it is important to solve two basic problems: (1) Ovules must be found that are suitable for fertilization by their own and foreign pollen grains; (2) conditions must be found that allow induced growth of arrested hybrid proembryos that are not able to grow further. In general, especially among species belonging to the same family, pollen grains often germinate on foreign stigmas but, because of various obstacles, their tubes are not able to reach the ovules. There is very little information about the germination of pollen on stigmas in comparison with that available about their growth on naked ovules. Guzowska (1971) selected several species belonging to the Caryophyllaceae and Cruciferae. Stigmas *in vivo* and ovules *in vitro* were self- and cross pollinated. In the latter case, the pollen used belonged to species of both families. Altogether, 174 combinations were obtained, on the basis of which it was ascertained that, in the Caryophyllaceae, pollen grains germinated easily and in a higher percentage on ovules than on stigmas. These results encourage the continuation of such experiments for obtaining hybrids *in vitro.*

The most satisfactory development of hybrid embryos in test tubes was found when ovules of *M. album* were pollinated with pollen grains of various species belonging to the same or other families (Table II). Also, some interesting results were obtained when ovules of *N. tabacum* were cross pollinated with species of the same family. Ovaries of both species are rich in well-developed placentas covered with several hundred ovules each (Fig. 2A). When ovules of *Melandrium* were cross pollinated pollen grains germinated profusely, producing long tubes that penetrated the ovules, and fertilization often occurred within a few hours. The development of proembryos concomitantly with endosperm, of endosperm only, or of proembryos only occurred in various combinations. The best development of hybrid embryos and endosperm took place in ovule cultures pollinated with pollen of *M. rubrum* (Fig. 2C), *Viscaria vulgaris* (Fig. 2B,D), *Silene schafta* (Fig. 2E), *Silene friwaldskyana, Silene alpina, Cucubalaus baccifer, Lychnis coronaria, Minuartia laricifolia, Vaccaria pyramidata,* and *Dianthus serotinus.* In all the above species, fertilization and the early stages of embryo development occurred. However, the percentage of ovules producing embryos varied depending on the species

TABLE II

Fertilization and the Formation of Hybrid Embryos after Cross Pollination of Placentas with Attached Ovules

Species	Seeds containing viable embryos	Seeds containing immature embryos	References
Melandrium album × *Melandrium rubrum*	+		Zenkteler (1967)
Melandrium album × *Viscaria vulgaris*	+		Zenkteler (1969)
Melandrium album × *Silene schafta*	+		Zenkteler (1967, 1969)
Melandrium album × *Silene friwaldskyana*		+	Zenkteler *et al.* (1975)
Melandrium album × *Silene alpina*		+	Zenkteler *et al.* (1975)
Melandrium album × *Silene tatarica*		+	Zenkteler (1967, 1969)
Melandrium album × *Minuartia laricifolia*		+	Zenkteler *et al.* (1975)
Melandrium album × *Vaccaria pyramidata*		+	Zenkteler *et al.* (1975)
Melandrium album × *Dianthus serotinus*		+	Zenkteler *et al.* (1975)
Melandrium album × *Dianthus carthusianorum*		+	Zenkteler (1967)
Melandrium album × *Campanula persicifolia*		+	Zenkteler *et al.* (1975)
Melandrium album × *Cucubalus baccifer*		+	Zenkteler (1969)
Melandrium rubrum × *Melandrium album*	+		Zenkteler (1969)
Melandrium tabacum × *Petunia hybrida*		+	Zenkteler and Melchers (1978)
Nicotiana tabacum × *Hyoscyamus niger*		+	Zenkteler and Melchers (1978)
Nicotiana tabacum × *Physosclaina praealta*		+	Zenkteler *et al.* (1981)
Nicotiana tabacum × *Withamia somnifera*		+	Zenkteler *et al.* (1981)
Nicotiana alata × *Nicotiana debneyi*	+		Zenkteler *et al.* (1981)
Zea mays × *Zea mexicana*	+		Dhaliwal and King (1978)
Melandrium album × *Cerastium arvense*	Only endosperm formation		Zenkteler *et al.* (1975)
Melandrium album × *Datura stramonium*	Only endosperm formation		Zenkteler (1971)
Melandrium album × *Petunia hybrida*	Only fertilization		Zenkteler and Melchers (1978)
Melandrium album × *Mathiola indica*	Only fertilization		Zenkteler (unpublished observations)

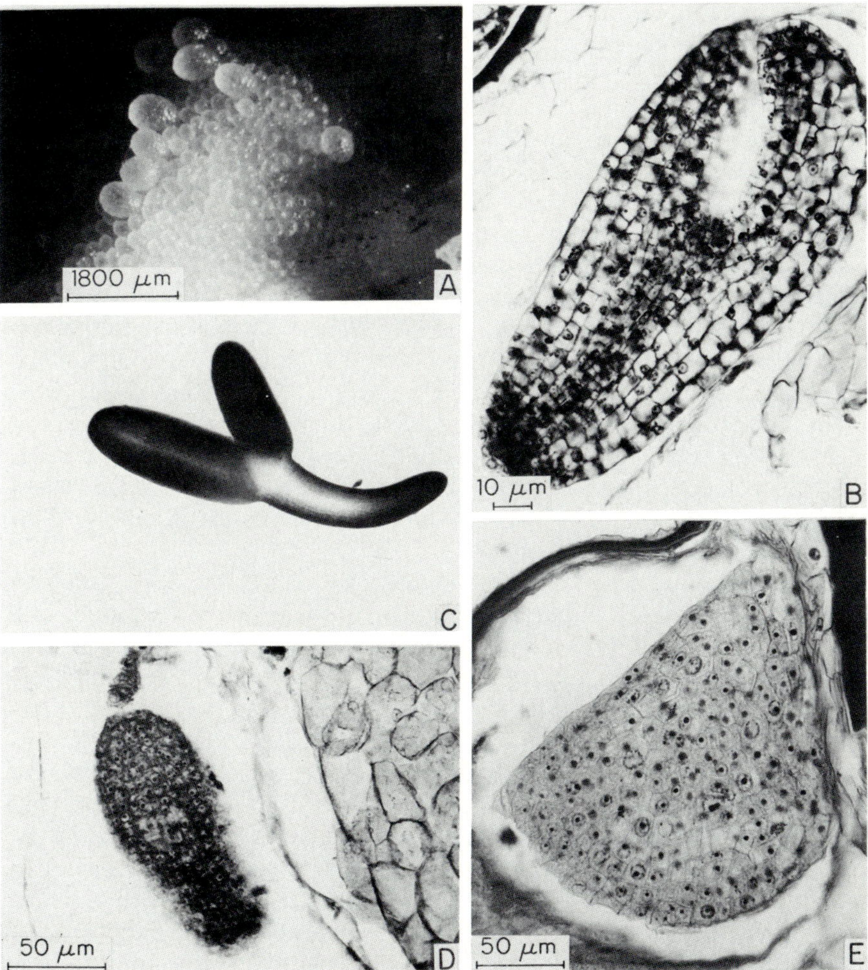

FIG. 2. (A) Enlarged ovules of *Nicotiana tabacum* situated on the placenta 4 days after pollination *in vitro* with pollen grains of *Hyoscyamus* niger. (B,D) Hybrid embryos of *Melandrium album* × *Viscaria vulgaris* 10 days after test tube pollination (ttp). (C) Fully developed hybrid embryo of *M. album* × *M. rubum* 12 days after ttp (× 30). (E) Hybrid embryo of *M. album* × *Silene schafta* 10 days after ttp.

used as pollinators. The early growth of proembryos was more or less similar in all the crossed species, but differences in size and structure of the embryos became marked as the embryos reached older stages. It was especially visible when *Melandrium* was pollinated in test tubes with pollen of *S. schafta* and *V. vulgaris*. These are the only cases in which intergeneric hybrid embryos and healthy plants developed (Fig. 3F). Otherwise, the immature embryos collapsed after reaching the globular stage. Most of the intergeneric hybrid embryos failed to persist beyond the eighth day. Several abnormalities were observed in the hybrid embryos and endosperm. In some crosses endosperm nuclei failed to divide normally and exhibited chromosomal bridges and other abnormalities. They often formed scattered masses of densely stained nuclei accumulated in various regions of embryo sacs, and, in the cross *M. album* × *Cucubalus baccifer,* endosperm did not develop at all and the small globular proembryos were situated inside the embryo sac cavity (Fig. 3A). Interesting results were obtained in the crosses *M. album* × *Campanula persicifolia, M. album* × *Datura stramonium,* and *M. album* × *P. hybrida.* In the first cross, only globular proembryos were formed without the development of endosperm. In the cross *Melandrium* × *Datura,* the process of fertilization of the secondary nucleus was traced but no fertilization was observed inside the egg cell. Consequently, a dividing egg cell or zygote was never noticed; only endosperm composed of up to 30 nuclei was seen. When *Melandrium* was pollinated with *Petunia,* a male gamete in mitotic division was noticed inside the egg cell.

Quite promising results were obtained by Zenkteler and Melchers (1978) in the cross *N. tabacum* × *Hyoscyamus niger* where after 7–8 days of culture the hybrid embryos reached such a large size that it was possible to isolate them and transfer them to a fresh liquid medium (Fig. 3B). In this case hybrid embryos or whole ovules were excised and transferred to a liquid medium (Murashige and Skoog, 1962) containing various additives. The cultures were agitated on a rotary shaker for 20 days. On the basic medium containing kinetin (2 mg/liter), 2,4-D (4 mg/liter), and sucrose, the cultured globular embryos developed further and reached the preheart stage in 7 days. After another week they started to produce callus. Unfortunately, this medium did not support the good growth of callus cells and the further growth of embryos was arrested. These experiments, which were partly successful with only one medium, show that there is some hope of success if the right medium can be found.

Hybrid embryos and plants were also obtained by Zenkteler (unpublished observations) in the cross *Nicotiana alata* × *Nicotiana debneyi.* Here the whole process of fertilization, embryogenesis, and seed development proceeded in test tubes. Seeds started to germinate 5 weeks following pollination of ovules. The hybrid plants (Fig. 3C–E) were completely male sterile. In some cases the female gametophyte developed normally.

There is only one paper concerning successful cross fertilization of ovules in monocotyledons. Dhaliwal and King (1978) used direct pollination of ovules of corn × corn and corn × teosinte and obtained well-developed embryos with endosperm. It must be pointed out that the valu-

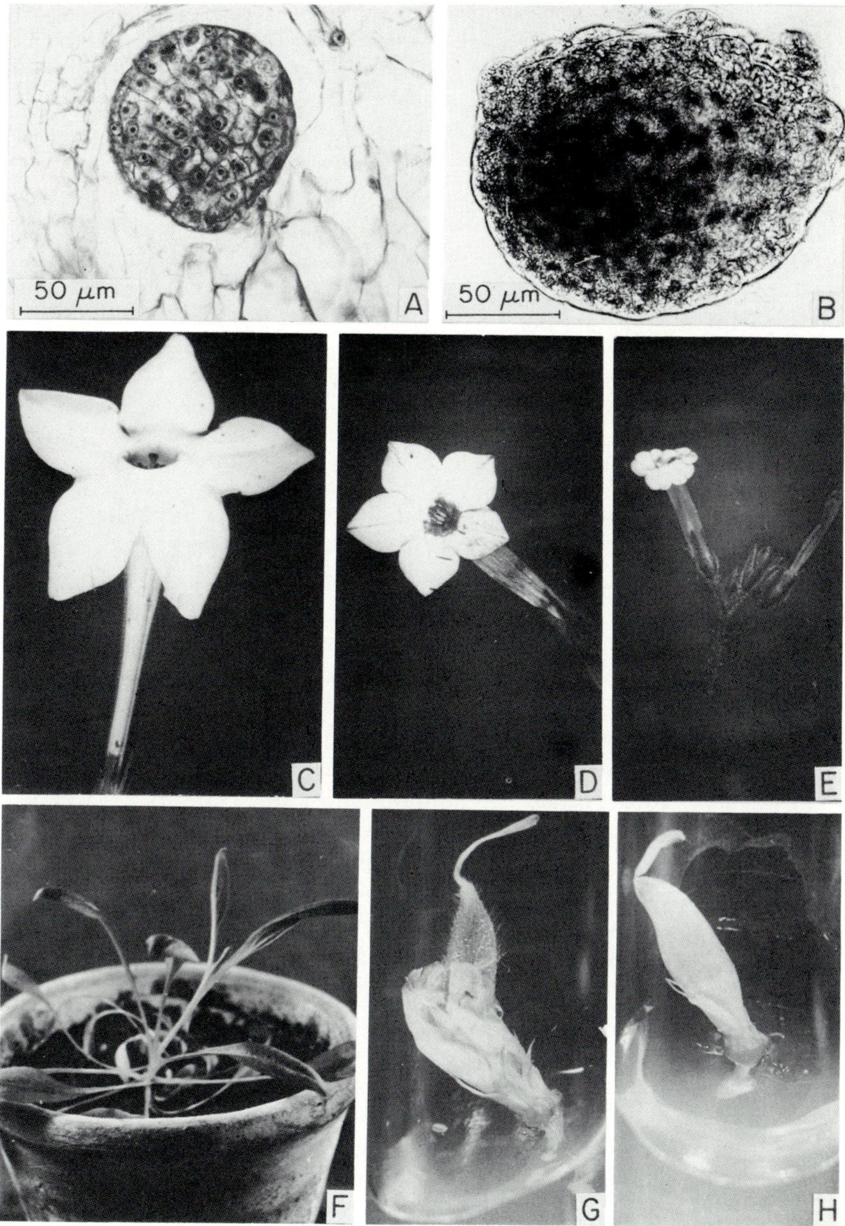

FIG. 3. (A) Hybrid embryo of *Melandrium album* × *Cucubalus baccifer* 7 days after test tube pollination (ttp); note lack of endosperm and perisperm. (B) Hybrid embryo in the pre-heart stage of *Nicotiana tabacum* × *Hyoscyamus niger* 8 days after ttp and 7 days after isolation from the ovule and culture in liquid medium. (C) Flower of *Nicotiana alata*, (D) flower

able experiments carried out on corn were possible partly because the authors used parts of cobs, which contained several ovules each. Therefore, corn with cobs containing many ovules is much more suitable for test tube fertilization than other grasses, which usually possess only one small and solitary ovule.

As can be seen, the technique of cross pollination of ovules *in vitro* enables us to culture both ovules and foreign pollen, thus making it feasible to fertilize the embryo sac. Depending on the species the fertilized egg cell may or may not develop further. Usually it is able to divide and, consequently, small embryos are formed. However, the growth responses of fertilized ovules have to be studied in detail in order to obtain a better understanding of embryogenesis and the nutritional requirements of embryos.

<h3 align="center">IV. Application of in Vitro Pollination in Overcoming
Self-Incompatibility and in Inducing the Formation
of Haploid Embryos</h3>

The effect of the stigma on the germination of pollen grains is well marked in self-sterile plants. In some plants successful self-pollination can be performed in the bud stage of the flower or when the flower is close to wilting. It is supposed that some factors that inhibit the germination of pollen grains just after natural self-pollination are not active in the bud stage or just before wilting. As bud pollination does not always overcome self-sterility, other devices have been worked out. Among them the technique of *in vitro* pollination and fertilization has been, in some cases, successful. The most effective experiments in overcoming prefertilization barriers have been carried out in a self-incompatible *Petunia axillaris* by Rangaswamy and Shivanna (1967, 1971a,b). These authors cultured the entire ovaries with the mass of ovules on the intact placentas. Subsequent to placental pollination many ovules were fertilized and during the next 24 days seeds containing dicotyledonous adult embryos and normal endosperm were formed. From the diploid seeds, fertile and self-incompatible plants were raised. Additional experiments have been performed on the plants to clarify whether stigmatic self-pollination and placental self-pollination could both be performed on the same pistil, and, if so, then

---

of a hybrid plant of *N. alata* × *Nicotiana debneyi,* and (E) flower of *N. debneyi,* (all × 0.5) (F) Hybrid plant of *Melandrium album* × *Viscaria vulgaris* (reduced about 3 ×). (G) Pistil of *Lathyrus odoratus* and (H) pistil of *Pisum sativum* 10 days after pollinating stigmas *in vitro.*

whether stigmatic self-pollination nullifies placental pollination. Rangaswamy and Shivanna applied the method of two-site pollinations in pistils cultured *in vitro*. The ovary wall was peeled to expose only one of the two placentas and the pistils were cultured. The exposed placentas and the stigmas were both used for *in vitro* pollinations. After they had self-pollinated stigmas and exposed placentas, the pollen tubes were inhibited in the style, but seeds developed from the ovules of the exposed placentas. These experiments demonstrated that stigmatic self-pollination is incompatible, whereas placental self-pollination is invariably successful.

The method of test tube fertilization was also applied with success in overcoming self-incompatibility in *P. hybrida* (Niimi, 1970). Here seeds were developed by culturing the whole mass of ovules with pollen grains. Seedlings were obtained by subculturing the developed seeds onto medium.

Since the studies of Blakeslee *et al.* (1922) on *Datura,* it has been possible in some species to induce parthenogenesis by applying inactivated pollen grains. In the earliest experiments, as well as in those being done at present, the pollen is placed on the stigmas. As the technique of *in vitro* pollination has succeeded in respect to several species, it seems to be very important to investigate the inducement of parthenogenesis in ovules cultured *in vitro*. The advantage of the method would be that pollen grains, when put directly on ovules, might play a much more effective role in inducing parthenogenesis than in the case of normal pollination. It is logical that the active substances of pollen grains responsible for the division of egg cell or synergide may influence the female gametophyte much more effectively when in direct contact with the ovule. Unfortunately, there is nearly a complete lack of information in this field. The work of Hess and Wagner (1974) on obtaining haploids of *Mimulus luteus* by pollinating *in vitro* ovules with pollen grains of *Torenia fournieri* is an exceptional one. The exposed ovules of *M. luteus* were overlaid with pollen from *Torenia* and 2–3 weeks after pollination seedlike development appeared in about 3% of the carpels. One third of the carpels produced viable seeds that later developed into plantlets, among which 1% represented haploids. That these were true haploids is proved by the following features: (1) approximately haploid chromosome number in root tip squashes, (2) sterility of the flowers, (3) reduced size of the plants, and (4) presence of nearly half of the amount of anthocyanins. In spite of the cytological analysis however, there is, no available information on the origin of the embryos. It is not clear whether they developed from an unfertilized female gametophyte or whether fertilization took place and only later, during embryogenesis, the chromosomes of the male partner were eliminated.

V. *In Vitro* Pollination of Pistils

The failure of test tube fertilization of naked ovules is still common for many species. This is partly because of physiological factors but also is partly because of unsuitable techniques. It is probable that information obtained on the culture of whole or parts of pistils will, in some cases, aid in obtaining seeds that cannot be produced by the direct pollination of ovules. There are some successful experiments on *in vitro* pollination of pistils and the production of viable seeds. The compatible aseptic pollen is deposited on the stigmas of the cultured pistils. For example, Rao (1965) pollinated pistils of *Nicotiana rustica;* Shivanna (1965), *Petunia violacea;* Usha (1965), *Antirrhinum majus;* and Dulieu (1963, 1966), *N. tabacum.* The normally viable seeds were developed following fertilization *in vitro.* In the case of *Pisum sativum* and *Lathyrus odoratus* (Zenkteler, unpublished observations; Fig. 3G,H), pollination *in vitro* of pistils led to successful fertilization and the formation of few seeds containing only immature embryos.

Much more interesting and valuable results were obtained by applying the technique of *in vitro* pollination of pistils of *Zea mays* (Sladky and Havel, 1976). Maize is one of the species in which fertilization of ovules *in vitro* has failed. Therefore, the authors verified the conditions for pollen germination and elaborated suitable media for the cultivation of ovaries. Excised ovaries along with cob segments were inoculated into the nutrient media in Petri dishes of small diameter placed in larger Petri dishes. The styles overhung into the larger dishes and only here were the stigmas pollinated. Pollen tubes reached the embryo sac within 3–4 days and, therefore, the styles were released from the ovaries 5 days following pollination. Later, the ovaries or segments of cobs were gradually transferred into solid nutrient media in Petri dishes. From the total number of pollinated ovules approximately 5% kernels developed to the stage of milk ripeness. Most embryos were normally developed and some even germinated *in situ* on the cob.

Intact maize ovaries were also cultured *in vitro* by Gengenbach (1977) and an average of 46% of them grew after pollination, which, according to the author, was a result of fertilization. About 20% of developed kernels were comprised of embryos and endosperm. If the embryos were excised and cultured on medium they germinated and produced seedlings.

The successful results with corn encourage the undertaking of experiments on other species, especially within the Gramineae, owing to the considerable economic importance of these crop plants. Finding a suitable method allowing the performance of the full process of pollination, fertilization, and embryogenesis *in vitro* in Gramineae would be of great value in breeding programs.

VI. Concluding Remarks

From this brief account, it is obvious that our knowledge about the processes that follow pollination and fertilization in test tubes is still scanty. Only a limited number of species have been used in test tube pollination, and only in very few species has sexual reproduction been carried out in culture. There are two main obstacles hindering sexual reproduction *in vitro* of plants belonging to the Angiospermae: (1) difficulties in performing double fertilization in directly pollinated ovules and (2) the inability of the zygote to undergo a normal process of embryogenesis. It cannot be overlooked that embryogenesis and seed formation must proceed in harmony; otherwise the embryo collapses and soon dies. It is well known that the developing embryo must absorb the endosperm and that the nutrients must flow from the vegetative tissues surrounding the embryo sac. The situation can alter when one tries to carry out sexual reproduction in completely sterile conditions. Probably, the isolation of placenta from the ovary evokes such a great physiological disturbance in the metabolic pathways in tissues surrounding the female gametophyte that even when fertilization takes place the zygote is not able to divide and shortly dies. This is a very critical and dangerous obstacle responsible for the degeneration of proembryos and endosperm in test tubes. The development of embryos in test tubes proceeded only in those species where the placenta was rich. It can be presumed that the amount of nutrients present in such placentas may be sufficient for the early development of embryos. From the literature to date, we have no information on whether nutrients present in the medium can be absorbed by the placenta, and, if so, then whether they are later transferred to the ovules, thus nourishing the developing embryos.

Theoretically the *in vitro* technique allows the creation of conditions that allow normal embryo growth and differentiation. However, it is astonishing that the research activities on zygotic embryos formed *in vitro,* and even *in vivo,* are limited and have declined during the past few years. This can be clearly seen when the lack of knowledge about the inducement of proembryos *in vitro* is compared with the flourishing literature concerning the culture of single cells and protoplasts. It is also astonishing that nowadays we are able to find conditions for stimulating the regeneration of protoplasts but concomitantly are not able to find suitable environments to induce zygotic proembryos to differentiate. That is probably the main reason why the technique of test tube fertilization has succeeded with only very few species in spite of the fact that its prospects are wide open. Attention must be given to the various *in vitro* techniques for the inducement of small proembryos. Only then, with more knowledge about the nutritional requirements of proembryos in test tubes, will it be possible to produce interspecific hybrids by means of embryo cultures. New methods enabling *in vitro* fertilization have to be used more extensively, followed by excision of the developing embryo before it breaks down. Artificially made culture media are not satisfactory for the culture of proembryos. Natural extracts of liquid endosperm must be used to get more suitable conditions for embryogenesis. Fertilization should also be accomplished by chemical treatment of the inoculated ovules. The concentration of hormones or the osmolarity of the medium are not the only important factors in embryogenesis. It is possible that, especially in the case of inducement of the development of hybrid proembryos, ovules should be directly treated by immunosuppressants. The response of ovules to environmental factors that may affect the process of embryogenesis increases the probability that placental pollination can help in overcoming the barriers of crossability.

The culture of pollinated placentas may also constitute an important tool in inducing parthenogenesis and the development of haploid embryos. Why should only immature pollen grains (microspores) respond to environmental conditions and undergo androgenesis (see Chapter 8, Part A)? The same may happen with immature female gametophytes when they are put in a suitable nutrient medium. The female gametophyte should react to the environmental factors in a way similar to microspores. Theoretically it ought to be the case that the female gametophyte, which consists of several cells endowed with a rich cytoplasm, is much more capable of dividing parthenogenetically than the small and cytoplasm-poor pollen grains. It can also be presumed that the foreign pollen grains, when put on young ovules, stimulate the immature female gametophyte and induce it to follow a different course of development. Here again, the interaction of the female gametophyte and the surrounding tissues, especially the distribution of nutrients, may play a substantial role in inducing the haploid cells to divide. Whatever other methodological difficulties there are in carrying out these types of experiments, it is worthwhile to attack them. It is obvious that male gametophytes in direct contact with the ovules affect the female gametophyte much more strongly than when merely on the stigma and style. The *in vitro* pollination method also offers conditions for the entrance of pollen tube into the immature female gametophyte. It is difficult to say now whether there is a chemotactic activity of such ovules, and, if there is, then whether the entering pollen would burst. It cannot be excluded that a pollen tube enters and bursts and that the presence of male gametes inside the immature embryo sac alters the process of megagametogenesis and evokes events that as yet have not been discovered by experimental embryologists.

Whatever other problems can be handled by the method of intraovarian and placental pollination, it must be stressed again that research in this field is very limited as compared to that in other branches of tissue culture. Therefore, our attention must be given to the area of *in vitro* methods, which would help to raise the proembryos in test tubes, and only later on to producing hybrid and haploid plants.

REFERENCES

Bajaj, Y. P. S. (1964). *Proc. Inst. Sci. India.* **30B,** 175–185.
Balatkova, V., and Tupy, J. (1972). *Biol. Plant.* **14,** 82–88.
Blakeslee, A. F., Belling, J., Farnham, M. F., and Bergner, A. D. (1922). *Science* **55,** 646–647.
Bosio, M. G. (1940). *Nuovo G. Bot. ital.* **47,** 591–598.
Chopra, R. N., and Sabharwal. P. S. (1963). *Plant Tissue Organ Cult., Symp. Int. Soc. Plant Morphol. Delhi* pp. 257–264.
Dahlrgen, K. V. O. (1926). *Sven. Bot. Tidskr.* **20,** 97.
Dhaliwal, S., and King, P. J. (1978). *Theor. Appl. Genet.* **53,** 43–46.
Dulieu, H. L. (1963). *C. R. Acad. Sci., Ser. D* **265,** 3344–3346.
Dulieu, H. L. (1966). *Phytomorphology* **16,** 69–75.
Gengenbach, B. G. (1977). *Planta* **134,** 91–93.
Green, C. E., and Phillips, R. L. (1975). *Crop Sci.* **15,** 417–421.
Guzowska, I. (1971). *Genet. Pol.* **12,** 261–265.
Hess, D., and Wagner, G. (1974). *Z. Pflanzenphysiol.* **72,** 466–468.
Jensen, W. A., and Fisher, D. B. (1968). *Planta* **78,** 158–183.
Johri, B. M., and Vasil, I. K. (1961). *Bot. Rev.* **27,** 325–381.
Joshi, P. C. (1962). *Plant Embryol., Symp. CSIR, New Delhi* pp. 199–204.
Kameya, T., Hinata, K., and Mizushima, U. (1966). *Proc. Jpn. Acad.* **42,** 165–167.
Kanta, K. (1960). *Nature (London)* **188,** 683–684.
Kanta, K., and Maheshwari, P. (1963). *Phytomorphology* **13,** 230–237.
Kanta, K., Rangaswamy, N. S., and Maheshwari, P. (1962). *Nature (London)* **194,** 1214–1217.
Kapoor, M. (1959). *Phytomorphology* **9,** 313–315.
Linsmaier, E. M., and Skoog, F. (1965). *Physiol. Plant.* **18,** 100–127.
Maheshwari, N., and Lal, M. (1961). *Phytomorphology* **11,** 307–314.
Mascarenhas, J. P. (1975). *Bot. Rev.* **41,** 259–314.
Murashige, T., and Skoog, F. (1962). *Physiol. Plant.* **15,** 473–497.
Nakajima, T., and Wakizuka, T. (1977). *Interspecif. Hybrid. Plant Breed.,Proc. Eucarpia Congr. 8th, Madrid* pp. 253–258.
Nakajima, T., Doyama, Y., and Matsumoto, H. (1969). *Jpn. J. Breed.* **19,** 373–379.
Niemirowicz-Szczyt, K., and Kubicki, B. (1979). *Genet. Pol.* **20,** 117–125.
Niimi, Y. (1970). *J. Jpn. Hortic. Sci.* **39,** 345–352.
Niimi, Y. (1976). *J. Jpn. Hortic. Sci.* **45**(2), 168–172.
Nitsch, J. P. (1951). *Am. J. Bot.* **38,** 566–577.
Nitsch, J. P. (1969). *Phytomorphology* **10,** 389–404.
Rangaswamy, N. S. (1961). *Phytomorphology* **11,** 109–127.
Rangaswamy, N. S., and Shivanna, K. R. (1967). *Nature (London)* **216,** 937–939.
Rangaswamy, N. S., and Shivanna, K. R. (1969). *Curr. Sci.* **38,** 257–259.
Rangaswamy, N. S., and Shivanna, K. R. (1971a). *J. Indian Bot. Soc.* **50A,** 286–296.

Rangaswamy, N. S., and Shivanna, K. R. (1971b). *Phytomorphology* **21**, 284–289.

Rao, P. S. (1965). *Phyton (Buenos Aires)* **22**, 165–167.

Shivanna, K. R. (1965). *Phytomorphology* **15**, 183–185.

Sladky, Z., and Havel, L. (1976). *Biol. Plant.* **18**, 469–472.

Strasburger, E. (1886). *Jahrb. Wiss. Bot.* **17**, 50–98.

Usha, S. V. (1965). *Curr. Sci.* **34**, 511–513.

Vasil, I. K. (1973). *Naturwissenschaften* **60**, 247–253.

Vasil, I. K. (1974). *In* "Fertilization in Higher Plants" (H. F. Linskens, ed.), pp. 115–118. North Holland Publ., Amsterdam.

White, P. R. (1963). "The Cultivation of Animal and Plant Cells." New York.

Zenkteler, M. (1965). *Naturwissenschaften* **23**, 645–646.

Zenkteler, M. (1967). *Experientia* **23**, 775–777.

Zenkteler, M. (1969). *PTPN Pr. Kom. Biol.* **32**(2), 1–71.

Zenkteler, M. (1971). *Experientia* **26**, 661–662.

Zenkteler, M., Slusarkiewicz-Jarzina, A., and Woźna, J. (1981). *Bull. Soc. Amis Sci. Lett. Poznań Ser. D,* Livraison 21.

Zenkteler, M., and Melchers, G. (1978). *Theor. Appl. Genet.* **52**, 81–90.

Zenkteler, M., Misiura, E., and Guzowska, I. (1975). *In* "Form, Structure and Function in Plants," B. M. Johri Commemoration Vol., pp. 180–187. Sarita Prakashan, Meerut, India.

Zubkova, M., and Sladky, Z. (1975). *Biol. Plant.* **17**, 276–280.

Chapter 16

Endosperm Culture

B. M. JOHRI, P. S. SRIVASTAVA,[1] AND A. P. RASTE[2]

Department of Botany, University of Delhi, Delhi, India

I. Introduction

In any sexually reproducing system nutrition of the embryo is an important and vital aspect of the life cycle. The female gametophyte, a haploid prefertilization product, fulfils this function in gymnosperms; whereas in angiosperms,[3] it is accomplished by a new structure, the endosperm, produced by an unique event of double fertilization. Endosperm is formed in most cases (over 81% of families) as the result of a fusion of three haploid nuclei, namely, the two polar nuclei and one of the male gametes. The endosperm, therefore, contains a triploid number of chromosomes. It is the main nutritive tissue for the embryo, and its role in embryonic growth has been emphasized by various investigators (Sargant, 1900; Van Over-

---

[1] Permanent address: Department of Botany, S.G.T.B. Khalsa College, University of Delhi, Delhi 110007, India.

[2] Permanent address: Department of Botany, Deshbandhu College, University of Delhi, Kalkaji, New Delhi 110019, India.

[3] Orchidaceae, Podostemaceae, and Trapaceae lack endosperm.

157

TABLE I

MORPHOGENIC RESPONSE OF CULTURED ENDOSPERM OF AUTOTROPHIC ANGIOSPERMS

Plant species	Family	Inoculum	Medium[a]	Response	References
Zea mays	Gramineae	Immature endosperm	BM + potato extract or extract of young corn kernels	Slight proliferation of cells adjacent to the embryo	Lampe and Mills (1933, cited in LaRue, 1936)
			BM + YE	Proliferation of endosperm, producing callus of unlimited growth	LaRue (1949)
			BM + asparagine	Proliferation of endosperm and growth of callus as good as on BM + YE	Straus and LaRue (1954)
			BM + YE	Proliferation of endosperm; the callus became nodular after transfer to BM + CH	Sehgal (1969)
Asimina triloba	Annonaceae	Mature seed piece	BM + sucrose 2%	Proliferation of endosperm	Lampton (1952)
Lolium perenne	Gramineae	Immature endosperm (9–10 days after pollination)	BM + YE	Proliferation of endosperm and establishment of tissue cultures	Norstog (1956)
Cucumis sativus	Cucurbitaceae	Immature endosperm (10 days after pollination)	BM + IAA + DPU + CH	Proliferation of endosperm, producing callus; differentiation of meristematic zones in callus	Nakajima (1962)
Ricinus communis	Euphorbiaceae	Decoated mature seed	BM + 2,4-D + KN + YE	Proliferation of endosperm and establishment of tissue cultures; differentiation of tracheidal cells in callus	Satsangi and Mohan Ram (1965)

		Endosperm from germinated seed	BM + 2,4-D + KN + YE	Proliferation of endosperm (endosperm from dried seeds failed to grow)	Brown *et al.* (1970)
		Fresh decoated seed	BM + 2,4-D + KN + YE	Proliferation of endosperm, producing callus that could be continuously subcultured	Johri and Srivastava (1972)
Croton bonplandianum	Euphorbiaceae	Decoated seed	BM + 2,4-D + KN + YE	Proliferation of endosperm; differentiation of tracheidal cells and embryo-like cell aggregates from endosperm callus	Bhojwani (1966)
		Endosperm callus subcultured	BM or BM + CH	Differentiation of roots from callus	Bhojwani and Johri (1971)
Jatropha panduraefolia	Euphorbiaceae	Decoated mature seed	BM + 2,4-D + KN + YE	Proliferation of endosperm producing profusely growing callus with differentiated tracheidal cells	Srivastava (1971b)
		Endosperm callus subcultured	BM + NAA + KN + CH	Differentiation of roots and shoots with high frequency; NAA promoted root formation, whereas KN favored shoot formation (CH being common to both)	
Putranjiva roxburghii	Euphorbiaceae	Decoated mature seed	BM + IAA + KN + CH	Proliferation of endosperm producing profusely growing callus; differentiation of roots and shoots and, subsequently, plantlets	Srivastava (1973); Srivastava and Johri (1978)

(continued)

TABLE I (*continued*)

Plant species	Family	Inoculum	Medium[a]	Response	References
Codiaeum variegatum	Euphorbiaceae	Entire seeds or decoated seeds	BM + 2,4-D + KN + CM + CH	Proliferation of endosperm; differentiation of roots and shoots from endosperm callus	Chikkannaiah and Gayatri (1974)
Hordeum vulgare	Gramineae	Immature endosperm (8 days after proliferation)	BM + CH + IAA	Proliferation of endosperm, forming slow-growing callus	Sehgal (1974)
		Endosperm callus transferred	BM + CM + 2,4-D	Callus grew profusely	
Triticum aestivum	Gramineae	Immature endosperm (8 days after pollination)	BM + CM + IAA or BM + CM + KN + 2,4-D	Proliferation of endosperm forming actively-growing callus	Sehgal (1974)
Oryza sativa	Gramineae	Immature endosperm (4–7 days after pollination)	BM + 2,4-D + YE	Proliferation of endosperm leading to establishment of tissue cultures; elimination of 2,4-D resulted in the differentiation of many roots and a few plantlets	Nakano *et al.* (1975)

160

Species	Family	Explant	Medium	Response	Reference
Nigella damascena	Ranunculaceae	Mature seed	BM + 2,4-D	Differentiation of embryoids (up to globular stage)	Sethi and Rangaswamy (1976)
Vitis	Vitidaceae	Seed	BM + 2,4-D + CH	Proliferation of endosperm	Mu *et al.* (1977a)
Malus pumila	Rosaceae	Immature endosperm (without embryo)	BM + KN + 2,4-D	Proliferation of endosperm and differentiation of buds	Mu *et al.* (1977b)
Petroselinum hortense	Umbelliferae	Seed	BM	Proliferation of endosperm; differentiation of embryoids	Masuda *et al.* (1977)
Citrus grandis, C. sinensis	Rutaceae	Young endosperm cellular (with or without embryo)	BM + 2,4-D + BAP + CH	Proliferation of endosperm forming callus	Wang and Chang (1978)
Citrus grandis		Endosperm callus subcultured	BM (with high mineral concentration) + GA	Differentiation of embryoids developing eventually into triploid plantlets	Wang and Chang (1978)
Coffea arabica	Rubiaceae	Seed	BM + CH + malt extract	Studies were made with regard to caffeine synthesis in endosperm tissue and no attention was paid to morphogenetic response	Keller *et al.* (1972); Monaco *et al.* (1977)

^a BM, basal medium: The basal medium generally contained mineral salts, and a carbon energy source such as glucose or sucrose. The references cited are recommended for further details of the composition. YE, yeast extract, IAA, indoleacetic acid; DPU, 1,3-diphenylurea; CH, casein hydrolysate; 2,4-D, 2,4-dichlorophenoxyacetic acid; KN, kinetin; NAA, naphthaleneacetic acid; CM, coconut milk; BAP, 6-benzylaminopurine; GA, gibberellin.

162

TABLE II

Morphogenic Response of Cultured Endosperm of Parasitic Angiosperms

Plant species	Family	Inoculum	Medium[a]	Response	References
Dendrophthoe falcata	Loranthaceae	Mature seed	BM + IAA + KN + CH	Proliferation of endosperm, and differentiation of shoot buds and haustoria	Nag and Johri (1971)
Exocarpus cupressiformis	Santalaceae	Mature seed	BM + IAA + KN + CH	Differentiation of shoot buds directly from endosperm	Johri and Bhojwani (1965)
Leptomeria acida	Santalaceae	Mature seed	BM + IAA + KN	Proliferation of endosperm and differentiation of shoot buds	Nag and Johri (1971)
Nuytsia floribunda	Loranthaceae	Mature seed	BM + IBA + KN + CH	Proliferation of endosperm; differentiation of tracheidal elements in callus	Nag and Johri (1971)
Phoradendron tomentosum	Loranthaceae	Decoated mature seed	BM + 2,4-D + KN + YE	Proliferation of endosperm; differentiation of papilla-like outgrowths	Bajaj (1970)
Santalum album	Santalaceae	Mature seed (presence of embryo obligatory)	BM + 2,4-D + KN + YE	Proliferation of endosperm	Rangaswamy and Rao (1963)
Scurrula pulverulenta	Loranthaceae	Mature seed	BM + cytokinin[b]	Differentiation of shoot buds directly from endosperm	Bhojwani and Johri (1970)

| *Taxilus vestitus* | Loranthaceae | Mature seed or excised endosperm (half-split) | BM + cytokinin[c] | Differentiation of shoot buds directly from endosperm; presence of embryo suppressed shoot bud differentiation (implied from much better response of endosperm without the embryo); differentiation of haustoria occurred when IBA and CH were added to KN medium | Johri and Nag (1970); Nag and Johri (1971) |
| *Taxillus cuneatus* | Loranthaceae | Mature seed or excised endosperm | BM + cytokinin | Similar to *T. vestitus*; shoot buds differentiated subsequent to proliferation of endosperm, haustoria differentiated in a medium with KN alone | Nag and Johri (1971) |

[a] Abbreviations as in Table I. The basal medium generally contained mineral salts and a carbon energy source, such as glucose or sucrose. The references cited are recommended for further details of the composition.

[b] Sequence of decreasing effectiveness: 6-(γ,γ-dimethylallylamino-)purine, zeatin, kinetin, benzyladenine, and triacanthine.

[c] Sequence of decreasing effectiveness: 6-(γ,γ-dimethylallylamino-)purine, 6-benzylaminopurine, 6-furfurylaminopurine (kinetin), 6-isoamylaminopurine, 6-furfurylamino-8-azapurine.

beek *et al.*, 1942; Brink and Cooper, 1947; Raghavan, 1976). During the maturation of seed, the endosperm may be completely consumed (nonendospermous seed) or may persist as a massive tissue (endospermous seed) containing reserve food in the form of starch, fat, or proteins. This food material is utilized during germination and postgermination development until the seedling becomes autotrophic.

The morphological nature of the endosperm of angiosperms has been a subject of discussion since the middle of the nineteenth century (Maheshwari, 1950). It had earlier been referred to as a "second embryo," modified to serve as food tissue for the zygotic embryo, or a "maimed" embryo. The early theories reflected in these terms arose because of the different morphogenetic fate of the primary endosperm nucleus in relation to the other product of double fertilization, the zygote. Also, the endosperm is the only ovular tissue that does not form nonzygotic embryos; all the other ovular tissues do so in nature. Does this mean, therefore, that the endosperm lacks the capacity to undergo organogenic differentiation? To grow the excised endosperm *in vitro* by the techniques of plant tissue culture has proved to be an ideal approach to resolving this problem.

In this chapter we shall first deal with growth and differentiation in endosperm cultures and then discuss the effect of various physical and chemical factors on the growth of endosperm callus. The results of histological and cytological studies of endosperm cultures also are described. At the end attention is drawn to the applicability of endosperm cultures in various areas, including raising of triploid plants in relation to plant improvement.

Attempts to grow endosperm tissue in cultures began in the 1930s, and immature and mature endosperm of various angiosperm taxa (autotrophic as well as parasitic) have been successfully cultured (see Tables I and II).

In 1933 Lampe and Mills (cited in LaRue, 1936) first attempted the culture of immature endosperm of *Zea mays* on a medium supplemented with an extract of potato or young corn kernels. Although they used an undefined medium, they observed slight proliferation of cells in the vicinity of the embryo.

During the 1940s, LaRue made extensive investigations (LaRue, 1947, 1949) to establish endosperm cultures. LaRue (1947) reported that "in *Zea mays*, a few specimens—less than one in a thousand developed roots and one formed a root–shoot axis and miniature leaves." He eventually succeeded in establishing a continuously growing tissue in culture from immature maize endosperm (LaRue, 1949). Five years earlier LaRue (1944) had also reported organogenic differentiation in endosperm cultures of castor bean. However, extensive investigations by subsequent workers on maize and castor bean could not confirm

LaRue's claim of achieving organogenesis. Several investigators since then have cultured maize endosperm but invariably have failed to induce organogenesis (Fig. 1; see Sehgal, 1969). Sehgal (1974) cultured the immature endosperm of *Hordeum* and *Triticum* and obtained only proliferation; all efforts to induce organogenesis failed. These repeated failures to induce organogenesis may perhaps result from a poor response, in general, of monocots. However, in *Oryza sativa* organogenesis was successfully induced in cultures of immature endosperm by Nakano and his co-workers (Nakano *et al.*, 1975). The only other instance of proliferation and organogenesis in cultures of immature endosperm is that of apple (Mu *et al.*, 1977b).

Induction of cell division with consequent proliferation of mature endosperm was initially considered to be difficult. Continued efforts by various investigators brought success, and not only proliferation but even organogenesis occurred in the cultures of mature endosperm of several autotrophic (Table I) as well as parasitic (Table II) angiosperm taxa. The first successful establishments of tissue cultures from mature endosperm were achieved by Rangaswamy and Rao (1963) in a parasitic species, *Santalum album,* and in an autotrophic species, *Ricinus communis,* by Satsangi and

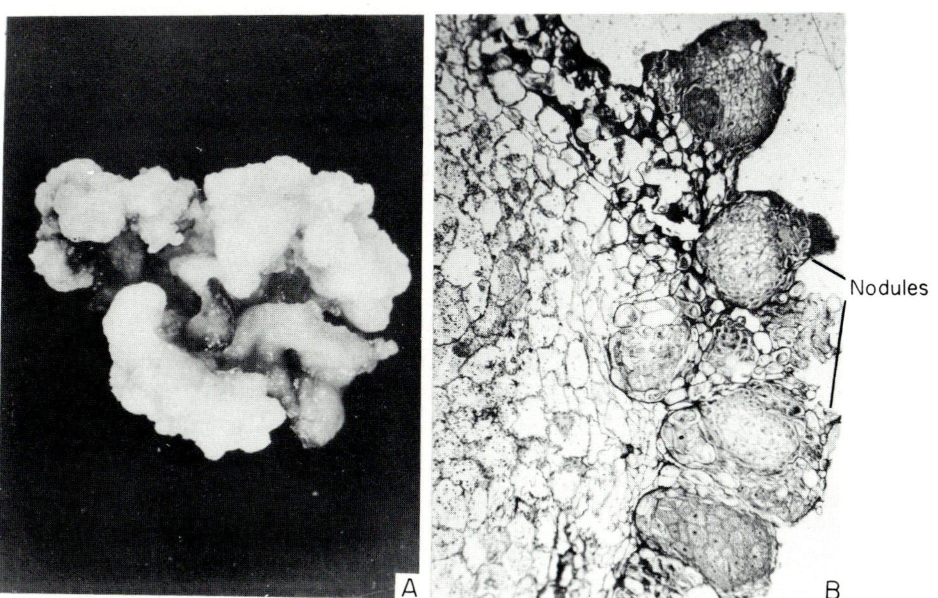

FIG. 1. *Zea mays*. (A) Four-week-old endosperm culture on BM + YE. (B) Section of callused endosperm 4 weeks after planting on BM + CH; note nodules at the periphery. (After Sehgal, 1969.)

Mohan Ram (1965). Organogenesis, however, has not been achieved so far in either of these. In all plant species tested so far, the frequency of organogenesis is much higher in endosperm cultures of parasites than in those of autotrophs.

II. Growth Response

A. CALLUS FORMATION

The endosperm tissues of the species cultured so far have shown more or less similar nutrient requirements, although some differences have been noticed in the cultures of immature and mature endosperm tissues.

Of the waxy, starchy, and sugary varieties of maize, only the sugary variety has been reported to yield tissue cultures. Maize endosperm was cultured by LaRue (1947) on media supplemented with tomato juice (TJ), extracts of grape, green corn kernels, coconut milk (CM), and yeast extract (YE). Twenty percent TJ induced best growth of the tissue and showed an 18-fold increase within 3 weeks. Straus (1960a,b) found that the active component in TJ was similar to asparagine; this result was later substantiated when better callus growth was obtained with asparagine than with TJ, YE, or casein hydrolysate (CH). Aspartic acid and glutamic acid could also sustain growth, but a combination of aspartic acid and asparagine proved better. For endosperm of *Triticum* (Sehgal, 1974), *Vitis* (Mu *et al.*, 1977a), and *Codiaeum* (Chikkannaiah and Gayatri, 1974), CM was an important additive for inducing cell proliferation.

Nakajima (1962) obtained proliferation of endosperm of *Cucumis* on a medium containing indoleacetic acid (IAA), 1,3-diphenylurea (DPU), and CH. The endosperm callus exhibited meristematic zones. The reports on culture of immature endosperm clearly indicate that the tissue requires such substances as YE or CH and cell division factors in addition to basal medium.

The age of the endosperm at culture is critical for its growth *in vitro*. For example, in maize, wheat, and barley endosperm tissue younger than 8 or older than 12 days after pollination did not grow in cultures. Likewise, the endosperm of *Lolium* (Norstog, 1956) and *Cucumis* (Nakajima, 1962) can be grown only when excised 7–10 days after pollination. Nakano *et al.* (1975) pointed out that in rice only cellular endosperm, cultured 4–7 days after pollination, shows proliferation. Coenocytic endosperm reared just 3–4 days after pollination does not form callus. In maize the endosperm becomes cellular 3–5 days after pollination and until 8 days the cells remain meristematic, after which mitosis is restricted to the outermost layer. Finally, after 22 days mitosis ceases. However, this does not explain the inability of younger or older endosperm tissue of

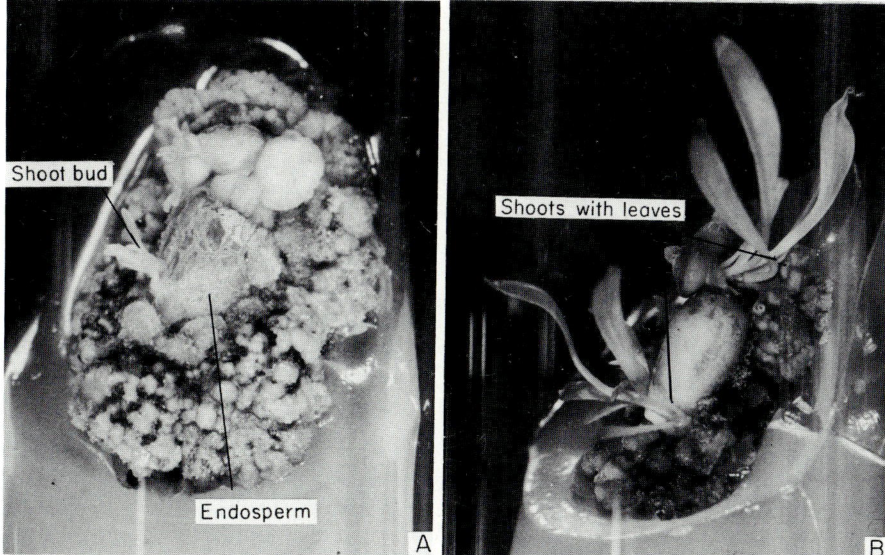

FIG. 2. *Dendrophthoe falcata*. (A) Ten-week-old endosperm culture on BM + IAA + KN + CH, showing callusing and a shoot bud on the uncallused surface. (B) Fourteen-week-old culture showing shoots (with leaves) developed from callused endosperm. (After Nag, 1970.)

kinin could overcome this decrease in the potentiality. In contrast to *Exocarpus,* the endosperm tissue of *Dendrophthoe* exhibited an enhancement in the percentage of cultures showing shoot bud formation on CH-fortified medium. Shoot bud differentiation was observed in 46% culture with CH (2000 ppm) as compared to 32% without it. The buds were mostly confined to the distal end of endosperm. Induction of buds did not occur on media having naphthaleneacetic acid (NAA),2,4-D, or 2,4,5-trichlorophenoxyacetic acid (2,4,5-T), and with indolepropionic acid (IPA) only deformed buds differentiated. The endosperm of *T. cuneatus* showed buds in 85% cultures, and haustoria in 40% after 10 weeks, on IAA + KN + CH medium. On IBA medium bud formation occurred in 55% and haustoria in 60% of cultures. In *T. vestitus* there was no bud formation on IBA + KN + CH medium. This is in contrast to *Dendrophthoe* and *T. cuneatus.*

Mature endosperm of only 13 autotrophic taxa [*Antigonon, Mimusops, Sapium* (unpublished), *Ricinus, Vitis, Croton, Jatropha, Putranjiva, Petroselinum, Nigella, Codiaeum,* apple,[4] and *Citrus*] have been cultured and induced to form callus, and only the last eight have exhibited organ differentiation (see Table I). The endosperm callus in these species becomes compact upon continuous subculture, and organs differentiate only after transfer to a fresh medium. However, the endosperm cultures of *Pe-*

---

[4] In this case the inoculum was immature endosperm.

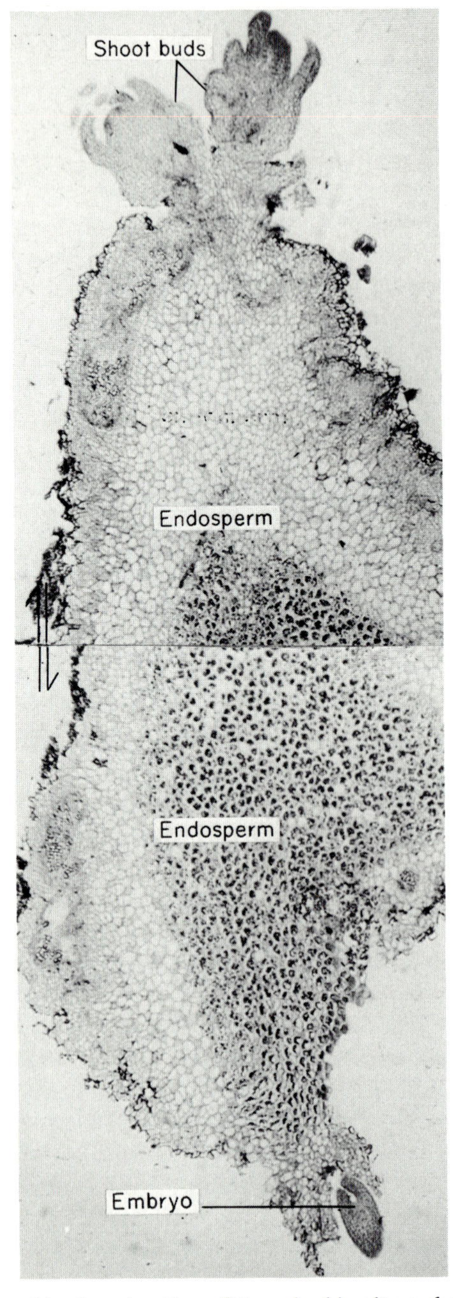

FIG. 3. *Leptomeria acida*. Longisection of 8-week-old culture showing shoot buds developed from endosperm. Note the arrested intact embryo. (After Nag and Johri. 1971.)

troselinum and *Codiaeum* produce callus and differentiate organs on the same medium. The compact callus of *Jatropha* obtained on 2,4-D + KN + YE medium, upon transfer to KN alone, grew vigorously, turned green, and produced some nodules (Srivastava and Johri, 1974). Addition of CH to the KN medium resulted in a friable callus and, later, small, green shoot buds developed in 22% of cultures. These buds elongated up to 2 cm but failed to unfold and develop into normal leaves (Fig. 4A–C). Auxin did not influence the differentiation of buds. Root differentiation also occurred in endosperm callus of *Jatropha* (Srivastava, 1971b; Srivastava and Johri, 1974) upon transfer to a medium with NAA + CH, in 43% cultures (Fig. 4D). Bhojwani and Johri (1971) also reported the formation of roots in the mature endosperm callus of *Croton* transferred to CH or basal medium alone from 2,4-D + KN + CH medium. This happened both in liquid and in semisolid medium. The number of cultures showing root formation increased with lower concentrations of auxins, IBA being most effective.

The nodulated endosperm callus of *Putranjiva* differentiated buds in 80% of cultures on IAA + KN + CH after 8 weeks. After another 3 weeks the buds grew into shoots that later produced three to four pairs of leaves. Indolepropionic acid was much more effective than either IAA or IBA. Among the cytokinins, 6-(γ,γ-dimethylallylamino)purine induced shoot bud differentiation in the maximum number of cultures. As in *Exocarpus*, the least effective was triacanthine.

Experiments with elimination of different inorganic salts present in the basal medium showed that, without potassium nitrate, calcium nitrate, or potassium dihydrogen phosphate, the growth of callus and differentiation of shoot buds was very poor in *Putranjiva*.

In *Putranjiva*, although the differentiation of shoots was quite common, roots appeared only in a few cultures. Even the shoots which attained a length of about 15 cm failed to differentiate roots. When the shoots were excised from the callus and planted on a different medium, root differentiation was noticed in some cultures. In 20% of cultures roots differentiated opposite the shoot end of the callus and, eventually, complete plantlets (Fig. 5) with root–shoot axes were obtained. The plantlets were comparable to the normal young plants obtained on basal medium. These plants were triploid (Srivastava, 1973).

Contrary to the promotion of shoot bud differentiation with elimination of CH in the endosperm cultures of *Exocarpus*, CH was indispensable for the differentiation of shoot buds from the endosperm callus of *Putranjiva*. As in *Dendrophthoe*, in *Putranjiva* the growth of callus was much better with CH (2000 ppm). On 3000 ppm CH shoot growth was inhibited. The effect of CH could partly be replaced in *Putranjiva* by valine, one of the

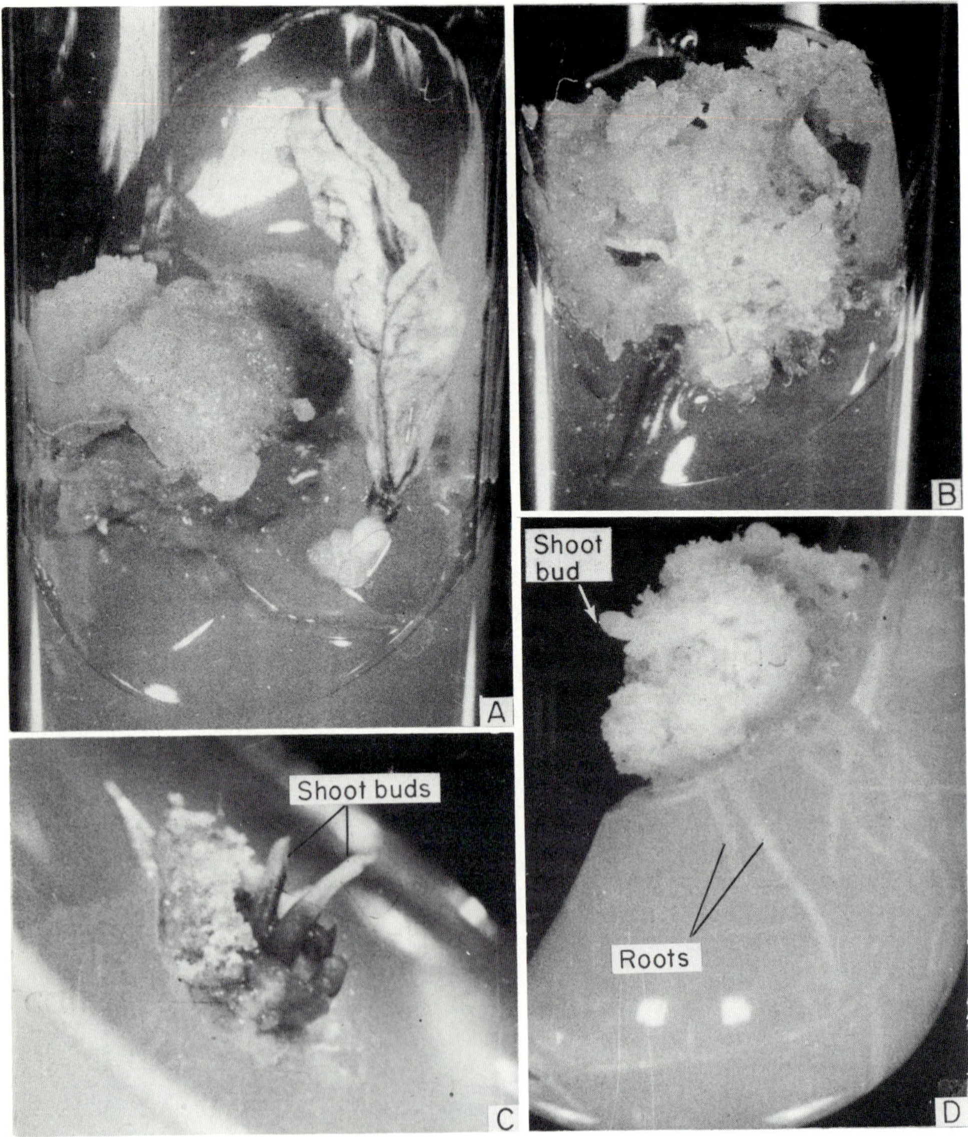

FIG. 4. *Jatropha panduraefolia*. (A) Six-week-old culture on BM + 2,4-D + KN + YE; note the callused endosperm and inhibition of germination. (B) Endosperm callus after 10 weeks. (C) A 26-week-old culture showing shoot differentiation on BM + KN + CH. (D) Eight-week-old endosperm callus on BM + NAA + CH; note well-differentiated roots and a small shoot bud. (After Srivastava and Johri, 1974.)

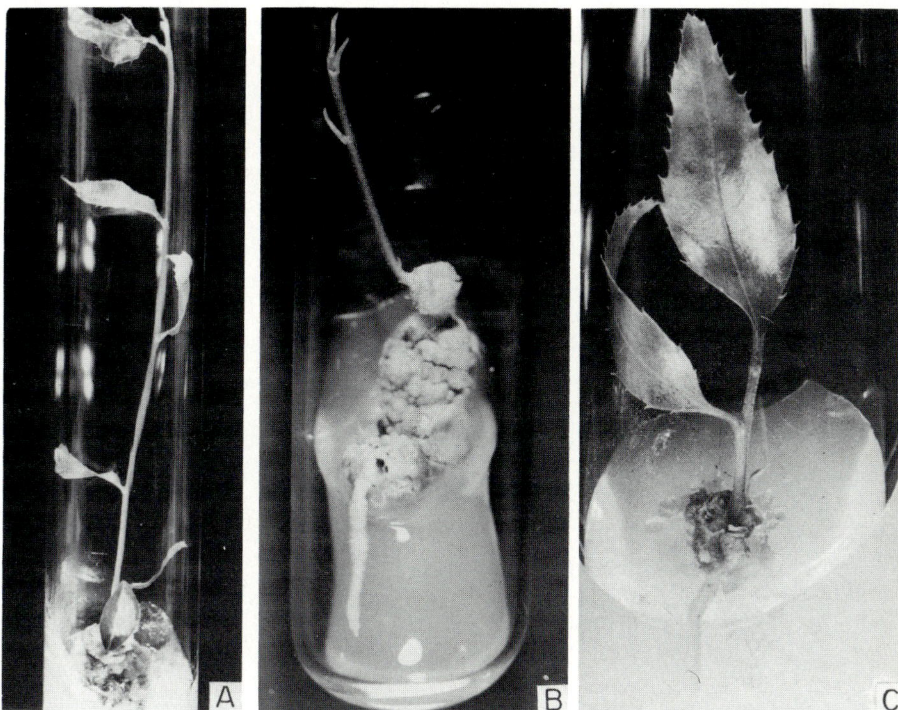

FIG. 5. *Putranjiva roxburghii.* (A) A 25-week-old shoot developed from endosperm callus on BM + IAA + KN + CH. (B) A 21-week-old culture showing a plantlet. (C) Differentiation of root from endosperm callus at the opposite end of shoot. (A, unpublished; B and C, after Johri and Srivastava, 1973.)

constituent amino acids of CH. When all the 13 amino acids present in CH were tried together, not only was the growth of callus slow but the cultures also showed poor development of shoots. With leucine and valine, shoot differentiation occurred in 65% cultures but the growth of callus was poor.

There are very few reports of endosperm callus showing embryogenesis. In *Ricinus* (Satsangi and Mohan Ram, 1965), *Croton* (Bhojwani, 1966), and *Nigella* (Sethi and Rangaswamy, 1976), cell assemblages resembling early stages of embryogenesis were observed. The embryoids differentiated only up to globular stage. In *Petroselinum,* however, fully developed embryoids leading to the differentiation of complete plantlets have been observed (Masuda *et al.*, 1977), but their origin has not been shown with certainty and they may as well have arisen from the zygotic callus. The germinated seeds rupture and, 2 weeks later, small masses of cells appear on basal medium. Once the meristematic activity had been initiated, not only callus proliferation but also embryogenesis continued for a prolonged period without subculturing. The first indication of embryogenesis was observed within 15 days after inoculation. Besides the embryos, twisted rootlike structures and irregularly shaped leaflets were also scattered along the upper periphery of the callus. If the callus was retained on the same medium for a longer period without subculturing, numerous shoots and roots and, eventually, plantlets differentiated after 50 days.

Immature endosperm of only two taxa is reported to show organ differentiation. In *Oryza* immature cellular endosperm produced callus on 2,-4-D + YE medium and, upon transfer to IAA + YE, differentiation of plantlets was noticed (Nakano *et al.*, 1975). Apple endosperm exhibited differentiation of plantlets on benzylaminopurine + CH medium (Mu *et al.*, 1977b).

The position of half-split endosperm on the surface of medium had a significant effect on the differentiation and distribution of buds. The site of bud formation was more or less predictable. In *T. vestitus* (Johri and Nag, 1970) when the half-split endosperm was planted with the cut surface in contact with the medium (containing KN), 12–18 shoot buds developed in all the cultures directly (i.e., without callusing) in about 10 weeks and, if the cut surface was away from the medium, only one to three deformed buds developed in 30% cultures (Fig. 6). When the half-

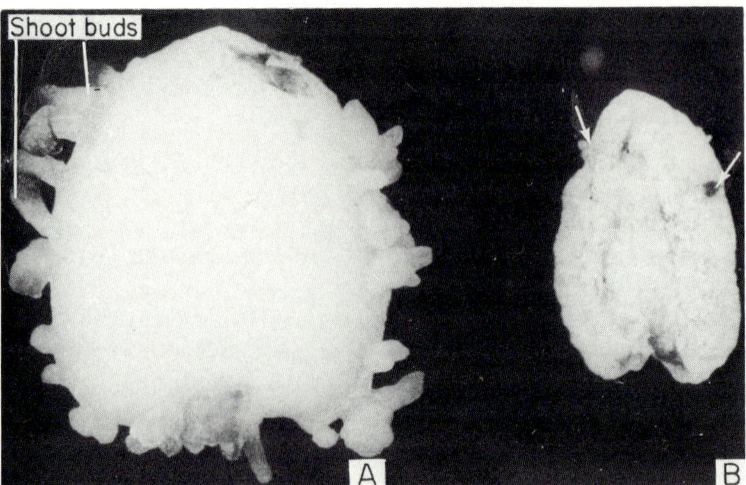

FIG. 6.   *Taxillus vestitus*. (A) Ten-week-old culture showing differentiated shoot buds from an endosperm-half with its cut surface in contact with the medium, on BM + KN. (B) Only two deformed buds (arrow) developed when the cut surface was away from the medium. (After Johri and Nag, 1970.)

split endosperm was planted vertically, or horizontally, the first two buds invariably differentiated from the epidermal cells near the cut end situated immediately above the level of medium. Most of the buds were restricted to the injured portions of the endosperm in direct contact with the medium. Occasionally, one to three buds also developed around the injured surface that was not in direct contact with the medium.

Buds were also formed when the half-split endosperm was soaked in KN (0.025 ppm) for 4 hours and planted on basal medium. These buds, however, were no longer confined to the cut margins in contact with the medium but appeared all over the surface. A few haustoria also developed at the base of the buds. A similar response was noticed when transverse pieces of seeds were soaked in KN solution and planted on basal medium.

III. Physical and Chemical Factors and Callus Growth

The data on the growth requirements of endosperm callus are quite meager, and very few workers have paid attention to the influence of physical and chemical factors.

A. pH

The effect of pH on the growth of endosperm callus has been studied in maize, *Asimina,* and *Ricinus.* The pH of the nutrient media has been usually maintained between 4.5 and 6.3. In maize, on a fresh weight basis, best growth occurred at pH 7.0 (Straus and LaRue, 1954), whereas on dry weight basis pH 6.1 appeared optimal. At pH 4.0 a 100% increase in the growth of endosperm callus of *Asimina* was observed, whereas pH 5.0 resulted in only 95% increase in growth (Lampton, 1952). In *Ricinus* best growth of endosperm callus occurred at pH 5.0, whereas pH 5.6 proved satisfactory for the endosperm of *Jatropha* and *Putranjiva* (Johri and Srivastava, 1972; see also Srivastava, 1971c). Apple endosperm produced callus and differentiated organs at pH 6.0–6.2

B. Light and Temperature

There was at least a 50% decrease in growth of maize endosperm at 30°C, but at 25°C a four-fold increase (Straus and LaRue, 1954) was observed. In *Jatropha* and *Ricinus* 24–26°C supported best growth of the endosperm callus (Srivastava, 1971c).

Maize endosperm callus showed satisfactory growth only when maintained in the dark. Continuous light supported best growth of the endosperm callus of *Ricinus* (Johri and Srivastava, 1972). Light did not play any significant role in the growth of endosperm callus of *Lolium* (Norstog, 1956).

C. SUGARS

As a carbon energy source, sucrose has been commonly used in endosperm tissue cultures. With higher concentrations of sucrose there is generally a decrease in growth. Of the different sugars tested for maize endosperm, sucrose, glucose, and fructose proved to be the best for growth, whereas lactose, arabinose, galactose, and rhamnose proved inhibitory (Straus and LaRue, 1954); even starch supported growth. Similar results have been achieved with *Lolium multiflorum* (Smith and Stone, 1973) and *Ricinus* (Srivastava, 1971c). The endosperm of *Ricinus* was unable to utilize galactose, lactose, arabinose, and raffinose as carbon sources (Johri and Srivastava, 1972).

Lampton (1952) obtained satisfactory growth of endosperm tissue on starch medium. However, the endosperm tissue of *Jatropha, Ricinus,* and *Putranjiva* did not respond to starch-containing medium.

D. TOMATO JUICE, YEAST EXTRACT, AND CASEIN HYDROLYSATE

Tomato juice promoted the growth of endosperm of corn (LaRue, 1949; Sternheimer, 1954), but the growth was erratic and unpredictable because the kinin-like activity of TJ changes correspondingly with the age of the fruit from which it is obtained. Although the addition of such undefined substances as tomato juice (TJ), coconut milk (CM), yeast extract (YE), and casein hydrolysate (CH), and extracts of many seeds adds to the activity of the medium, the nature of the stimulatory substances contained therein has not yet been entirely elucidated. Straus and LaRue (1954) and Tamaoki and Ullstrup (1958) could fairly replace TJ with YE for the growth of corn endosperm callus. Yeast extract is now known to be essential for the satisfactory growth of mature endosperm callus of a number of taxa: *Croton* (Bhojwani, 1966), *Cucumis* (Nakajima, 1962), *Lolium* (Norstog, 1956), *Santalum* (Rangaswamy and Rao, 1963), and *Ricinus* (Srivastava, 1971a). Casein hydrolysate, which also supports the growth of endosperm, is not comparable to YE with regard to universality for callus growth and differentiation.

E. AMINO ACIDS

The addition of 1.5×10^{-2} *M* asparagine to basal medium resulted in better growth of maize endosperm (Straus and LaRue, 1954; Straus,

1960a,b); arginine was ineffective. However, glutamine, glutamic acid, and aspartic acid ($1.5 \times 10^{-2} M$) supported appreciable growth. In *Ricinus* glutamic acid proved most effective for the growth of endosperm callus (Johri and Srivastava, 1972).

In the cultures of maize endosperm, aspartic acid and cysteine were best promoters of anthocyanin synthesis, whereas methionine, asparagine, glutamine, and valine inhibited pigmentation of endosperm callus tissue (Straus, 1960a). In *Ricinus* endosperm callus pigment synthesis failed to occur on any of the amino acid-enriched media.

IV. Histological Studies

In *Scurrula* the differentiation of shoot buds occurred as a result of meristematic activity in the epidermal as well as the hypodermal cells of the endosperm. The buds developed from the peripheral cells in *Exocarpus*. Likewise, in *Taxillus* and *Dendrophthoe* proliferation of the endosperm started from the cells that were in direct contact with the medium. Nag (1970) often noticed anticlinal divisions in the epidermal cells of cultured endosperm tissue. Occasionally, in *Leptomeria*, division of cells in a row was observed. In the parasitic taxa tested so far, differentiation of shoot buds in endosperm cultures was achieved either by direct divisions of the epidermal cells or after the formation of callus.

The pattern of initial growth of endosperm, especially in *Putranjiva*, is similar to that of carrot root segments in cultures. Numerous nodules or growth centers are formed after 30 days of culture. The endosperm mostly proliferates after laying down a cambium-like layer or by cell divisions followed by enlargement. Mitoses begin in the peripheral cells. In *Osyris* localized peripheral meristems lead to the formation of surface outgrowths in which large patches of tracheidal cells are formed (Johri and Bhojwani, 1965). In *Jatropha, Putranjiva,* and *Ricinus* the growth pattern is similar to that in *Osyris*. The initial growth of cultured maize endosperm tissue occurs as a result of divisions in the aleurone cells and the underlying layers.

Anatomical studies of callused endosperm of *Jatropha* and *Putranjiva* revealed that the cotyledons of the associated embryo expand and show some proliferation along with the radicle. Soon the endosperm overtakes them and proliferates more vigorously. The endosperm callus is therefore transferred to fresh medium to avoid contamination by the embryonal callus.

In squash preparations and microtome sections, 4-week-old endosperm callus of *Croton, Jatropha,* and *Putranjiva* showed only parenchymatous cells. After another 3–5 weeks the tracheidal cells differentiated.

The thickening of cells of the callus tissue and the differentiation and organization of tracheidal cells varied a great deal on different media. The disposition of thick-walled cells in *Jatropha, Putranjiva,* and *Ricinus* endosperm callus was at first irregular but later became more pronounced than that in the cambium-like zones. The organization of tracheidal cells and the cambium-like layer resembled a vascular bundle in 5-month-old endosperm cultures of *Ricinus*. The phloroglucinol–HCl test for lignin, both in squash preparations and in freehand sections, was positive. Maximum tracheidal cells differentiated in the endosperm callus of *Ricinus* with 2–4% sucrose. Culture medium without any sugar inhibited callus growth as well as the differentiation of tracheidal cells. Also, tracheidal cells did not differentiate in *Ricinus* endosperm callus on a medium with arabinose, lactose, mannose, and sorbose. When the callus obtained on sucrose medium was transferred to a sucrose-free medium, there was a gradual decrease in the growth rate of callus and in the frequency of tracheidal cells.

Cell groups with thickened walls were observed in 9-week-old callus of *Jatropha* (Srivastava and Johri, 1974). After 12 weeks these groups developed reticulate thickenings. The endosperm callus of *Jatropha* and *Ricinus* showed actively dividing cells on YE-supplemented medium besides multicellular filaments and cell clusters with hypertrophied nuclei and fragmented nucleoli. Occasionally, cell assemblages simulating stages of embryogeny were observed. In *Jatropha* and *Putranjiva* differentiation of tracheidal cells accompanied organogenesis. In *Putranjiva* a periderm-like layer was observed in 6-month-old endosperm callus.

The roots developed from the endosperm callus exhibited calyptrogen, dermatogen, periblem, and plerome in *Croton, Putranjiva,* and *Jatropha*. The mature root in *Croton* possessed a root cap. Transverse sections of roots showed the epidermal layer, followed by cortex, pericycle, and endodermis. The xylem was poorly developed.

V. Cytological Studies

In prolonged cultures endosperm callus exhibits cells of different ploidy. Occurrence of aneuploid and polyploid cells was observed in maize endosperm by Straus and LaRue (1954), and in *Lolium* by Norstog (1956) and Norstog *et al.* (1969). In the endosperm cultures of *Croton* and *Jatropha,* the callus also exhibited cells of different ploidy. However, in no case did the organs differentiate from cells of higher ploidy than the normal endosperm. In *Dendrophthoe* and *Taxillus* there was neither a change in chromosome number nor a loss in organ-forming capacity of the callus. *Petroselinum* endosperm callus, maintained for more than half a year through several passages, also did not show any deterioration in its embryogenic potentiality. The organ-forming capacity of the mature endosperm callus of *Putranjiva* was not lost even after 24 months of continuous culture on the same medium. It may be attributed to the fact that the callus retained its constancy in chromosome number and thus maintained external as well as internal stability (Partanen, 1965). Cytological studies of endosperm callus have not been conducted in other taxa.

In the chromosome counts of *in vitro,* differentiated, diploid (from embryonal callus) and triploid (from endosperm callus) leaf cells of *Dendrophthoe* there was a 2:3 ratio in the size of the nucleus, but there was no difference in the size of chromosomes (Johri and Nag, 1974). Nag (1970) observed that there was no effect of either the nutrients in the medium or the position of the explant on the chromosomal constitution of the cells.

VI. "Embryo Factor" and Growth Response

Johri and Nag (1970) observed that in *T. vestitus,* after 9 weeks on KN (10 ppm) medium, the "seed" (endosperm with embryo) showed only three buds per culture in 40% cultures, whereas the half-split endosperm (without embryo) showed 13 buds per culture in almost all the cultures. The delayed appearance of shoot buds from the endosperm in "seed" cultures was seemingly a result of the presence of the embryo. The inhibitory effect of the embryo was also evident from the fact that in half-split "seeds" only six buds, as against 13 in half-split endosperm, developed in each culture.

The mature endosperm of some species fails to proliferate if cultured without the embryo. Perhaps, the embryo releases certain substances (embryo factor) for the activation of the endosperm (Paleg, 1960; Ingle and Hageman, 1965; Brown *et al.,* 1970). Bhojwani (1966) in *Croton,* and Srivastava (1971c) in *Putranjiva* and *Ricinus,* attempted to replace the "embryo factor." The endosperm pieces were presoaked in different concentrations of GA_3, IAA, or KN solutions for 12, 24, 36, and 48 hours. Explants soaked in IAA (2 ppm) for 24 hours, when cultured, enlarged considerably but did not proliferate. Better response was observed with *Putranjiva* and *Ricinus* endosperm pieces soaked in GA_3 (2 ppm) solution for 36 hours. Such endosperm pieces proliferated when cultured on IAA + KN + CH (*Putranjiva*) or on 2,4-D + KN + YE medium (*Ricinus*). In *Putranjiva* the cut margins (after 4 weeks) formed small, green buds. Therefore, the "embryo factor" could be replaced by GA_3 in *Cro-*

ton, Ricinus, and *Putranjiva.* Mu and co-workers (1977a,b) observed that the association of embryo with the endosperm at culture is not necessary for the proliferation of endosperm and subsequent differentiation of organs. In *Citrus* (Wang and Chang, 1978), also, the presence of embryo was not obligatory for the proliferation and differentiation of plantlet from endosperm. However, Wang and Chang (1978) did use gibberellin in the medium. Therefore, what the "embryo factor" is, and what role it plays in endosperm cultures, remains to be elucidated.

VII. Concluding Remarks

The endosperm tissue is of much significance, as it totally lacks (*in vivo*) any degree of differentiation. Earlier attempts to induce (*in vitro*) organogenesis in endosperm failed, although it was achieved in other plant parts. In *Ricinus, Zea mays,* and *Cucumis* proliferation resulting in a callus tissue of unlimited growth was obtained, but organogenesis did not occur even though endosperm callus showed groups of meristematic cells. The induction of organogenesis in endosperm cultures, therefore, has always been a challenging problem and, after many failures, has been achieved in only a few taxa. Extensive investigations therefore must be undertaken to examine the recalcitrant nature of endosperm for organogenesis, in general, and for embryogenesis, in particular.

Endosperm is an excellent experimental system for physiological and morphogenetic studies. Tissue cultures that are morphologically homogeneous and that have the capacity of unlimited growth *in vitro* show great promise in the study of problems on metabolism and differentiation.

Endosperm lacks vascular elements, and buds differentiate directly (without callusing) from the epidermal cells. Therefore, the differentiation of buds cannot be correlated with the presence of tracheidal elements. However, in autotrophic members, vascular tissues are discernible in the callus prior to shoot differentiation.

The distribution of shoot buds in half-split endosperm cultures indicates a remarkable influence of implanting position and injury on organogenesis. Restriction of buds adjacent to the cut surface in direct contact with the medium suggests that along the injured region the cells acquire meristematic activity much earlier than the other cells. In the seed cultures of *D. falcata,* however, the buds mostly develop from the cells in the exposed portion of endosperm, that is, away from the medium.

Investigations provide ample evidence that triploid plants can be raised through endosperm culture. The conventional method for raising triploids is through a cross between a tetraploid and diploid. This cross, however, may not be successful because of reasons including failure of endosperm development with subsequent abortion of the embryo. Moreover, repeated production of triploids is a cumbersome process. In these cases the culture of endosperm may prove to be of advantage. Presently, there are a number of plants, such as several varieties of banana, apple, beet, tea, mulberry, and watermelon, in which triploids are in commercial use. In *Populus tremuloides* it has been shown by Winton (1970) that triploid plants (as compared to diploid counterparts) are superior for pulp wood.

In addition, in those hybrids where growth and development of embryo is affected because of poor development of endosperm, endosperm culture should prove quite profitable. Floris (1969) was able to grow wheat embryos on a different endosperm of the same plant. Ziebur and Brink (1951) succeeded in growing *Raphanus sativus* and *Capsella bursa-pastoris* embryos on barley endosperm and endosperm extracts. Likewise, Krüse (1973, 1974) was able to culture 9- to 12-day-old hybrid embryos of *Hordeum* × *Triticum*, *Hordeum* × *Agropyrum*, and *Hordeum* × *Secale* on a medium that had endosperm of *Hordeum* on its upper surface. Using *Hordeum* endosperm as the nurse tissue, young embryos of these hybrids can be induced to germinate and form normal hybrid plants.

The production of triploid plantlets from the endosperm of *Citrus* (Wang and Chang, 1978) is a very significant advancement. It is desirable that plant breeders apply these techniques to economically important plants where conventional breeding methods prove futile.

REFERENCES

Bajaj, Y. P. S. (1970). *Z. Pflanzenphysiol.* **63,** 408–415.
Bhojwani, S. S. (1966). *Phytomorphology* **16,** 349–353.
Bhojwani, S. S. (1968). Ph.D. Thesis, Univ. of Delhi, Delhi.
Bhojwani, S. S., and Johri, B. M. (1970). *Z. Pflanzenphysiol.* **63,** 269–275.
Bhojwani, S. S., and Johri, B. M. (1971). *New Phytol.* **70,** 761–766.
Brink, R. A., and Cooper, D. C. (1947). *Bot. Rev.* **13,** 423–541.
Brown, D. J., Canvin, D. T., and Zilkey, B. F. (1970). *Can. J. Bot.* **48,** 2323–2331.
Chikkannaiah, P. S., and Gayatri, M. C. (1974). *Curr. Sci.* **43,** 23–24.
Floris, C. (1969). *G. Bot. Ital.* **103,** 59–64.
Ingle, J., and Hageman, R. H. (1965). *Plant Physiol.* **40,** 672–675.
Johri, B. M., and Bhojwani, S. S. (1965). *Nature (London)* **208,** 1345–1347.
Johri, B. M., and Nag, K. K. (1970). *Curr. Sci.* **39,** 177–179.
Johri, B. M., and Nag, K. K. (1974). *Cytologia* **39,** 801–813.
Johri, B. M., and Srivastava, P. S. (1972). *In* "Advances in Plant Morphology" V. Puri Commemoration Vol. (Y. S. Murty, B. M. Johri, H. Y. Mohan Ram, and T. M. Verghese, eds.), pp. 339–358. Sarita Prakashan, Meerut, India.
Johri, B. M., and Srivastava, P. S. (1973). *Z. Pflanzenphysiol.* **70,** 285–304.
Keller, H., Wanner, H., and Baumann, T. W. (1972). *Planta* **108,** 339–350.
Krüse, A. (1973). *Hereditas* **73,** 157–161.
Krüse, A. (1974). *Hereditas* **77,** 219–224.
Lampton, R. K. (1952). Ph.D. Thesis, Univ. of Michigan, Ann Arbor.
LaRue, C. D. (1936). *Bull. Torrey Bot. Club* **63,** 365–382.

LaRue, C. D. (1944). *Am. J. Bot.* **31,** 45 (Abstr.)

LaRue, C. D. (1947). *Am. J. Bot.* **34,** 585–586.

LaRue, C. D. (1949). *Am. J. Bot.* **36,** 798.

Maheshwari, P. (1950). "An Introduction to the Embryology of Angiosperms." McGraw-Hill, New York.

Masuda, K., Koda, Y., and Okazawa, Y. (1977). *Physiol. Plant.* **41,** 135–138.

Monaco, L. C., Sondahl, M. R., Carvalho, A., Crocomo, O. J., and Sharp, W. R. (1977). *In* "Applied and Fundamental Aspects of Plant Cell, Tissue and Organ Culture" (J. Reinert and Y. P. S. Bajaj, eds.), pp. 109–126. Springer-Verlag, Berlin and New York.

Mu, S.-K., Kwei, Y.-L., Liu, S.-K., Chang, F.-C., Lo, F.-M., Yang, M.-Y., and Wang, F.-H. (1977a). *Chih Wu Hsueh Pao* **19,** 93–94.

Mu, S.-K., Liu, S.-Q., Zhou, Y.-K., Qian, N. F., Zhang, P., Zie, H.-X., Zhang, F. S., and Yan, Z.-L. (1977b). *Sci. Sin.* **20,** 370–376.

Nag, K. K. (1970). Ph.D. Thesis, Univ. of Delhi, Delhi.

Nag, K. K., and Johri, B. M. (1971). *Phytomorphology* **21,** 202–218.

Nakajima, T. (1962). *Bull. Univ. Osaka Prefect., Ser. B.* **13,** 13–48.

Nakano, H., Tashiro, T., and Maeda, E. (1975). *Z. Pflanzenphysiol.* **76,** 444–449.

Norstog, K. (1956). *Bot. Gaz. (Chicago)* **117,** 253–259.

Norstog, K., Wall, W. E., and Howland, G. P. (1969). *Bot. Gaz. (Chicago)* **130,** 83–86.

Paleg, L. G. (1960). *Plant Physiol.* **35,** 902–906.

Partanen, C. R. (1965). *Am. J. Bot.* **52,** 204–209.

Raghavan, V. (1976). "Experimental Embryogenesis in Vascular Plants." Academic Press, New York.

Rangaswamy, N. S., and Rao, P. S. (1963). *Phytomorphology* **13,** 450–454.

Sargant, E. (1900). *Ann. Bot. (London)* **14,** 689–712.

Satsangi, A., and Mohan Ram, H. Y. (1965). *Phytomorphology* **15,** 20–30.

Sehgal, C. B. (1969). *Beitr. Biol. Pflanz.* **46,** 233–238.

Sehgal, C. B. (1974). *Curr. Sci.* **43,** 38–40.

Sethi, M., and Rangaswamy, N. S. (1976). *Curr. Sci.* **45,** 109–111.

Smith, M. M., and Stone, B. A. (1973). *Aust. J. Biol. Sci.* **26,** 123–133.

Srivastava, P. S. (1971a). *Curr. Sci.* **40,** 337–339.

Srivastava, P. S. (1971b). *Z. Pflanzenphysiol.* **66,** 93–96.

Srivastava, P. S. (1971c). Ph.D. Thesis, Univ. of Delhi, Delhi.

Srivastava, P. S. (1973). *Z. Pflanzenphysiol.* **69,** 270–273.

Srivastava, P. S., and Johri, B. M. (1974). *Beitr. Biol. Pflanz.* **50,** 255–268.

Srivastava, P. S., and Johri, B. M. (1978). *Beitr. Biol. Pflanz.* **54,** 381–397.

Sternheimer, E. P. (1954). *Bull. Torrey Bot. Club* **81,** 111–113.

Straus, J. (1960a). *Plant Physiol.* **35,** 645–650.

Straus, J. (1960b). *Am. J. Bot.* **47,** 641–647.

Straus, J., and LaRue, C. D. (1954). *Am. J. Bot.* **41,** 687–694.

Tamaoki, T., and Ullstrup, A. J. (1958). *Bull. Torrey Bot. Club* **85,** 260–272.

Van Overbeek, J., Conklin, M. E., and Blakeslee, A. P. (1942). *Am. J. Bot.* **29,** 472–477.

Wang, T., and Chang, C. (1978). *Proc. Symp. Plant Tissue Cult., Peking,* pp. 463–468.

Winton, L. L. (1970). *Am. J. Bot.* **57,** 904–909.

Ziebur, N. K., and Brink, R. A. (1951). *Am. J. Bot.* **38,** 253–256.

Chapter 17

The Formation of Secondary Metabolites in Plant Tissue and Cell Cultures

H. BÖHM

Institute of Plant Biochemistry of the Academy of Sciences of the GDR, Halle/Salle, German Democratic Republic

I. Introduction

Whether or not a given natural compound is a secondary metabolite may be interpreted differently by different biochemists or physiologists. This indicates that the criteria of defining secondary metabolites are not absolutely clear. Nevertheless, the existence of a broad metabolic area outside the primary metabolic processes is evident. This so-called secondary metabolism results in such characteristic compounds as flavonoids, carotenoids, alkaloids, cardiac glycosides, antibiotics, and volatile oils. It is known in microorganisms, animals, and especially in plants. Generally, one plant forms secondary metabolites belonging to different groups of natural compounds (Luckner, 1971).

In contrast, cell cultures derived from plants can be completely or partially inactive in secondary metabolism. This phenomenon has been explained by the lack of differentiation processes that precede secondary metabolite formation in the course of the realization of differentiation programs in whole plants (Luckner *et al.*, 1977). However, the presence of genes responsible for the biosynthesis of secondary substances has been confirmed in unproductive cell cultures (Hirotani and Furuya, 1977). It also has been shown that one secondary compound is not synthesized although the appropriate precursor is available in the cells (Bauch and Leistner, 1978a). Finally, the absence of accumulating spaces like those present in intact plants does not necessarily prevent the formation of secondary metabolites in plant cell cultures (Lang and Höchster, 1977). Consequently, the lack of gene expressions that lead to specific enzymes of secondary metabolism seems to be the main reason for the unproductivity of cell cultures in this area. The differentiation processes must be initiated by the cell cultures themselves. No essential influence of explants has been observed. This corresponds to the fact that highly productive cell cultures can be derived from organs that are unable to form secondary compounds characteristic for the whole plant (Böhm, 1977).

Cell cultures capable of synthesizing secondary metabolites are the subject of this chapter. Because of their properties they are of significance not only for basic experiments (Böhm, 1977; Butcher, 1977; Overton and Picken, 1977) but also for industrial processes (Staba, 1977; Alfermann and Reinhard, 1978; Zenk, 1978). The following is a selection of results and problems that mark the growth in understanding phenomena of secondary metabolite formation in plant tissue and cell cultures during the last years.

II. The Biosynthesis of Secondary Metabolites

A. Intermediates and Enzymes

1. *Sesquiterpenoids*

Using callus cultures of *Andrographis paniculata* Overton and co-workers (Overton and Picken, 1976, 1977) have directed their investigation to the biosynthesis of paniculides A, B, and C (Fig. 1). These sesquiterpenoid lactones are completely absent in the respective intact plants, which are characterized by a group of diterpenoids (Butcher and Connolly, 1971). They are structurally based on γ-bisabolene. Feeding experiments with ^{14}C-labeled farnesol isomers have indicated 2-*cis*,6-*trans*-farnesol pyrophosphate as a natural intermediate that is transformed into γ-bisabolene by simple cyclization (Fig. 1). Other studies have revealed that *Andrographis paniculata* callus cultures synthesize the Z-isomer of the latter compound. Obviously, Z-γ-bisabolene is converted into the paniculides without rearrangement (Overton and Picken, 1976).

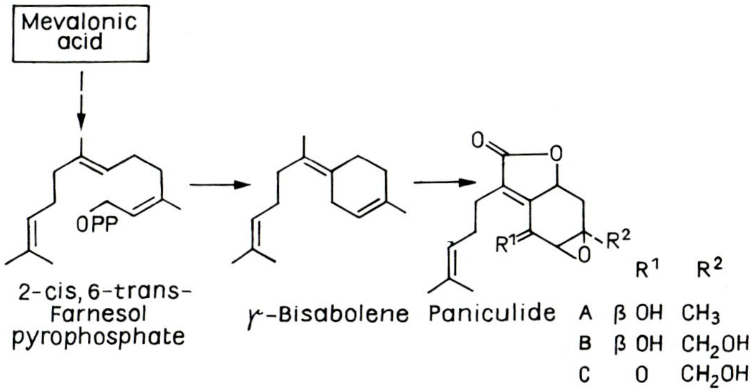

FIG. 1. The biosynthesis of paniculide B, a representative of sesquiterpenoids, characteristic for *Androgrphis paniculata* callus cultures. (From Overton and Picken, 1976; Butcher and Connolly, 1971.)

2. *Corynanthe-type Alkaloids*

Experiments with enzyme preparations obtained from *Catharanthus roseus* cell cultures (T, S)[1] have remarkably improved knowledge of the biosynthetic pathway of *Corynanthe*-type alkaloids. Scott and Lee (1975) demonstrated the *in vitro* transformation of tryptamine and secologanin into geissochizine and ajmalicine (Fig. 2). Confirming this result in principle, Stöckigt and associates (1976) have found that the overall reaction absolutely depends on the presence of a reduced nicotine pyridine nucleotide (NADPH or NADH). If such a component is absent in the incubation mixture, an intermediate of the ajmalicine biosynthesis accumulates. It was identified as 20,21-didehydroajmalicine (cathenamine, Fig. 2) and should be converted into ajmalicine and its isomers by a stereochemically controlled reduction (Stöckigt *et al.*, 1977). According to time-dependent incorporation experiments with enzyme extracts from *C. roseus* tissue cultures, geissoschizine is an alternative intermediate to cathenamine (Lee *et al.*, 1979) (Fig. 2). In contrast, Stöckigt (1978) has concluded from his investigation that this alkaloid is indirectly involved in the biosynthesis of ajmalicine and related compounds.

Partial purification of the enzyme system from *C. roseus* cell cultures revealed that a β-D-glucosidase is involved in the formation of *Corynanthe*-type alkaloids (Scott *et al.*, 1977b). Consequently, δ-D-gluconolactone, known as an inhibitor of plant glucosidases, inhibits the *in vitro* synthesis of ajmalicine and its isomers (Treimer and Zenk, 1978). Under its influence a substance accumulates that has been identified as strictosidine

---

[1] In discussing cell culture here, the types of culture originally used are symbolized by capital letters: C, callus culture; T, tissue culture; S, cell suspension culture.

(=isovincoside, Fig. 2). It results from the condensation of tryptamine with secologanin and must be considered the key intermediate in the biosynthesis of monoterpenoid indole alkaloids with 3α and 3β configurations, instead of the epimeric vincoside (Stöckigt and Zenk, 1977a,b; Rueffer *et al.*, 1978) that previously had been postulated in this position. Strictosidine synthetase could be detected in cell cultures derived from plants belonging to different genera of the family Apocynaceae. In this screening the enzyme has shown a pH optimum between 6.0 and 7.0 (Treimer and Zenk, 1979), in accord with the pH optimum of the enzyme system catalyzing the overall reaction found at 6.5 (Treimer and Zenk, 1978). Isolated from *C. roseus* cell suspension cultures, strictosidine synthetase has been purified 740-fold and characterized in detail (Mizukami *et al.*, 1979).

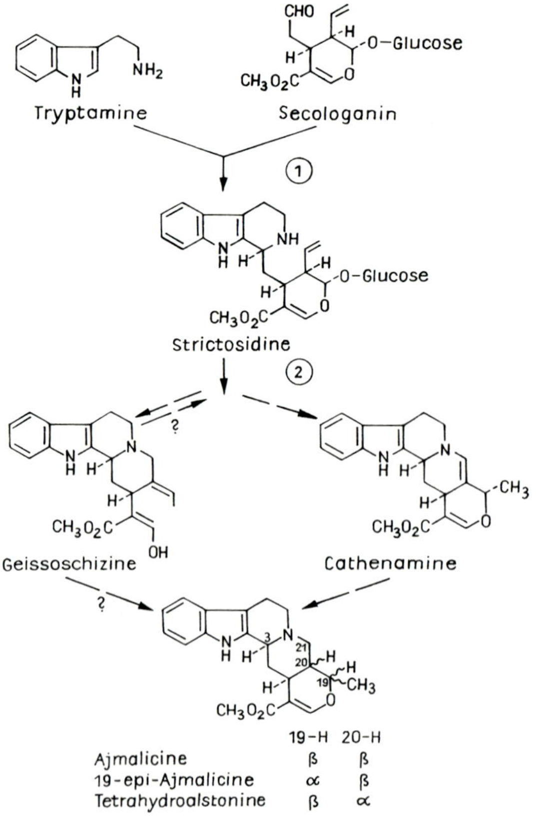

	19-H	20-H
Ajmalicine	β	β
19-epi-Ajmalicine	α	β
Tetrahydroalstonine	β	α

FIG. 2.   Intermediates and enzymes. (1) Strictosidine synthetase; (2) β-ᴅ-glucosidase, involved in the biosynthesis of ajmalicine and its isomers in cell cultures (T, S) of *Catharanthus roseus*. (From Scott *et al.*, 1977b; Stöckigt *et al.*, 1977; Treimer and Zenk, 1979; Stöckigt, 1978; Lee *et al.*, 1979.)

The fact that strictosidine and not vincoside is the first specific intermediate of the biosynthesis of *Corynanthe*-type alkaloids also has been demonstrated by the following experiments: Separate addition of radioactively labeled samples of both compounds to incubation mixtures prepared with enzyme extracts from *C. roseus* callus cultures has revealed end product formation from strictosidine only (Scott *et al.*, 1977a). 4,21-Dehydrocorynantheine aldehyde has been indirectly identified as an intermediate in the course of strictosidine transformation (Stöckigt *et al.*, 1978).

3. *Flavonoids*

In the field of flavonoid glycoside biosynthesis extensive investigations with both intact plants and plant cell cultures have led to a detailed knowledge of intermediates and several enzymes (Grisebach and Hahlbrock, 1974). Cell cultures have been used to study enzymes engaged in the formation of flavanones and preceding reactions as well as in glycosidations and *O*-methylations of 15C-compounds.

From illuminated cell suspension cultures of *Petroselinum hortense* a CoA ligase was isolated and partially purified which shows a broad specificity toward substituted cinnamic acids. Of all substrates tested, *p*-coumarate is the most efficient. According to further experiments, only its trans isomer is activated. Evidently, the enzyme from parsley cell cultures represents a *p*-coumarate:CoA ligase involved in the flavonoid biosynthesis by providing flavanone synthase with the phenylpropane substrate. This *p*-coumarate:CoA ligase could not be separated into isomers. Its molecular weight was estimated at about 67,000 (Knobloch and Hahlbrock, 1977).

The behavior of acetyl-CoA carboxylase in suspended parsley cells after illumination or dilution of cultures indicates the cooperation of this enzyme in the biosynthesis of flavonoid glycosides (see also Section II,B,2). Acetyl-CoA carboxylase is well known from fatty acid metabolism as the enzyme responsible for the synthesis of malonyl-CoA. This molecule is used as substrate for the *in vitro* formation of the A ring of flavonoids catalyzed by flavanone synthase (Kreuzaler *et al.*, 1978) (Fig. 3). Whether one type of acetyl-CoA carboxylase acts both in fatty acid and flavonoid biosynthesis or at least two types of the enzyme are separately involved in these metabolic areas is still an open question (Ebel and Hahlbrock, 1977).

In order to study the substrate specificity of flavanone synthase toward differently substituted cinnamic acids, this enzyme was extensively purified after isolation from illuminated cell suspension cultures of *P. hortense*. At pH 8.0 the addition of *p*-coumaroyl-CoA to the incubation mix-

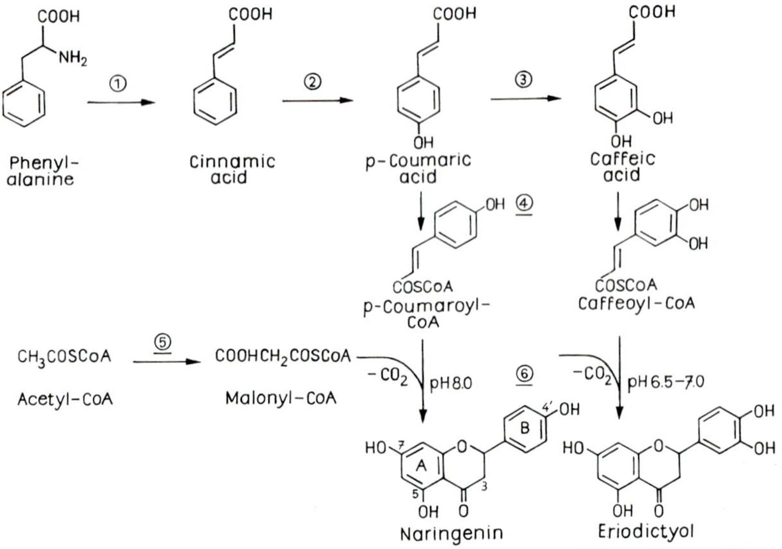

FIG. 3. Flavanone synthesis catalyzed by enzymes extracted from *Petroselinum hortense* cell suspension cultures. (1) Phenylalanine ammonia-lyase; (2) cinnamate 4-hydroxylase; (3) *p*-coumarate 3-hydroxylase; (4) *p*-coumarate:CoA ligase; (5) acetyl-CoA carboxylase; (6) flavanone synthase. Underlined enzyme symbols refer to recent experiments. (From Ebel and Hahlbrock, 1977; Kreuzaler *et al.*, 1978; Saleh *et al.*, 1978.)

ture containing also malonyl-CoA has led to an intensive formation of naringenin (Fig. 3). Under otherwise identical conditions the substrate caffeoyl-CoA has given rise to only a very small amount of eriodictyol (5,7,3′,4′-tetrahydroxyflavanone). Feruloyl-CoA has not been used as substrate for flavanone formation. Experiments with flavanone synthase from *Haplopappus gracilis* cell suspension cultures confirmed these results. However, when experimental conditions were varied it was shown that caffeoyl-CoA was an efficient substrate for eriodictyol synthesis at pH values of 6.5–7.0. According to reinvestigations the same is true for the *in vitro* system derived from parsley cell cultures. Obviously, *p*-coumaroyl-CoA and caffeoyl-CoA are native substrates of the flavanone synthase in *P. hortense* and *H. gracillis*.

The mechanism by which they are combined with six carbon atoms in each case to form the related flavanone molecules might be signaled by compounds that occur in addition to the main products of *in vitro* reactions. Irrespective of their special structures, these compounds show an aromatic ring with a side chain containing one, two, or four C atoms in addition to three C atoms coming from the phenylpropane moiety. They have been named "release products" (Fig. 4) because of the assumption

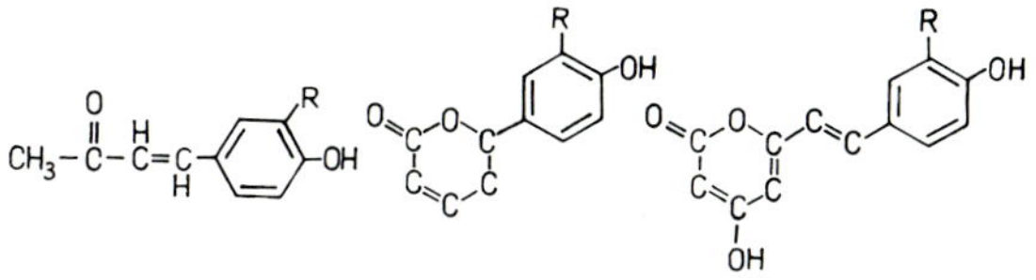

FIG. 4. Release products of *in vitro* transformation of *p*-coumaroyl-CoA (R = H), caffeoyl-CoA (R = OH), and feruloyl-CoA (R = OCH₃) catalyzed by flavanone synthase from parsley cell suspension cultures. (Redrawn from Hrazdina *et al.*, 1976.)

that flavanone synthase is a multifunctional enzyme that performs a stepwise condensation of cinnamic acid derivatives with malonate molecules at different binding sites. In the course of this process, unfinished products can be released if their translocation from one binding site to another is blocked (Hrazdina *et al.*, 1976; Saleh *et al.*, 1978).

The finding that the same flavanone synthase accepts *p*-coumaroyl-CoA as well as caffeoyl-CoA as substrate (Saleh *et al.*, 1978) is important with respect to the question of when in the course of the flavonoid glycoside biosynthesis the substitution pattern of the B ring is attached. Although manifold experimental data are available it has not yet been clarified whether this pattern is determined at the cinnamic acid stage or completed after the formation of the 15C-compound (Hess, 1968; Hrazdina *et al.*, 1976). A condition postulated by the so-called "cinnamic acid start hypothesis" has been demonstrated on principle by the results of Saleh and associates (1978). Remarkably, the first flavanone synthase isolated from the tissue of intact plants—tulip anthers—exhibits the substrate specificity known for that from plant cell cultures and can additionally transform feruloyl-CoA into homoeriodictyol (5,7,4'-trihydroxy-3'-methoxyflavanone) with high efficiency (Sütfeld *et al.*, 1978).

Another problem connected with the function of flavanone synthase seems to remain: Are the immediate products of *in vivo* reactions catalyzed by this type of enzymes flavanones or chalcones? Evidently, *in vitro* reactions always result in flavanone structures (Kreuzaler and Hahlbrock, 1975) even if the plant material from which the flavanone synthase has been extracted exclusively accumulates chalcones (Sütfeld *et al.*, 1978). The *in vivo* situation, however, should be considered also in light of the following findings: Certain mutants of *Callistephus chinensis* and *Petunia hybrida* are unable to synthesize flavonols and anthocyanins in flower parts. They show no or very weak chalcone:flavanone isomerase activity and accumulate chalcones (Kuhn *et al.*, 1978; Forkmann and Kuhn, 1979) (see Note Added in Proof).

Some enzymes responsible for the glycosylation of 15C-compounds have been isolated from plant cell cultures. One of these enzymes, extracted from cell suspension cultures of *Glycine max,* was purified about 5- to 10-fold. It predominantly glucosylates the 3 position of flavonols and uses UDP-glucose as glucosyl donor. The transferase exhibits no activity toward cinnamic acids and simple phenols (Poulton and Kauer, 1977). From cell suspension cultures of *Haplopappus gracilis* an enzyme has been isolated that transfers glucosyl from UDP-glucose to the 3 position of cyanidin, other anthocyanidins, and even flavonols. The *in vitro* glycosylation has its pH optimum at about 8.0 (Saleh *et al.,* 1976).

O-Methylation in position 3 of the B ring of several flavones and flavonoles is catalyzed by an enzyme extracted from soybean cell suspension cultures. In this reaction *S*-adenosyl-L-methionine acts as methyl donor. The methyltransferase has been purified 38-fold. It can also use a flavone glucoside (luteolin 7-*O*-glucoside) as substrate. This is of interest with respect to the sequence of glycosylation and *O*-methylation in the course of the variation of 15C-compounds (Poulton *et al.,* 1976a, 1977). The methylation of hydroxy groups in the A ring of various flavonoids has been demonstrated with cell-free extracts from *Citrus mitis* callus cultures. Attempts to purify the methyltransferase(s) have been unsuccessful, so far, because of the instability of crude enzyme preparations (Brunet *et al.,* 1978; Brunet and Ibrahim, 1978).

4. *Lignin Monomers*

Besides flavonoid glycosides lignin also belongs to the large group of phenylpropanoid compounds. As generally known, aldehydes and especially alcohols derived from *p*-coumaric, ferulic, and sinapic acids are the essential building units of this polymer. The investigation of their biosynthesis has been remarkably improved by experiments at the enzymatic level for some years (Grisebach, 1977; G. G. Gross, 1977). Because the steps up to the activation of substituted cinnamic acids are in principle identical in lignin and flavonoid biosynthesis, the following part of the pathway leading to lignin monomers has been of special interest (Fig. 5). Cell suspension cultures of *Glycine max,* which form lignin, have contributed to the investigation of this sequence.

Cinnamoyl-CoA:NADPH reductase catalyzes the NADPH-dependent reduction of substituted cinnamoyl-CoA esters to the related cinnamaldehydes (Fig. 5). It shows the highest degree of specificity toward feruloyl-CoA. Also 5-hydroxyferuloyl-CoA and *p*-coumaroyl-CoA are efficient substrates. The molecular weight of the 1660-fold purified enzyme is about 38,000. The reaction has its pH optimum between 6.0 and 6.2 (Wengenmayer *et al.,* 1976). A cinnamyl alcohol dehydrogenase seems to be responsible for the last step of the biosynthesis of lignin monomers. Ob-

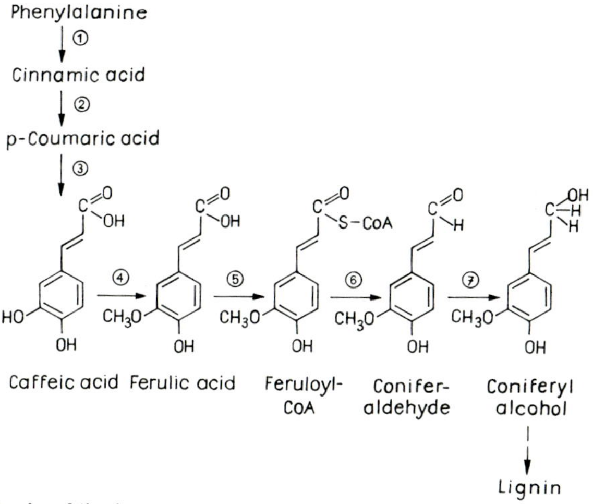

Phenylalanine
↓ ①
Cinnamic acid
↓ ②
p-Coumaric acid
↓ ③

Caffeic acid — ④ → Ferulic acid — ⑤ → Feruloyl-CoA — ⑥ → Conifer-aldehyde — ⑦ → Coniferyl alcohol

↓

Lignin

FIG. 5. Synthesis of lignin monomers catalyzed by enzymes isolated from cell suspension cultures of *Glycine max*. (1) Phenylalanine ammonia-lyase; (2) cinnamate 4-hydroxylase; (3) p-coumarate 3-hydroxylase; (4) O-methyltransferase; (5) isoenzyme 1 of a p-coumarate:CoA ligase; (6) cinnamoyl-CoA:NADPH reductase; (7) cinnamyl alcohol dehydrogenase. Formulas mark reaction steps recently investigated. (From Poulton *et al.*, 1976a,b; Knobloch and Hahlbrock, 1975; Wengenmayer *et al.*, 1976; Wyrambik and Grisebach, 1975.)

viously, the reverse reaction of this oxidoreductase is of physiological significance for the transformation of substituted cinnamaldehydes into the related cinnamyl alcohols (Fig. 5). After isolation from soybean cells the enzyme could be separated into two isoenzymes. The only substrate accepted by isoenzyme 1 is coniferyl alcohol. Isoenzyme 2 shows a broad specificity toward several substituted cinnamyl alcohols and cinnamaldehydes. The latter compounds are more efficient substrates than the former ones. In each group the ferulic acid derivative—coniferyl alcohol and coniferaldehyde, respectively—shows the highest efficiency. The isoenzymes have been partially purified. Their molecular weights are 43,000 (isoenzyme 1) and 69,000 (isoenzyme 2). The following optimal pH values have been observed for the reactions catalyzed by isoenzyme 2: 8.8 for coniferyl alcohol oxidation and 6.5 for coniferaldehyde reduction (Wyrambik and Grisebach, 1975).

With respect to the important role of methoxylated cinnamic acid derivatives in lignin biosynthesis, two further enzymes are considered to be specifically involved in this metabolic pathway. Both have been isolated from cell suspension cultures of *Glycine max*. An O-methylating enzyme exclusively transfers the methyl group from S-adenosyl-L-methionine at the meta positions of hydroxyl cinnamic acids. On several substrates tested, caffeic and 5-hydroxyferulic acids are transformed with high efficiency (Poulton *et al.*, 1976a,b) (Fig. 5).

The other enzyme that is noteworthy in this connection is the isoenzyme 1 of a *p*-coumarate:CoA ligase. Although isoenzyme 2 cannot use methoxycinnamic acids as substrates, isoenzyme 1 preferentially activates these intermediates. In a range of substituted cinnamic acids tested as substrates, ferulic, sinapic, and 5-hydroxyferulic acids show the highest efficiency (Knobloch and Hahlbrock, 1975) (Fig. 5).

The selective functions of methyltransferases and ligases have been extensively discussed by Hahlbrock (1977) as a branch point from which different intermediates are introduced into different specific pathways of phenylpropanoid biosynthesis. Especially the function of isoenzyme 1 of *p*-coumarate:CoA ligase has been estimated as the selection of efficient substrates for the reducing enzymes of the biosynthesis of lignin monomers. Although the cited interpretation is most obvious, it should be considered in light of the properties of flavanone synthase that have since been reported (Saleh *et al.*, 1978; Sütfeld *et al.*, 1978).

B. REGULATORY ASPECTS

The several principles and mechanisms in regulating metabolic processes include such different phenomena as the substrate specificity of enzymes and the existence of isoenzymes, discussed in Section II,A,4, as well as compartmentation by cellular structures (see Section V). This section is directed to the change of specific enzyme activity, a very important kind of regulation. This phenomenon may be caused by the formation, activation, degradation, and inhibition of enzyme molecules.

1. Triggering Factors

Light and auxins are well-known factors triggering the formation of flavonoids and other pigments in plant cell cultures (Böhm, 1977). As an exception, Alfermann and associates (1975) have selected strains of *Daucus carota* callus cultures that form cyanidin glycosides in the dark on media without auxins. To explain this interesting fact one can assume that another factor instead of light or 2,4-dichlorophenoxyacetic acid (2,4-D) triggers anthocyanin biosynthesis or that the dependence of this metabolic pathway on effectors is removed. Each of these situations might be caused by mutation. Anthocyanin formation in dark-grown cell suspension cultures of *Haplopappus gracillis* is markedly stimulated only by continuous irradiation with high doses of ultraviolet (UV) light below 345 nm. No linear correlation between UV dose and the intensity of flavonoid formation could be observed. Before anthocyanin accumulates, the activities of phenylalanine ammonia-lyase (PAL), chalcone:flavanone isomerase, and flavanone synthase are drastically increased (Wellmann *et al.*, 1976).

Obviously, elicitors represent a special group of triggering factors. These substances are isolated from microorganisms and can trigger or drastically stimulate the formation of phytoalexins in plant tissues. Although the structures of most elicitors are unknown, the effector obtained from *Phytophtora megasperma* var. sojae (Pms), for instance, can be characterized as a polysaccharide built up by modified β-1,3-glucan molecules. Phytoalexins are comprised of different groups of natural compounds, especially isoflavonoids, terpenoids, polyacetylenes, and dihydrophenanthrenes. Their formation by plant tissues in response to the influence of various environmental factors is discussed as part of a plant defense system (Grisebach and Ebel, 1978; D. Gross, 1977). For some years cell cultures derived from plants of several genera of the family Leguminosae have been used to study problems of phytoalexin formation. In this section only experiments with more or less physicochemically definable triggering factors are considered.

If soybean cell suspension cultures growing in the dark are treated with solutions of the Pms elicitor, glyceollin, an isoflavonoid, appears 12–15 hours later and accumulates further. Before this time a rapid increase of PAL activity starts, reaching its maximum level about 20 hours after the addition of the effector. These processes can also be observed under the influence of nigerian instead of the Pms elicitor. Other commercially available polysaccharides are uneffective. Untreated soybean cell cultures show no glyceollin formation. It is remarkable that the addition of the effectors also results in the halting of cell culture growth (Ebel *et al.*, 1976). The "model inducer" ribonuclease can trigger the accumulation of the isoflavonoid phaseollin in cell suspension cultures of *Phaseolus vulgaris*. This process is accompanied by a drastic increase of the specific activity of PAL, the first enzyme of phenylpropanoid biosynthesis. It has been demonstrated by application of comparative density labeling technique that the *de novo* synthesis of PAL at least partially causes the increase of enzyme activity (Lamb and Dixon, 1978). Surprisingly, cell suspension cultures of *P. vulgaris* accumulate phytoalexins also after treatment with an aqueous extract of bean hypocotyls. Besides phaseollin as the main component, phaseollidin, phaseollinisoflavan, and kievitone are detectable at lower concentrations. Traces of these compounds are the activities of phenylalanine ammonia-lyase, (PAL), chalcone:flavalexin formation depends on the growth phase of subcultured cells. Phaseollin reaches its maximum amount 30–40 hours after the addition of bean hypocotyl extract to cell cultures. At least up to this time cell growth is prevented. Whereas phytoalexin concentration subsequently decreases, the growth rate of cell cultures becomes similar to that of controls. The nature of triggering factors is still unknown (Hargraeves and Selby, 1978).

From the stopping or delay of cell growth in experiments with elicitors and other effectors the question arises of whether or not these substances trigger phytoalexin formation directly or trigger instead a change in primary metabolism leading to phytoalexin accumulation. This circumstance and the high unspecificity of the correlation between triggering factor and cell response seem to restrict the significance of the system "triggering of phytoalexin formation in plant cell cultures" for investigations directed to fundamental problems of regulation.

Enzymes involved in lignin biosynthesis (PAL, cinnamic acid 4-hydroxylase, *O*-methyltransferase, peroxidase) show maxima of their activities 2 days after the transfer of tobacco cells into fresh liquid media containing 2,4-D or 6-benzyladenine as phytohormones. In both auxin- and cytokinin-containing media enzyme activities subsequently decrease. However, although they remain at a low level in the absence of 6-benzyladenine, they reach further maxima under the influence of this cytokinin between the tenth and twelfth day of subculture. Only after this time can lignification be observed. Consequently, *Nicotiana tabacum* cell cultures in an auxin medium do not form lignin and tracheids. In a cytokinin medium large cell clusters show higher activities of enzymes involved in lignin biosynthesis than smaller ones do (Yamada and Kuboi, 1976; Kuboi and Yamada, 1976, 1978). Experiments with callus cultures of *Phaseolus vulgaris* have also confirmed the dependence of lignin biosynthesis on cytokinins and have demonstrated changes of PAL and *O*-methyltransferase activities under various experimental conditions (Haddon and Northcote, 1976).

2. *Enzyme Induction*

Continuing their research program Hahlbrock and co-workers have elaborated detailed information on coordinated changes of enzyme activities and on related processes at the epigenetic level. The subject of all these experiments has been flavonoid glycoside biosynthesis in cell suspension cultures of *Petroselinum hortense*. As noted in Section II,B,1, it depends on the influence of light.

When parsley cell cultures grown in the dark for 10 days are illuminated, the enzymes involved in the biosynthesis of flavonoid glycosides show different behaviors. The activities of PAL, cinnamic acid 4-hydroxylase, and *p*-coumarate:CoA ligase (enzymes of group 1) increase after a lag phase of about 2 hours and reach their maxima about 17, 22, and 23 hours, respectively, after transfer to light (Fig. 6). In contrast, enzymes of

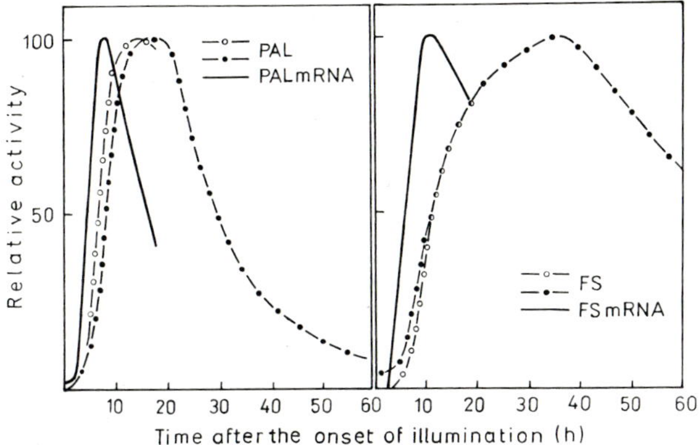

FIG. 6.   Activities of phenylalanine ammonia-lyase (PAL), flavanone synthase (FS), and their messenger ribobucleic acids (mRNA) after illumination of *Petroselinum hortense* cell suspension cultures previously grown in the dark for 7 (–O–, – – –) or 10 (—●—) days. (From Hahlbrock *et al.*, 1976; Schröder *et al.*, 1979.)

another group (group 2), for instance flavanone synthase, acetyl-CoA carboxylase, and specific methyl and glycosyl transferases, increase in activities after a lag phase of about 4 hours and are characterized by peaks of their activities between about 25 and 40 hours after the start of illumination (Fig. 6). Enzymes that are engaged in primary metabolism do not respond to the influence of light in these experiments. Obviously, the effect of illumination is highly selective. The different responses of enzymes involved in flavonoid glycoside biosynthesis indicate coordinated regulation of the members of each group and the existence of two regulatory systems. Interestingly, they correspond to the functions of these enzymes in different areas of phenylpropanoid metabolism: Whereas the enzymes of group 1 participate in the formation of all phenylpropane derivatives, the enzymes of group 2 are engaged exclusively in flavonoid glycoside biosynthesis (Hahlbrock *et al.*, 1976; Ebel and Hahlbrock, 1977).

Other results give further evidence of the *de novo* synthesis of enzymes following the exposure of cell cultures to light. From illuminated parsley cell cultures the messenger ribonucleic acids (mRNAs) of PAL and flavanone synthase, as representatives of both enzyme groups, could be isolated. They are characterized by their incubation in wheat germ extracts (Schröder *et al.*, 1976) or rabbit reticulocyte lysates (Ragg *et al.*, 1977; Schröder, 1977; Schröder *et al.*, 1977) and the treatment of translation products with specific antisera. In order to label proteins formed *in vitro* besides 19 other amino acids, [35S]methionine is added to the incubation mixture. Finally, immunoprecipitates are analyzed by gel electrophoresis. The application of this method also has revealed that parsley cells, after cultures are transferred to distilled water in the dark, contain the mRNA of PAL but not of flavanone synthase. The stimulating effect of dilution is generally restricted to the enzymes of group 1 (Böhm, 1977). Untreated cell cultures of *P. hortense* show neither PAL mRNA nor flavanone synthetase mRNA activity.

Continuous illumination of parsley cell cultures previously grown in the dark for 7 days causes a rapid increase of both PAL mRNA and flavanone synthase mRNA activity. These processes are detectable after a lag phase of about 2 hours and reach maximum levels in 9 and 11 hours, respectively, after the onset of light (Fig. 6). Obviously, in each case the increase and optimum mRNA activity precede the increase and optimum of the related enzyme activity. These data are in agreement with what is generally known of the sequence of mRNA and protein appearance in the course of gene expression. They permit the explanation of changes of enzyme activities under the influence of light by the changes of mRNA activities under the same conditions. Remarkably, in consequence of the coordinated induction of both mRNAs discussed here, the distance between the maximum of flavanone synthase mRNA activity and that of flavanone synthase activity is relatively wide. To what extent this finding and others, for instance, data concerning the decrease of mRNA activity, reflect the *in vivo* situations has been intensively investigated (Schröder *et al.*, 1979).

Evidently, light-induced changes of enzyme activity do not result only from the synthesis but also from the degradation of enzyme molecules. Experiments with continuously or only briefly illuminated parsley cell suspension cultures have shown that degradation is approximately constant in the case of PAL induction. The half-life of this enzyme very much depends on the growth phase of cells used for irradiation. It has been determined in cells growing in the linear phase of cultures to be about 10 hours but in those from the stationary phase it is only about 5 hours (Hahlbrock, 1976; Betz *et al.*, 1978). Later experiments have on principle confirmed these half-lives of PAL. They also have provided information about the half-life of flavanone synthase: Measuring enzyme activity in briefly illuminated parsley cells from cultures growing in the linear phase has revealed a value of 6–7 hours. In contrast, when cell extracts from the same cultures have been analyzed by immunoprecipitation, a half-life of about 20 hours has been obtained. This discrepancy points to the possibility that flavanone synthase loses its catalyzing activity before the degradation process begins (Schröder *et al.*, 1979).

3. *Inhibition*

Isoenzyme 2 but not isoenzyme 1 of cinnamyl alcohol dehydrogenase isolated from soybean cell suspension cultures is inhibited by the substrates of the forward and reverse reactions. Coniferyl alcohol and coniferaldehyde are effective above concentrations of 150 μM and 40 μM, respectively (Wyrambik and Grisebach, 1975). Cyanidin strongly inhibits the activity of UDP-glucose:cyanidin 3-O-glucosyltransferase from *Haplopappus gracilis* cell cultures if its concentration exceeds 250 μM. This enzyme is also inhibited by the product cyanidin-3-glucoside up to 30% (Saleh *et al.*, 1976). The activity of flavanone synthase is drastically reduced under the influence of naringenin or eriodictyol in a concentration of about 18 μM (Saleh *et al.*, 1978). Product inhibition has also been observed with cinnamoyl-CoA:NADPH reductase from soybean cell suspension cultures in the presence of CoASH or NADP (Wengenmayer *et al.*, 1976).

It has repeatedly been reported that auxins and gibberellins can inhibit the formation of secondary metabolites in plant cell cultures. For instance, *Daucus carota* callus cultures cannot produce anthocyanins on a medium containing 3×10^{-6} M gibberellic acid (GA_3), whereas they exhibit a typical blue-red color on the same medium without GA_3. The half-maximum dose of inhibition has been found at 10^{-8} M GA_3. Colorless, anthocyanin-free cultures show only 15% of the PAL activity detectable in the anthocyanin-producing cultures. Despite this, the total amounts of cinnamic acid derivatives are similar in both cell lines. Gibberellic acid does not inhibit the reaction catalyzed by PAL *in vitro*. The assumption that the low PAL activity in anthocyanin-free cell cultures results from low enzyme formation could be confirmed experimentally.

The specific activity of hydroxycinnamate:CoA ligase, moreover, is lower in colorless than in colored cell cultures of *D. carota*. This enzyme exhibits different substrate specificities in the two cell lines. Although *p*-coumaric acid is most efficiently used in cells containing anthocyanin, it remains nearly inactive in anthocyanin-free ones. Hydroxycinnamate:CoA ligase extracted from colored cells could be separated into three peaks of activity but gave rise to only two peaks after isolation from colorless cells. The GA_3-mediated variation of the phenylpropanoid metabolism is caused by a change of isoenzyme patterns in *D. carota* cell cultures (Braun and Seitz, 1975; Heinzmann and Seitz, 1977; Heinzmann *et al.*, 1977).

Because of low PAL activity, on the one hand, and large amounts of phenylpropanoid compounds, on the other, PAL cannot be considered a key enzyme in anthocyanin-free cell cultures of *D. carota* (Braun and Seitz, 1975). This corresponds to results obtained from experiments with anthocyanin-producing *D. carota* cell cultures (Stärk *et al.*, 1976) and tissue cultures from tea plants (Shipilova *et al.*, 1978). The role of PAL in phenylpropanoid metabolism has been critically considered by Margna (1977).

III. The Spectra of Secondary Metabolites

Cell cultures derived from various plant organs are generally identical in their spectra of a certain group of secondary metabolites. This result of earlier experiments has been confirmed by recent findings. It is indeed remarkable because not only are some plant parts biosynthetically inactive in the synthesis of a certain class of compounds (see Section I), but also because productive plant organs can differ in their spectra of a certain group of secondary metabolites. Obviously, the relation between the organ-specific volatile oil composition in *Ruta graveolens* roots and the volatile oil spectrum of cell cultures derived from this plant part (Nagel and Reinhard, 1975) remains an exception.

According to the foregoing consideration the spectra of secondary metabolites in plant cell cultures can be compared with the compositions of corresponding secondary compounds in the whole plant. Doing this, Delfel and Rothfus (1977) observed a "fortunate similarity." Cephalotaxine and its antitumor esters, harringtonine, isoharringtonine, homoharringtonine, and desoxyharringtonine, characteristic of *Cephalotaxus harringtonia,* occur in callus cultures derived from this tree. Their relative amounts are very similar to those in the intact plant. Frequently, the quantitative proportion of secondary compounds known from plants is changed in the respective cell cultures. For instance, although thebaine is the main alkaloid of *Papaver bracteatum,* its concentration is very low in cell cultures (C, S) derived from this poppy. In contrast, protopine, occurring in traces in the intact plant, predominates in the cell culture spectrum (Kamimura and Nishikawa, 1976; Kamimura *et al.*, 1976). Nevertheless, in this case the typical component of the original plant is present in the derived cell culture. Cell cultures of 13 other species of the family Papaveraceae has been found to form simple, uniform alkaloid spectra without any species-specific compound (Ikuta *et al.*, 1974; Franke and Böhm, 1978). Moreover, onion tissue cultures fail to produce the main flavor component of *Allium cepa* plants, *S*-propenyl cysteine sulfoxide. Only the three less important flavor compounds, *S*-allyl, *S*-methyl, and *S*-propyl cysteine sulfoxides, are detectable at about 10% of the level found in the intact plants (Turnbull and Collin, 1978).

A further phenomenon concerning spectra of secondary metabolites in cell cultures is the occurrence of components unknown in the original plant. Some reported examples are 19-*epi*-ajmalicine in *Catharanthus roseus* cell suspension cultures (Stöckigt *et al.*, 1976), aromorine in callus cultures of *Stephanis cepharantha* (Akasu *et al.*, 1976), and homodeoxyharringtonine in *Cephalotaxus harringtonia* callus cultures (Delfel and Rothfus, 1977).

Light is a definable factor influencing the alkaloid spectrum in tissue cultures of *Ruta graveolens*. It triggers the formation of arborinine, which is lacking in dark-grown tissue cultures (Szendrei *et al.*, 1976). The occurrence of some aromatic metabolites in cells of *Galium mollugo* depends on their culture in media characterized by special phytohormone compositions (Bauch and Leistner, 1978a). In this connection it should be mentioned that the formation of certain organs by cell cultures often influences spectra of secondary metabolites that differ from those of the respective plants. At least the regeneration of intact plants from cell cultures is accompanied by the restitution of the species-specific patterns of secondary metabolites. Illustrating this general experience, root formation by onion tissue cultures only leads to an increase in the amounts of less important flavor compounds, whereas in the course of shoot or plantlet formation the main flavor component predominates (Turnbull and Collin, 1978). Remarkably, abnormal alkaloid spectra have been found to remain in plants regenerated from callus cultures of *Datura innoxia* and in their descendants, disappearing only after a few generations (Tabata *et al.*, 1977).

Certainly, there are several reasons for the differences between secondary metabolite spectra in cell cultures and those in the corresponding plants. However, they can generally be explained by the lack of specific enzymes, the changed localization of substrate and enzyme or the transformation of accumulated intermediates into new secondary compounds in plant cell cultures. Further aspects of these interesting phenomena have been discussed by Böhm (1977, 1978).

IV. The Contents of Secondary Metabolites

A. BIOSYNTHETIC CAPACITY

The biosynthetic capacity of a plant cell culture is expressed by the amount of secondary metabolites formed in a biomass unit during a certain time. This feature is of special significance for the industrial production by plant cell cultures of secondary metabolites used in medicine or technology. In order to determine it correctly, at least two facts, exemplified by the following findings, must be considered.

Callus cultures obtained from *Digitalis purpurea* explants contain detectable amounts of typical glycosides. However, at least after the fifth subculture of these cells on static media the secondary metabolites disappear or their contents decrease drastically. Cells of the eighteenth subculture exhibit no traces of glycosides (Kartnig *et al.*, 1976; Kartnig, 1977;

Wichtl *et al.*, 1978). Interestingly, in these experiments the cardenolide glycoside formation in callus cultures derived from leaves quantitatively —and qualitatively—differs from that in root callus cultures (Kartnig *et al.*, 1979). The other fact noted above concerns the occurrence of secondary metabolites in media. For instance, paniculides, the sesquiterpenoid lactones characteristic for *Andrographis paniculata* cell cultures (S), are almost exclusively detectable only in the liquid media (Butcher and Connolly, 1971). About half of the amount of glyceollin, an isoflavonoid formed by soybean cell cultures (S) in response to the Pms elicitor, has been measured in culture filtrates (Ebel *et al.*, 1976). Calluses derived fom *Coffea arabica* plants excrete up to 90% of their caffeine into the static media (Frischknecht *et al.*, 1977), and the new alkaloid deoxyharringtonine occurs exclusively in the culture medium of *Cephalotaxus harringtonia* cells (Delfel and Rothfus, 1977).

Consequently, data representative for the biosynthetic capacity of a plant cell culture in the area of secondary metabolism can be obtained only by the analysis of cells after a long time of cultivation and by the proof of corresponding media. Moreover, it should be considered in this connection that the detectable amount of a given secondary metabolite is not necessarily equal to the quantity formed. Degradation processes may change the ratio of both values. Bauch and Leistner (1978b) believe that the lack of success in attempts to incorporate different precursors into two naphthols synthesized by *Galium mollugo* cell cultures (S) is caused by an extremely high turnover of these secondary compounds. The native degradation of secondary metabolites is difficult to investigate (Barz, 1977).

Up to the middle of the 1970s only a few cell cultures were reported to show biosynthetic capacities like those of the respective intact plants. However, since that time the concentrations of secondary metabolites in cell cultures have been repeatedly shown to reach or exceed levels known for the same compounds from original plants. Examples of unselected plant cell cultures with high biosynthetic capacities are cell suspension cultures of *Morinda citrifolia* (Zenk *et al.*, 1975) and *Galium mollugo* (Bauch and Leistner, 1978a) and *Cassia tora* callus cultures (Tabata *et al.*, 1975), all of which produce anthraquinones. Callus cultures of *Stephania cepharantha* show three times the amount of biscoclaurine alkaloids than the intact plant (Akasu *et al.*, 1976). Other alkaloids are formed in the same concentrations as in the original plants: serpentine in *Catharanthus roseus* callus cultures (Döller *et al.*, 1976) and caffein in callus cultures of *Coffea arabica* (Frischknecht *et al.*, 1977). Rosmarinic acid, the caffeoyl ester of 3,4-dihydroxyphenyllactic acid, is accumulated up to 15% of dry weight by *Coleus blumei* cell suspension cultures (Razzaque and Ellis,

1977; Zenk *et al.*, 1977b). No higher concentration than this has been reported for a single secondary metabolite in a plant cell culture.

It is an interesting fact of practical relevance that the cells of a plant cell culture differ in their capacities to snythesize certain groups of secondary metabolites. Because of this circumstance, isolated cell clusters or even single cells generally give rise to cell culture strains, the biosynthetic features of which are different from those of the original cell culture. This has been demonstrated by the selection of highly productive variant strains from cell suspension cultures of *Catharanthus roseus* (Zenk *et al.*, 1977a) and *Nicotiana tabacum* (Ogino *et al.*, 1978). In these experiments small cell aggregates and single cells were separated without information on their serpentine/ajmalicine and nicotine contents, respectively. Only in cell colonies developed from the starting material could alkaloid amounts be determined by radioimmunoassay (Zenk *et al.*, 1977a) or the cell squash method (Ogino *et al.*, 1978). They have varied in percentage of cell dry weight from zero to 1.4 for serpentine, from zero to 0.8 for ajmalicine, and from about zero to 2.2 for nicotine, illustrating to what extent the cells of the original cell cultures differ in their biosynthetic capacities. According to a proposal by Tabata and associates (1978), this phenomenon may be explained by the assumption that cytoplasmic factors differently influence or control the expression of the uniform genetic information located in cell nuclei. The selection of dark and light colored pieces of cell cultures (T,C) from *Macleaya microcarpa* (Koblitz *et al.*, 1975) and *Lithospermum erythrorhizon* (Mizukami *et al.*, 1978) led to strains with high and low contents of protopine-type alkaloids and shikonin derivatives, respectively. Although the red color of the latter compounds allows the estimation of their amounts in single cells or cell groups, the protopine-type alkaloids, as colorless substances, fail to give such information. Obviously, their contents are indicated by the accompanying yellow colored alkaloid sanguinarine or even by secondary metabolites without alkaloidal properties characterized as carotenoids (Koblitz *et al.*, 1975; Böhm and Franke, 1978).

B. Influencing Factors and Conditions

Attempts to increase the contents of secondary metabolites in plant cell cultures by feeding related precursors and intermediates have been continued during the last few years. As in earlier experiments, the results are variable and often contradictory.

Shikimic acid, L-phenylalanine, or L-tyrosine added to the B5 medium of *Coleus blumei* cell suspension cultures in various amounts did not influence the formation of rosmarinic acid or inhibit its synthesis in higher concentrations (Razzaque and Ellis, 1977). However, cell suspension cultures of *C. blumei* have also been reported to produce about 100% more rosmarinic acid than the control when supplied with 500 mg L-phenylalanine per liter of B5 medium (Zenk *et al.*, 1977b). Although cinnamic acid, *p*-coumaric acid, caffeic acid, and dihydroxyphenylalanine (DOPA) considerably decreased the growth rates of this cell culture, only under a limited number of conditions were the concentrations of rosmarinic acid slightly increased (Razzaque and Ellis, 1977). Phenylalanine added to *Lithospermum erythrorhizon* cell cultures (C) during the fourth week of a subculture has shown a positive effect on the formation of shikonin. However, compounds biosynthetically derived from this amino acid have inhibited not only cell growth but also, at concentrations above $10^{-5}\,M$, secondary metabolite formation (Mizukami *et al.*, 1977). Anthraquinone production in cell suspension cultures of *Galium mollugo* can be increased by supplying *o*-succinylbenzoic acid (Bauch and Leistner, 1978a). When ornithine, putrescine, and nicotinic acid were separately fed to tobacco tissue cultures, only the first precursor caused a slight increase of nicotine, although the second one was applied in relatively high concentrations to a cell culture strain showing very weak alkaloid formation (Ohta *et al.*, 1978a,b).

Catharanthus roseus cell cultures (C, S, T) have repeatedly been the object of feeding experiments directed to the enhancement of alkaloid production above control levels. Tryptamine added in a concentration of 100 mg/liter during the second or third week of culture has stimulated cell growth and alkaloid metabolism (Krueger and Carew, 1978). When this precursor was present in the same concentration from the start of a subculture, it caused the inhibition of both cell growth and alkaloid formation (Zenk *et al.*, 1977a; Döller *et al.*, 1976; Döller, 1978). Supplying cell cultures with tryptophan can lead to a threefold increase of alkaloid yield compared with the control (Zenk *et al.*, 1977a; Deus, 1978) or can have a negative effect on alkaloid formation under some conditions (Döller *et al.*, 1976; Döller, 1978). In this situation it has been fruitful to examine the significance of high internal tryptophan concentrations for the biosynthesis of serpentine and/or ajmalicine in cell cultures of *C. roseus*. Cell lines producing 30 and 40 times the normal amounts of free tryptophan could be selected because of their resistance to 5-methyltryptophan. However, these resistant cell lines exhibit neither increased tryptamine contents nor higher alkaloid levels compared with sensitive cell cultures. They can even fail in alkaloid formation (Schallenberg and Berlin, 1979; Scott *et al.*, 1979). The assumption that the low activity of "tryptophan decarboxylase" prevents the transformation of the tryptophan pool into an adequate amount of alkaloids in resistent cell lines has been supported by initial experimental data (Schallenberg and Berlin, 1979). The feeding of loganin has led to an important increase of serpentine concentrations in *C. roseus* cell cultures. Other precursors of the monoterpene parts of indole alkaloids have negative or slightly positive effects on alkaloid formation (Döller *et al.*, 1976; Döller, 1978; Zenk *et al.*, 1977a; Krueger and Carew, 1978).

Among the nutritional factors of plant cell cultures sucrose may be of special significance for secondary metabolite formation. Concentrations of this medium component above 2% can enhance the yield of different compounds not only by stimulating cell growth but also by increasing biosynthetic activity. This has been demonstrated with cell cultures producing alkaloids (Döller, 1978; Ohta *et al.*, 1978b), anthraquinones (Zenk *et al.*, 1975; Bauch and Leistner, 1978a), naphthoquinones (Mizukami *et al.*, 1977), and rosmarinic acid (Zenk *et al.*, 1977b).

Studying the effect of phytohormones on secondary metabolism in plant cell cultures, several research groups have confirmed earlier findings. Although relatively low concentrations of auxins and cytokinins can positively influence the formation of secondary compounds, higher concentrations generally inhibit this process. Moreover, the fact that auxins can remarkably differ in their physiological relevance has been discovered by several experiments (Brain and Lockwood, 1976; Döller *et al.*, 1976; Marshall and Staba, 1976; Shah *et al.*, 1976; Zenk *et al.*, 1977b; Deus, 1978; Ohta *et al.*, 1978b).

Under otherwise identical conditions, light causes an increase of serpentine concentrations in cell cultures (C, T) of *Catharanthus roseus* (Döller *et al.*, 1976; Roller, 1978) but an inhibition of nicotine formation in tobacco tissue cultures (Ohta and Yatazawa, 1978).

V. Secondary Metabolite Formation and
Cellular Structures

Undoubtedly, the realization of secondary metabolism in plant cell cultures as a part of biochemical differentiation is closely correlated with morphological differentiation (Böhm, 1977). For instance, the function of several enzymes may depend on the presence of certain membranes. However, compartmentation of enzymes, substrates, and end products by cellular structures is known as a mechanism for regulating metabolic processes (Luckner *et al.*, 1980).

A microsomal fraction obtained from cell suspension cultures of *Glycine max* shows PAL and cinnamic acid 4-hydroxylase activities. Subfractionation of this material reveals that both enzymes are exclusively connected with endoplasmic reticulum. Whereas cinnamic acid 4-hydroxylase is accepted as being a membrane-bound enzyme, a very loose association is assumed between PAL and endoplasmic reticulum *in vivo* (Postius and Kindl, 1978).

Enzymes involved in flavonoid and anthocyanin biosynthesis—flavanone synthase, chalcone:flavanone isomerase, and UDP-glucose:anthocyanidin-3-*O*-glucosyltransferase—have been found in cytosol fractions obtained from *Hippeastrum* and *Tulipa* protoplasts. In contrast to a proposal concerning sites of anthocyanin formation (Fritsch and Grisebach, 1975), no enzyme activity could be detected in any vacuole fractions. Obviously, the cited enzymes are not connected with tonoplasts *in vivo* (Hrazdina *et al.*, 1978).

According to electron microscopic investigations lamiaceous tannins consisting of cinnamic acid esters, for instance, rosmarinic acid, are located in the cytoplasm, at the tonoplast, and in vacuoles of *Osmium basilicum* cell cultures in liquid media. Surprisingly, deposits of these substances are surrounded by membranes only within vacuoles but not in the cytoplasm. The site of tannin formation could not be detected. Nevertheless, endoplasmic reticulum and microvesicles seem not to participate in this process (Lang *et al.*, 1977). In *Coleus blumei* cell suspension cultures rosmarinic acid is accumulated in intercellular spaces or in vesicles near cell walls (Peterson and Ellis, 1977).

The cells rich in alkaloids scattered through cell cultures (T, S) of *Macleaya microcarpa* obviously represent latex cells. Arguments that they are specialized for a storage function have found support: In suspension cultures the size of cell aggregates is generally reduced if the auxin concentration of media increases. Simultaneously, the number of accumulating cells decreases in relation to the total number of cells, and increasing amounts of alkaloids are excreted. Aggregates consisting of about 10 cells show no accumulating cells. Under this condition *M. microcarpa* cell suspension cultures can synthesize remarkable amounts of alkaloids. Therefore, cells rich in alkaloids should be considered not as of relevance for alkaloid formation but as specialized in the accumulation of alkaloids formed by surrounding cells (Böhm, 1978; Franke and Böhm, 1978).

Callus cultures of *Ruta graveolens* growing under continuous light are green and can develop leaves. In these organs, as well as toward the periphery of the callus, secretory glands occur. As demonstrated in a light and electron microscopic study they are initiated lysigenously or schizolysigenously (Peterson *et al.*, 1978).

VI. Concluding Remarks

For several years the initiation and maintenance of a desired cell culture have been generally successful because of enhanced experience and improved methods in this field. Recently it has been reported that from cell cultures showing no or very weak secondary metabolite formation, metabolically active cell culture strains can be derived by selecting single cells or cell groups. Considering this situation there seem to be at present two main problems concerning secondary metabolism in plant cell cultures. The first is represented by absolutely unproductive cell cultures. The second problem results from the comparison of the spectra of secondary metabolites in cell cultures with those of respective intact plants. Further understanding and overcoming of these difficulties will bring not only progress in this area of research but also further contributions to the discussion of fundamental questions in biology and biochemistry. In each case it must be considered whether or not a result obtained from experiments with plant cell cultures is caused by special conditions in these systems or is representative for the situation in whole plants.

ACKNOWLEDGMENTS

I am very grateful to the following colleagues for providing me with unpublished results and manuscripts: Dr. J. Berlin, Braunschweig-Stöckheim, German Federal Republic; Dr. G. Brunet, Montreal, Quebec, Canada; Prof. Dr. K. Hahlbrock, Freiburg, German Federal Republic; Prof. Dr. S. L. Lee, College Station, Texas; Prof. Dr. M. Luckner, Halle, German Democratic Republic; and Dr. U. Seitz, Tübingen, German Federal Republic.

REFERENCES

Akasu, M., Itokawa, H., and Fujita, M. (1976). *Phytochemistry* **15,** 471–473.
Alfermann, A. W., and Reinhard, E. (1978). *In* "Production of Natural Compounds by Cell Culture Methods" (A. W. Alfermann and E. Reinhard, eds.), BPT-Report, pp. 3–15. Gesellschaft für Strahlen und Umweltforschung, Munich.
Alfermann, A. W., Merz, D., and Reinhard, E. (1975). *Planta Med.* Suppl., 70–78.
Barz, W. (1977). *In* "Plant Tissue Culture and Its Bio-technological Application" (W. Barz, E. Reinhard, and M. H. Zenk, eds.), pp. 153–171. Springer-Verlag, Berlin and New York.
Bauch, H. J., and Leistner, E. (1978a). *Planta Med.* **33,** 105–123.
Bauch, H. J., and Leistner, E. (1978b). *Planta Med.* **33,** 124–127.
Betz, B., Schäfer, E., and Hahlbrock, K. (1978). *Arch. Biochem. Biophys.* **190,** 126–135.
Böhm, H. (1977). *In* "Secondary Metabolism and Cell Differentiation" (M. Luckner, L. Nover, and H. Böhm, eds.), pp. 103–123. Springer-Verlag, Berlin and New York.
Böhm, H. (1978). *In* "Frontiers of Plant Tissue Culture, 1978" (T. A. Thorpe, ed.), pp. 201–211. Univ. of Calgary, Offset Printing Services, Calgary, Alberta.
Böhm, H., and Franke, J. (1978). Unpublished observations.
Brain, K. R., and Lockwood, G. B. (1976). *Phytochemistry* **15,** 1651–1654.
Braun, G., and Seitz, U. (1975). *Biochem. Physiol. Pflanz.* **168,** 93–100.
Brunet, G., and Ibrahim, R. K. (1978). *Abstr. Int. Congr. Plant Tissue Cell Cult., 4th, Calgary* No. 1407.
Brunet, G., Saleh, N. A. M., and Ibrahim, R. K. (1978). *Z. Naturforsch., Teil C* **33,** 786–788.
Butcher, D. N. (1977). *In* "Applied and Fundamental Aspects of Plant Cell, Tissue, and Organ Culture" (J. Reinert and Y. P. S. Bajaj, eds.), pp. 668–693. Springer-Verlag, Berlin and New York.
Butcher, D. N., and Connolly, J. D. (1971). *J. Exp. Bot.* **22,** 314–322.
Delfel, N. E., and Rothfus, J. A. (1977). *Phytochemistry* **16,** 1595–1598.
Deus, B. (1978). *In* "Production of Natural Compounds by Cell Culture Methods" (A. W.

Alfermann and E. Reinhard, eds.), BPT-Report, pp. 118–123. Gesellschaft für Strahlen und Umweltforschung, Munich.

Döller, G. (1978). *In* "Production of Natural Compounds by Cell Culture Methods" (A. W. Alfermann and E. Reinhard, eds.), BPT-Report, pp. 109–116. Gesellschaft für Strahlen und Umweltforschung, Munich.

Döller, G., Alfermann, A. W., and Reinhard, E. (1976). *Planta Med.* **30,** 14–20.

Ebel, J., and Hahlbrock, K. (1977). *Eur. J. Biochem.* **75,** 201–209.

Ebel, J., Ayers, A. R., and Albersheim, P. (1976). *Plant Physiol.* **57,** 775–779.

Forkmann, G., and Kuhn, B. (1979). *Planta* **144,** 189–192.

Franke, J., and Böhm, H. (1978). Unpublished observations.

Frischknecht, P. M., Baumann, T. W., and Wanner, H. (1977). *Planta Med.* **31,** 344–350.

Fritsch, H., and Grisebach, H. (1975). *Phytochemistry.* **14,** 2437–2442.

Grisebach, H. (1977). *Naturwissenschaften.* **64,** 619–625.

Grisebach, H., and Ebel, J. (1978). *Angew. Chem.* **90,** 668–681.

Grisebach, H., and Hahlbrock, K. (1974). *In* "Metabolism and Regulation of Secondary Plant Products" (V. C. Runeckles and E. E. Conn, eds.), pp. 21–52. Academic Press, New York.

Gross, D. (1977). *Fortschr. Chem. Org. Naturst.* **34,** 187–247.

Gross, G. G. (1977). *In* "The Structure, Biosynthesis, and Degradation of Wood" (F. A. Loewus and V. C. Runeckles, eds.), pp. 141–184. Academic Press, New York.

Haddon, L., and Northcote, D. H. (1976). *Planta* **128,** 255–262.

Hahlbrock, K. (1976). *Eur. J. Biochem.* **63,** 137–145.

Hahlbrock, K. (1977). *In* "Plant Tissue Culture and Its Bio-technological Application" (W. Barz, E. Reinhard, and M. H. Zenk, eds.), pp. 95–111. Springer-Verlag, Berlin and New York.

Hahlbrock, K., Knobloch, K.-H., Kreuzaler, F., Potts, J. R. M., and Wellmann, E. (1976). *Eur. J. Biochem.* **61,** 199–206.

Hargraeves, J. A., and Selby, C. (1978). *Phytochemistry* **17,** 1099–1102.

Heinzmann, U., and Seitz, U. (1977). *Planta* **135,** 63–67.

Heinzmann, U., Seitz, U., and Seitz, U. (1977). *Planta* **135,** 313–318.

Hess, D. (1968). "Biochemische Genetik," pp. 89–104. Springer-Verlag, Berlin and New York.

Hirotani, M., and Furuya, T. (1977). *Phytochemistry* **16,** 610–611.

Hrazdina, G., Kreuzaler, F., Hahlbrock, K., and Grisebach, H. (1976). *Arch. Biochem. Biophys.* **175,** 392–399.

Hrazdina, G., Wagner, G. J., and Siegelman, H. W. (1978). *Phytochemistry* **17,** 53–56.

Ikuta, A., Syono, K., and Furuya, T. (1974). *Phytochemistry* **13,** 2175–2179.

Kamimura, S., and Nishikawa, M. (1976). *Agric. Biol. Chem.* **40,** 907–911.

Kamimura, S., Akutsu, M., and Nishikawa, M. (1976). *Agric. Biol. Chem.* **40,** 913–919.

Kartnig, T. (1977). *In* "Plant Tissue Culture and Its Bio-technological Application" (W. Barz, E. Reinhard, and M. H. Zenk, eds.), pp. 44–51. Springer-Verlag, Berlin and New York.

Kartnig, T., Russheim, U., and Maunz, B. (1976). *Planta Med.* **29,** 275–282.

Kartnig, T., Russheim, U., Trousil, G., and Maunz, B. (1979). *Planta Med.* **35,** 275–278.

Knobloch, K.-H., and Hahlbrock, K. (1975). *Eur. J. Biochem.* **52,** 311–320.

Knobloch, K.-H., and Hahlbrock, K. (1977). *Arch. Biochem. Biophys.* **184,** 237–248.

Koblitz, H., Schumann, U., Böhm, H., and Franke, J. (1975). *Experientia* **31,** 768–769.

Kreuzaler, F., and Hahlbrock, K. (1975). *Eur. J. Biochem.* **56,** 205–213.

Kreuzaler, F., Light, R. J., and Hahlbrock, K. (1978). *FEBS Lett.* **94,** 175–178.

Krueger, R. J., and Carew, D. P. (1978). *Lloydia* **41,** 327–331.

Kuboi, T., and Yamada, Y. (1976). *Phytochemistry* **15,** 397–400.

Kuboi, T., and Yamada, Y. (1978). *Biochim. Biophys. Acta* **542,** 181–190.

Kuhn, B., Forkmann, G., and Seyffert, W. (1978). *Planta* **138,** 199–203.

Lamb, C. J., and Dixon, R. A. (1978). *FEBS Lett.* **94,** 277–280.

Lang, E., and Hörster, H. (1977). *Planta Med.* **31,** 112–118.

Lang, E., Hörster, H., Friedrich, H., Themann, H., and Amelunxen, F. (1977). *Cytobiologie* **15,** 372–381.

Lee, S. L., Hirata, T., and Scott, A. I. (1979). *Tetrahedron Lett.* 691–694.

Luckner, M. (1971). *Pharmazie* **26,** 717–724.

Luckner, M., Nover, L., and Böhm, H. (1977). "Secondary Metabolism and Cell Differentiation." Springer-Verlag, Berlin and New York.

Luckner, M., Diettrich, B., and Lerbs, W. (1980). *Phytochemistry,* in press.

Margna, U. (1977). *Phytochemistry* **16,** 419–426.

Marshall, J. G., and Staba, E. J. (1976). *Phytochemistry* **15,** 53–55.

Mizukami, H., Konoshima, M., and Tabata, M. (1977). *Phytochemistry* **16,** 1183–1186.

Mizukami, H., Konoshima, M., and Tabata, M. (1978). *Phytochemistry* **17,** 95–97.

Mizukami, H., Nordloey, H., Lee, S. L., and Scott, A. I. (1979). Unpublished observations.

Nagel, M., and Reinhard, E. (1975). *Planta Med.* **27,** 264–271.

Ogino, T., Hiraoka, N., and Tabata, M. (1978). *Phytochemistry* **17,** 1907–1910.

Ohta, S., and Yatazawa, M. (1978). *Agric. Biol. Chem.* **42,** 873–877.

Ohta, S., Kojima, Y., and Yatazawa, M. (1978a). *Agric. Biol. Chem.* **42,** 1733–1738.

Ohta, S., Matsui, O., and Yatazawa, M. (1978b). *Agric. Biol. Chem.* **42,** 1245–1251.

Overton, K. H., and Picken, D. J. (1976). *J.C.S. Chem. Commun.* 105–106.

Overton, K. H., and Picken, D. J. (1977). *Fortschr. Chem. Org. Naturst.* **34,** 249–298.

Peterson, R. L., and Ellis, A. E. (1977). Unpublished observations.

Peterson, R. L., Scott, M. G., and Ellis, A. E. (1978). *Can. J. Bot.* **56,** 2717–2729.

Postius, C., and Kindl, H. (1978). *Z. Naturforsch., Teil* **33c,** 65–69.

Poulton, J. E., and Kauer, M. (1977). *Planta* **136,** 53–59.

Poulton, J. E., Grisebach, H., Ebel, J., Schaller-Hekeler, B., and Hahlbrock, K. (1976a). *Arch Biochem. Biophys.* **173,** 301–305.

Poulton, J. E., Hahlbrock, K., and Grisebach, H. (1976b). *Arch. Biochem. Biophys.* **176,** 449–456.

Poulton, J. E., Hahlbrock, K., and Grisebach, H. (1977). *Arch. Biochem. Biophys.* **180,** 543–549.

Ragg, H., Schröder, J., and Hahlbrock, K. (1977). *Biochim. Biophys. Acta* **474,** 226–233.

Razzaque, A., and Ellis, B. E. (1977). *Planta* **137,** 287–291.

Roller, U. (1978). *In* "Production of Natural Compounds by Cell Culture Methods" (A. W. Alfermann and E. Reinhard, eds.), BPT-Report, pp. 95–104. Gesellschaft für Strahlen und Umweltforschung, Munich.

Rueffer, M., Nagakura, N., and Zenk, M. H. (1978). *Tetrahedron Lett.* **18,** 1593–1596.

Saleh, N. A. M., Fritsch, H., Witkop, P., and Grisebach, H. (1976). *Planta* **133,** 41–45.

Saleh, N. A. M., Fritsch, H., Kreuzaler, F., and Grisebach, H. (1978). *Phytochemistry* **17,** 183–186.

Schallenberg, J., and Berlin, J. (1979). *Z. Naturforsch., Teil* **34c,** 541–545.

Schröder, J. (1977). *Arch. Biochem. Biophys.* **182,** 488–496.

Schröder, J., Betz, B., and Hahlbrock, K. (1976). *Eur. J. Biochem.* **67,** 527–541.

Schröder, J., Betz, B., and Hahlbrock, K. (1977). *Plant Physiol.* **60,** 440–445.

Schröder, J., Kreuzaler, F., Schäfer, E., and Hahlbrock, K. (1979). *J. Biol. Chem.* **254,** 57–65.

Scott, A. I., and Lee, S. L. (1975). *J. Am. Chem. Soc.* **97,** 6906.

Scott, A. I., Lee, S. L., de Capite, P., Culver, M. G., and Hutchinson, C. R. (1977a). *Heterocycles* **7,** 979–984.

Scott, A. I., Lee, S. L., and Wan, W. (1977b). *Biochem. Biophys. Res. Commun.* **75,** 1004–1009.

Scott, A. I., Mizukami, H., and Lee, S. L. (1979). *Phytochemistry* **18**, 795–798.
Shah, R. R., Subbaiah, K. V., and Mehta, A. R. (1976). *Can. J. Bot.* **54**, 1240–1245.
Shipilova, S. V., Koretzkaya, T. F., and Zaprometov, M. N. (1978). *Fiziol. Rast.* **25**, 552–555.
Staba, E. J. (1977). *In* "Applied and Fundamental Aspects of Plant Cell, Tissue, and Organ Culture" (J. Reinert and Y.P.S. Bajaj, eds.), pp. 694–702. Springer-Verlag, Berlin and New York.
Stärk, D., Alfermann, A. W., and Reinhard, E. (1976). *Planta Med.* **30**, 104–117.
Stöckigt, J. (1978). *J.C.S. Chem. Commun.* 1097–1099.
Stöckigt, J., and Zenk, M. H. (1977a). *FEBS Lett.* **79**, 233–237.
Stöckigt, J., and Zenk, M. H. (1977b). *J.C.S. Chem. Commun.* 646–648.
Stöckigt, J., Treimer, J., and Zenk, M. H. (1976). *FEBS Lett.* **70**, 267–270.
Stöckigt, J., Husson, H. P., Kan-Fan, C., and Zenk, M. H. (1977). *J.C.S. Chem. Commun.* 164–166.
Stöckigt, J., Rueffer, M., Zenk, M. H., and Hoyer, G. A. (1978).*Planta Med.* **33**, 188–192.
Sütfeld, R., Kehrel, B., and Wiermann, R. (1978). *Z. Naturforsch., Teil C,* 841–846.
Szendrei, K., Rozsa, Z., Reisch, J., Novak,I., Kusowkina, N., and Minker, E. (1976). *Herba Hung.* **15**(2), 23–29.
Tabata, M., Hiraoka, N., Ikenoue, M., Sano, Y., and Konoshima, M. (1975). *Lloydia* **38**, 131–134.
Tabata, M., Hayashi, K., and Konoshima, M. (1977). *Yakugaku Zasshi* **97**, 746–752.
Tabata, M., Ogino, T., Yoshioka, K., Yoshikawa, N., and Hiraoka, N. (1978). *In* "Frontiers of Plant Tissue Culture, 1978" (T. A. Thorpe, ed.), pp. 213–222. Univ. of Calgary, Offset Printing Services, Calgary, Alberta.
Treimer, J. F., and Zenk, M. H. (1978). *Phytochemistry* **17**, 227–231.
Treimer, J. F., and Zenk, M. H. (1979). *FEBS Lett.* **97**, 159–162.
Turnbull, A., and Collin, H. A. (1978). *Abstr., Int. Congr. Plant Tissue Cell Cult., Calgary* No. 711.
Wellmann, E., Hrazdina, G., and Grisebach, H. (1976). *Phytochemistry* **15**, 913–915.
Wengenmayer, H., Ebel, J., and Grisebach, H. (1976). *Eur. J. Biochem.* **65**, 529–536.
Wichtl, M., Jentsch, K., and Rücker, W. (1978). *Pharmazie* **33**, 229–233.
Wyrambik, D., and Grisebach, H. (1975). *Eur. J. Biochem.* **59**, 9–15.
Yamada, Y., and Kuboi, T. (1976). *Phytochemistry* **15**, 395–396.
Zenk, M. H. (1978). *In* "Frontiers of Plant Tissue Culture, 1978" (T. A. Thorpe, ed.), pp. 1–13. Univ. of Calgary Press, Calgary, Alberta.
Zenk, M. H., El Shagi, H., and Schulte, U. (1975). *Planta Med.* Suppl., 79–101.
Zenk, M. H., El-Shagi, H., Arens, H., Stöckigt, J., Weiler, E. W., and Deus, B. (1977a). *In* "Plant Tissue Culture and Its Bio-technological Application" (W. Barz, E. Reinhard, and M. H. Zenk, eds.), pp. 27–43. Springer-Verlag, Berlin and New York.
Zenk, M. H., El-Shagi, H., and Ulbrich, B. (1977b). *Naturwissenschaften* **64**, 585–586.

NOTE ADDED IN PROOF. Strictosidine synthase was also characterized by J. F. Treimer and M. H. Zenk (1979. *Eur. J. Biochem.* **101**, 225–233). 4,21-Dehydrogeissoschizine has been reported to be an intermediate in the biosynthetic pathway of *Corynanthe*-type alkaloids that is transformed into cathenamine by the new enzyme cathenamine synthase (M. Rueffer, C. Kan-Fan, H.-P. Husson, J. Stöckigt, and M. H. Zenk). Experiments directed to the further characterization of flavanone synthase (F. Kreuzaler, H. Ragg, W. Heller, R. Tesch, J. Witt, D. Hammer, and K. Hahlbrock, 1979. *Eur. J. Biochem.* **99**, 89–96) have revealed that this enzyme can form a chalcone rather than a flavanone under certain conditions (W. Heller and K. Hahlbrock, 1980. *Arch. Biochem. Biophys.* **200**, in press).

Chapter 18

Embryo Culture

V. RAGHAVAN

Department of Botany, The Ohio State University, Columbus, Ohio

I. Introduction

The excision of embryos from ovules and seeds of higher plants and their culture in defined media allow one to investigate the factors that influence embryonic growth under controlled conditions and, by implication, to define in chemical terms the milieu of the embryo sac, which nurtures the developing embryos. For example, the progressive and orderly transition of the fertilized egg (zygote) through well-defined stages of embryogenesis to a mature embryo may be related to certain changes in the *in vivo* chemical environment replete with balanced nutritional or hormonal factors. Culture *in vitro* of embryos also facilitates experiments to determine the factors that regulate growth of the primordial organs of the seedling plant and to study the metabolic and biochemical aspects of germination, which is difficult to undertake in embryos enclosed within seeds without interference from accessory tissues. The aim of this chapter is to assess the progress that has been made toward realization of these fundamental potentials of embryo culture technique. From a practical point of view, information obtained by *in vitro* culture of embryos has opened the way to obtaining plants from inviable hybrids and to by-passing the tradi-

209

tional treatments to overcome dormancy and accelerate germination in certain types of seeds. This aspect of embryo culture has been adequately reviewed elsewhere (Raghavan, 1977a).

The work of Hannig (1904), who removed mature embryos from seeds of *Raphanus* and *Cochlearia* and nurtured them into plantlets in a medium containing mineral salts and sucrose, is generally accepted as marking the birth of plant embryo culture. Since this discovery, embryos excised from ovules and seeds of the most varied species of plants have been grown under controlled conditions and much useful information on their nutritional requirements, growth, and differentiation has accumulated. This information is documented in several review articles (Brink and Cooper, 1947; Rappaport, 1954; Sanders and Ziebur, 1963; Narayanaswami and Norstog, 1964; Maheshwari and Rangaswamy, 1965; Wardlaw, 1965; Degivry, 1966; Raghavan, 1966) and books (Wardlaw, 1968; Raghavan, 1976) published during the last three decades. Reference also should be made to the articles by Sanders and Ziebur (1963), Raghavan (1967, 1977b), and Torrey (1973), which describe the general methodology followed in the culture of embryos and composition of the varied kinds of media employed.

Culture of embryos may be divided into the following two categories: culture of relatively mature and differentiated seed embryos and culture of immature, early division phase proembryos. The culture of seed embryos aims at analyzing the various parameters of embryonic growth and the metabolic and biochemical aspects of dormancy and germination. The motivation in proembryo culture is to understand the controls of differentiation and the nutritional requirements of progressively smaller embryos. In certain instances, culture of ovules and seeds has drawn attention to the causal agents that control growth and morphogenesis of the enclosed embryos. Although it might seem logical to begin with an account of proembryo culture and then proceed to seed embryo culture, historically it was the latter that developed first and paved the way for developments in proembryo culture. Hence, seed embryo culture is described first here.

II. Culture of Seed Embryos

It is no accident that the first successful embryo cultures were raised from mature embryos excised from seeds and not from immature embryos isolated from ovules. The reasons for this are not difficult to envisage. With a few exceptions, the embryo contained in the seed is a fully developed bipolar structure consisting of a contrasting meristem at each pole — the primordial root, or the radicle, and the primordial shoot, or the plumule—and one or two lateral appendages, the cotyledons. Therefore, at the time of culture the embryo consists of cells poised on the threshold to undergo division, elongation, and differentiation leading to the progressive development of the root and shoot axes of the seedling plant. Exhibiting simple nutritional requirements, seed embryos divested of seed coats and surrounding endosperm, if any, are independent and autotrophic and are able to grow into plantlets when supplied with a limited diet consisting of certain conventional inorganic salts and sucrose in the medium. The relatively large size of embryos and the ease with which they can be isolated without recourse to micromanipulative techniques have also contributed in a great measure to their use in the pioneering studies.

A. Nutrition and Metabolism

1. *Carbohydrate Nutrition*

It is the general experience that growth and survival of cultured embryos is markedly enhanced by supplementation of the medium with a carbon energy source. The superiority of sucrose in this pivotal role has been established for embryos of a few species of *Zea* (Andronescu, 1919), *Datura stramonium* (van Overbeek *et al.*, 1944), *Pinus nigra* (Radforth and Pegoraro, 1955), and *Capsella bursa-pastoris* (Rijven, 1952). From work with excised embryos of rice (*Oryza sativa*), Amemiya *et al.* (1956a,b) claimed that fructose, glucose, sucrose, and maltose were superior to xylose, galactose, and mannose. In other cases, such as for embryos of *Carex lurida* (Lee, 1952) and those excised from seeds of plants belonging to the family Rosaceae (Tukey, 1938), addition of glucose was found to satisfy the carbohydrate requirement as well as or better than sucrose. Some workers (Stingl, 1907; Dieterich, 1924; Esenbeck and Suessenguth, 1925) have reported that certain graminaceous embryos can metabolize substances such as starch, that are not ordinarily utilized by cultured embryos of other plants. These observations point strongly to the conclusion that embryos *in vitro* are able to utilize a wider range of carbohydrates than other explanted organs.

In recognition of the role of carbohydrates in enhancing embryonic growth and survival, it has been established that they invariably promote growth of the root (Buckner and Kastle, 1917; Rietsema *et al.*, 1953a; Honma, 1955; Ozsan and Cameron, 1963) or of both shoot and leaf primordia (Augusten, 1956; Buffard-Morel, 1968). Addition of sugar is necessary for effective vernalization of isolated cereal embryos. Purvis (1944) has described in detail the effect of different carbon sources on the growth of primordial organs of isolated rye (*Secale cereale* var. Petkus) embryos during vernalization. When isolated embryos were grown at 1°C on a medium devoid of any carbohydrate, the coleoptile primordium more than doubled its length during a growth period of 6 weeks, whereas under the same conditions the primary root and the lateral seminal roots hardly elongated. Maximum growth of these embryonic organs occurred in a medium containing 2% sucrose. Compared to growth of the embryo on a medium containing 2% sucrose, varying degrees of coleoptile extension unaccompanied by root elongation occurred in media containing arabinose, fructosan, mannitol, mannose, pyruvic acid, ribose, xylose, glycerol, and galactose supplied at molecular concentrations equivalent to 2% sucrose. Raffinose, maltose, glucose, and fructose were intermediate in their growth-promoting effects and significantly increased the length of both coleoptile and root over no-sugar controls. The overall significance of these results in understanding the pattern of sugar utilization by cultured embryos is uncertain, because (1) cultures were routinely grown at 1°C for vernalization and (2) the varying degrees of utilization of carbohydrates for vernalization process may distort their real contribution for growth processes in the embryo.

2. Nitrogen Nutrition

Although the standard embryo culture medium is principally supplied with inorganic nitrogen in the form of nitrate, nitrite, or ammonia, addition of various amino acids, singly or in combination, has a profound influence on the growth and development of cultured embryos. This information has in turn led to an evaluation of the growth of embryos resulting from the addition of individual amino acids to the medium with a view to gaining some insight into the metabolism and biosynthesis of the compounds in the cultured organ. In general, the amide glutamine has been found to be the most efficient source of nitrogen for growth of embryos of a number of species. The growth-promoting effect of glutamine is especially striking in comparison to that of asparagine, as seen in the work of Rijven (1956). In short-term experiments using embryos excised from such widely unrelated families as Cruciferae (*Capsella bursa-pastoris, Arabidopsis thaliana, Sisymbrium orientale*), Resedaceae (*Reseda odorata*), Leguminosae (*Medicago tribuloides, Medicago orbicularis*), Primulaceae (*Anagallis arvensis*), Solanaceae (*Datura stramonium*), Capparidaceae (*Cleome viscosa*), Chenopodiaceae (*Chenopodium album*), Gramineae (*Hordeum vulgare*), and Liliaceae (*Allium cepa*), glutamine enhanced growth in length considerably more than asparagine. For growth of embryos of *C. bursa-pastoris, A. thaliana,* and *R. odorata,* asparagine was even inhibitory. In detailed experiments on embryos of *C. bursa-pastoris* (Rijven, 1955), it was found that glutamic acid was not as effective as glutamine in promoting growth, whereas asparatic acid was not as inhibitory as asparagine. These results as well as the additional finding that γ-amino-butyric acid and β-alanine, the decarboxylation products of glutamic acid and aspartic acid, respectively, could not substitute for glutamine and asparagine militate against any meaningful interpretation of the contrasting effects of the amides on the growth of embryos.

Studies on nitrogen nutrition of cultured embryos have been particularly useful in assessing the extent of mutual synergism and antagonism between different amino acids. Sanders and Burkholder (1948) set the stage for this line of work with the finding that a mixture of 20 amino acids in the proportions in which they occur in casein hydrolyzate was as effective as the latter in promoting growth of pre- and early heart-shaped embryos of *Datura stramonium* and *Datura innoxia*. The reality of interaction between amino acids was evident when they were added in small groups to the medium. Therefore, although both beneficial and inhibitory amino acids were included in the complete mixture, its favorable effects on embryos of *D. stramonium* were not reproduced by the beneficial compounds alone. However, embryos of *D. innoxia* grew as well on a medium containing favorable amino acids as in the complete medium. These results were extended and confirmed by Harris (1956), who also noted strong antagonistic effect between L-phenylalanine and L-tyrosine, L-leucine and DL-valine, DL-isoleucine and DL-valine, and L-arginine and L-lysine on the growth of the primary root of cultured oat (*Avena sativa*) embryos. According to Miflin (1969), the inhibitory effects of leucine and valine on barley (*Hordeum vulgare*) embryos were synergistic, the two compounds causing greater growth inhibition than the sum of their individual effects. Growth inhibition by valine was relieved by isoleucine and that of leucine by the addition of both isoleucine and valine. The interrelationship between valine, leucine, and isoleucine implied by these results may very well occur because of the action of a common transaminase or a single enzyme catalyzing the dehydration of precursors of both isoleucine and valine.

Wright and Srb (1950) found that canavanine inhibition of growth of corn (*Zea mays*) embryos was almost completely mitigated by the simultaneous application of arginine and to some extent by citrulline, ornithine, glutamic acid, and lysine. According to Stokes (1953), after-ripened embryos of *Heracleum sphondylium* differentiated into normal seedlings when grown in a medium containing arginine, but growth was poor when serine and alanine were supplied along with arginine.

Competitive interactions between individual amino acids described above may represent an important physiological consideration in the control of growth and differentiation in cultured embryos. It is of interest to note that although embryos of *D. innoxia* grew as well in a medium containing the beneficial amino acids as in the complete medium, embryos grown in the former were abnormal with characteristically long cotyledons. Normal embryos with short cotyledons in proportion to the hypocotyl were formed only in a medium containing the complete amino acid mixture or casein hydrolyzate (Sanders and Burkholder, 1948). Studies by Stokes (1953) on embryos of *H. sphondylium* have shown that individual amino acids may influence growth of specific organs of the embryo.

3. *Vitamin Nutrition*

As the empirical importance of vitamins in plant nutrition came to be recognized, they were incorporated into the various culture media employed for growth of seed embryos. However, vitamins should not be regarded as invariable constituents of such media because it is probable that seed embryos, being autotrophic, are able to meet their vitamin requirements by cellular biosynthesis. This was demonstrated by Bonner and Bonner (1938), who found that excised embryos of pea variety "Wrinkled Winner," which synthesized ascorbic acid in the shoot, did not respond to the added vitamin, whereas growth of embryos of "Perfection," which had low ascorbic acid content, was promoted by this compound. Differing rates of synthesis of vitamins and their derivatives by embryos may make their inclusion in the medium stimulatory for growth. For example, in parallel with its effects on isolated roots, thiamine has been shown to promote root elongation in cultured embryos of several plants (Kögl and Haagen-Smit, 1936; Bonner and Axtman, 1937; Lammerts, 1942; Helmkamp and Bonner, 1953; Sircar and Lahiri, 1956; Bartels, 1957). A similar effect of ascorbic acid has been reported in cultured embryos of jute (*Corchorus capsularis*) (Mitra and Datta, 1951). In contrast, biotin (Kögl and Haagen-Smit, 1936), pantothenic acid (Bonner and Axtman, 1937), and niacin (Bonner, 1938) had more marked effect on shoot growth than on root growth. Pyridoxine has been shown to inhibit shoot growth in pea embryos (Helmkamp and Bonner, 1953), although it promotes coleoptile and root elongation in rice embryos (Sircar and Lahiri, 1956).

4. *Miscellaneous Compounds*

The thought that has persistently guided the use of additives to embryo culture media has been that of securing growth in culture of the embryo to the same extent as growth *in vitro*. A medium containing mineral salts and a carbohydrate was devised on this principle as far as ions and carbon energy source were concerned. Even in this medium, which was later supplemented with a variety of organic nitrogen sources and vitamins, embryos of many plants failed to grow or grew only feebly. This led to the recognition that embryos of different plants had different nutritional requirements and in turn resulted in the use of various undefined substances and plant extracts as additives to the medium until highly complex artificial media were formulated.

van Overbeek *et al.* (1942) made an important contribution to embryo culture work when they showed that very small embryos of *D. stramonium* could be grown to maturity when the medium was supplemented with the liquid endosperm of coconut (*Cocos nucifera*) known in tissue culture lore as coconut milk. Since this discovery, one of the commonest additives to plant tissue culture medium has been coconut milk, which presumably makes up for deficiencies of certain sugars, amino acids, growth hormones, and other critical metabolites. In the culture of seed embryos, coconut milk has been shown to promote overall growth as well as growth of specific organs, such as root, shoot, or leaf, of cultured embryos (Gorter, 1955; Norstog, 1956; Ball, 1959; Berlyn, 1962; Colonna *et al*; 1971; Anagnostakis, 1977). Although coconut milk has been reported to promote growth *in vitro* of embryos of common varieties of coconut (Cutter and Wilson, 1954; Abraham and Thomas, 1962; Fisher and Tsai, 1978), it is inhibitory for growth of embryos of the "Makapuno" variety (De Guzman *et al.*, 1971). In embryos of the stem parasite *Scurrula pulverulenta*, which differentiate a haustorium in culture, addition of coconut milk to the medium has been shown to inhibit this process (Johri and Bhojwani, 1970). Endosperm extracts of other plants have also been occasionally employed as additives to embryo culture media and both stimulatory and inhibitory effects have been reported (Augusten, 1956; Bajaj, 1968).

Another substance that has found its way into embryo culture media is activated charcoal (Wang and Huang, 1976; Fisher and Tsai, 1978). By absorbing inhibitory substances present in trace amounts in the medium or toxic products released by the cultured organ, charcoal is thought to promote the growth of embryos.

Finally, although it appears that certain organic additives promote growth of cultured embryos, it must be emphasized that selection of a suitable basal medium from among a number of such media currently being used can go a long way toward eliminating some of the difficulties associated with poor or irregular growth of the organ in culture. A study by Randolph and Khan (1960) on the effectiveness of different media on the growth responses of embryos of *Iris* and wheat (*Triticum vulgare*) has shown that adaptation of embryos to arbitrarily chosen media should not always be taken for granted. Apart from various combinations of major salts, trace elements, and sources of iron that make up a medium, the question of adaptation of the embryo is also related to the agar concentration of the medium. Because agar restricts the availability of water to cultured embryos, a relatively low concentration of agar is reported to be best suited for embryo culture (Randolph and Khan, 1960; Stoltz, 1971).

B. HORMONAL EFFECTS

Much of the interest in the role of hormones in embryo growth stems from the dramatic modifications induced in the growth and morphogenesis of the primordial root and shoot by exogenous hormones, such as auxins, gibberellins, and cytokinins, separately and in combination. The approach to the study of hormonal effects on cultured embryos was influenced by parallel work on other cultured plant organs as well as by the erroneous concept that each hormone played a regulatory role in a specific phase of plant development, such as cell division or cell elongation.

1. *Auxins*

The effects of auxins on the growth of embryos have been studied since 1936, soon after indoleacetic acid (IAA) was identified as the natural auxin. The bulk of the evidence accumulated to date indicates that auxins are generally inhibitory for growth of embryos. In a study of the effects of a range of concentrations of IAA on the growth of embryos of *Datura stramonium,* Rietsema *et al.* (1953b) made the important point that despite its general inhibitory effects, application of the hormone at extremely low concentrations led to a significant promotion of growth of the root primordium. In later studies on embryos of other plants, promotion of growth of both root and shoot or coleoptile primordia by added auxins has been documented, although these contrasting organs of the embryo generally exhibited differential sensitivity to auxin (Guttenberg and Wiedow, 1952; Kruyt, 1954; Sircar and Lahiri, 1956; Furuya and Soma, 1957; Raghavan and Torrey, 1964; Fisher and Tsai, 1978). Embryos of several plants are also known to respond to the presence of relatively high concentrations of auxin in the medium by the production of callus (Carew and Schwarting, 1958; Ball, 1959; Mitra and Kaul, 1964; Raghavan and Torrey, 1964; Sankhla *et al.*, 1967; Bulard, 1967a; Button *et al.*, 1971; Mehra and Mehra, 1974; Nataraja *et al.*, 1973; Nataraja, 1975; Green and Phillips, 1975; Cummings *et al.*, 1976). In recalcitrant embryos of oil palm (*Elaeis guineensis*) a combination of an auxin and a cytokinin was essential for initiation of callus growth (Rabéchault *et al.*, 1970).

The involvement of auxin in the production of callus on cultured embryos is now firmly established. By manipulation of the hormonal constituents of the medium the callus can be cultured indefinitely or induced to differentiate roots, shoot buds, and leaves. With the demonstration that the callus formed on embryos of *Cuscuta reflexa* (Maheshwari and Baldev, 1961) grown in a medium containing casein hydrolyzate and IAA formed adventitious embryos (embryoids), considerable interest has been generated in the induction of embryoids from somatic cells of plants (see Chapter 6, Part A).

2. *Gibberellins*

So far there are only a few reports in the literature that draw attention to the effects of gibberellins on the growth and morphogenesis of cultured embryos. An early observation of Dure and Jensen (1957) showed that in cotton (*Gossypium hirsutum*) embryos gibberellic acid (GA) promoted growth of the axis by cell division and that of the cotyledons by cell enlargement. Because these workers also showed that GA treatment depressed carbohydrate and nitrogen contents of embryos, it may be that the hormone acts by diluting the cell constituents rather than by inducing the synthesis of new macromolecules. Although roots of intact plants are known to be least sensitive to the various gibberellins, addition of GA to the medium has been shown to be particularly effective in promoting root elongation in cultured embryos of *Capsella bursa-pastoris* (Veen, 1963; Raghavan and Torrey, 1964). In cultured coconut embryos, root primordia initiated under the influence of GA generally failed to elongate unless embryos were transferred to a medium without the hormone (De Guzman, 1969). According to Skene (1969), a tendency for postgerminal arrest of growth observed in embryos of bean (*Phaseolus vulgaris*) cultured in a medium enriched with casein hydrolyzate was largely overcome by GA, which induced normal symptoms of growth, including formation of lateral roots and elongation of the hypocotyl. A similar enhancement of growth resulting in precocious germination was achieved in light-grown barley embryos by supplying the medium with GA (Norstog, 1972). Furthermore, GA has also been shown to promote growth of the coleorhiza in barley embryos, although the interaction of other hormones, such as IAA and kinetin, is not excluded here (Norstog, 1969b). Such observations are clearly not incompatible with the notion that endogenous gibberellins might control the proportional growth of the different parts of the embryo.

When presenting nutrient substances to cultured embryos of certain plants, the margin between success and failure will depend upon the site of application of the nutrient. This seems to have been the case when Bulard (1967b) encountered a striking inhibition of growth of the epicotyl of seed embryos of *Ginkgo biloba* cultured with their cotyledons in contact with media containing GA_3, GA_4, and GA_7. In later work (Bulard and Le Page-Degivry, 1968) it was shown that the effect of applied hormone was dependent upon the age of embryos and the site of absorption of the hormone; when GA was applied to the epicotyl of partially germinating embryos or when the hormone was allowed to be translocated through the embryonic root, it was inconsequential in growth inhibition.

    V. RAGHAVAN

There exists a large literature on the effects of GA on embryos excised from dormant seeds, that is, seeds containing well-developed embryos that do not germinate even under favorable external conditions. These seeds are stimulated to germinate when they are given a specific temperature pretreatment or a particular wavelength of light deemed necessary to break dormancy. Temperature-sensitive seeds do not generally germinate until they are subjected to a spell of cold temperature, a treatment known as after-ripening. Although embryos excised from non-after-ripened seeds of many cold-requiring species of plants do not grow in culture, Barton (1956) first showed that the low-temperature requirement of embryos of *Malus arnoldiana* could be overcome by treating them with GA before transplantation into soil. Since this discovery, GA has been usefully employed to secure growth of embryos excised from non-after-ripened seeds of *Euonymus europaeus* (Monin, 1964), hazel (*Corylus avellana*) (Bradbeer and Pinfield, 1967), apple (*Pyrus malus*) (Côme and Durand, 1971), *Taxus baccata* (Le Page-Degivry, 1973a), and *Sorbus aucuparia* (Bianco and Bulard, 1977). Similarly, embryo dormancy in non-after-ripened whole seeds can also be overcome by GA (Wareing and Saunders, 1971). The effects of exogenous GA are somewhat equivocal in embryos that normally grow in culture upon excision from non-after-ripened seeds (Baskin and Baskin, 1970, 1971; Pinfield and Stobart, 1972). Significant increases noted in the endogenous gibberellin content of seeds during chilling may substantiate a role for this class of hormones in overcoming dormancy (Lewak and Rudnicki, 1977).

The site and mode of action of GA in breaking dormancy of embryos of hazel have been investigated by culture of embryo axes. Based on the rapid reponse of excised embryo axes to exogenous GA, Jarvis and Wilson (1977) concluded that the hormone directly affects this part of the embryo than the cotyledons. In further work (Jarvis *et al.*, 1978) it was also found that in a medium containing sucrose and inorganic salts, addition of GA caused only a transient increase in the growth of the embryo axes. From this it has been concluded that the hormone plays a dual role in overcoming embryo dormancy, involving an initial effect of the axis and a later effect on the cotyledons. These observations are consistent with the view (Ross and Bradbeer, 1968) that GA synthesis potentiated during cold treatment can subsequently proceed at a higher temperature favorable for germination.

Although light requirement for germination of certain seeds, such as those of lettuce (*Lactuca sativa*), can be overcome by GA, results on the effects of GA on isolated embryos have been inconclusive because of the interaction of other hormones (Bewley and Fountain, 1972). In the seeds of *Phacelia tenacetifolia,* whose germination was normally inhibited by

light, excised embryos became light insensitive. Light sensitivity rein-
stated in the embryos by holding them in an osmoticum, however, was
reversed by GA (Chen, 1970). According to Bouvinet and Rabéchault
(1965a,b), GA induced rapid growth and differentiation in cultured oil
palm embryos in a matter of days, whereas germination inside the seed
was spread over a period of several years. In the seeds of *Protea com-
pacta* (van Staden *et al.*, 1972a) and *Comptonia peregrina* (Del Tredici
and Torrey, 1976), where chemical inhibitors prevent germination, addition of GA to the medium has been shown to induce normal growth and
differentiation of excised embryos in culture. Because none of the above
studies has gone beyond the demonstration of the effects of GA in inducing growth of embryos, the mechanism of action of the hormone in these
systems remains unclear.

3. *Cytokinins and Related Compounds*

Another group of plant hormones that has received some attention with
respect to morphogenesis of the embryo and overcoming embryo dormancy in seeds is cytokinins. Veen (1963) reported the effects of a range
of concentrations of kinetin on the growth of torpedo-shaped embryos of
Capsella bursa-pastoris in liquid media. Measuring increase in length of
the embryo 7 days after culture, this investigator found that addition of
kinetin at 10^{-5} gm/ml caused complete inhibition of growth of the cotyledons and hypoctyl and led to necrosis of the root primordium. When embryos of the same age were grown on solid medium in long-term cultures,
presence of kinetin in the medium provoked precocious leaf expansion
and callus growth (Raghavan and Torrey, 1964). Callus growth was also
characteristically observed when embryos of *Ginkgo biloba* were grown
in media supplemented with kinetin or benzylaminopurine (Bulard,
1967a).

In cultured graminaceous embryos, the scutellum seems to be particularly susceptible to the action of cytokinins and related substances. According to Narayanaswami (1959), addition of 20.0–40.0 mg/liter adenine
to the medium induced callus growth on the scutellar part of embryos of
Pennisetum typhoideum, whereas the embryo axis was unaffected. Norstog (1969a) found that when barley embryos were grown in a medium
containing 0.1 mg/liter kinetin, greening of the scutellum and coleoptile
was accompanied by the formation of typical foliar hairs on their abaxial
epidermal surface. The fact that these controversial structures of the embryo possess the potential to express foliar characteristics under experimental conditions is considered as strong evidence in support of their homology to leaflike organs. Another characteristic effect of kinetin on this
sytem is the formation of embryo-like outgrowths from the region of the

epiblast (Norstog, 1970). Formation of shoot buds has been reported to occur on embryos of *Dendrobium falcatum* grown in media containing various cytokinins (Nag and Johri, 1976). Cytokinin effects here are infinitely complex because of their interaction with auxins and casein hydrolyzate; moreover, the effects are also dependent upon the age of embryos at excision.

The search for additional compounds to break dormancy of seeds has led to the use of cytokinins in embryo culture. However, the limited number of trials so far conducted on cold-requiring seeds indicate that cytokinins are less effective than GA or chilling pretreatment in inducing growth of embryos. For example, in the widely investigated apple (*Malus domestica*) seeds, cytokinins were only marginally effective in inducing growth of dormant or partially dormant embryos (Badizadegan and Carlson, 1967; Kamiński and Pieniazek, 1968). Curiously enough, both kinetin and benzyladenine modified the photosensitivity of embryos excised from dormant or partially after-ripened seeds, although the basis of this effect is not clear (Lewak and Bryzek, 1974). Tzou *et al.* (1973) found that zeatin- or zeatin ribonucleoside-induced increases in root elongation and fresh weight observed in non-after-ripened ash (*Fraxinus americana*) embryos were inferior to the increases noted in embryos excised from after-ripened seeds. Although embryos from freshly harvested seeds of *Acer platanoides* did not respond to the presence of kinetin in the medium, the compound was effective in inducing growth of embryos excised from stored seeds or from seeds given suboptimal chilling pretreatments (Pinfield *et al.*, 1974). In the related *Acer pseudoplatanus*, in which embryos of non-after-ripened seeds grow normally in culture, kinetin action was limited to a transient promotion of radicle growth (Pinfield and Stobart, 1972). Several investigators (Brown and van Staden, 1973; van Staden *et al.*, 1972b; Webb *et al.*, 1973) have analyzed the changes in cytokinin levels of seeds during chilling, and the results of these studies indicate that cytokinins along with other hormonal substances play a significant role in the metabolic processes of the embryo during after-ripening.

4. *Other Plant Hormones*

There have, in general, been insufficient investigations on the effects of other hormones on growth and morphogenesis of embryos, and this is particularly true of ethylene. Although ethylene is a gas at ordinary temperature, in recent years it has become clear that it is an endogenously produced hormone involved in normal control of certain aspects of plant growth and development. Moreover, the role of ethylene in breaking seed dormancy has been widely recognized and studied (Ketring, 1977). The only published work on the effects of ethylene on embryos is that by Kepczynski *et al*. (1977), who showed that the gas at 0.01–0.1% accelerated the growth in culture of embryos excised from partially after-ripened *Malus domestica* seeds. In confirmation of a requirement for ethylene for embryo growth it was also shown that analogs of ethylene biosynthesis, such as rhizobitoxine and hydroxyquinoline sulfate, or trapping the gas from the ambient culture medium by mercuric perchlorate inhibited growth of the embryos. Because excised *M. domestica* embryos gradually evolve ethylene during after-ripening, it is reasonable to conclude that ethylene biosynthesis is causally related to their ability to grow in culture.

Unlike other hormones, the almost universal response of plant cells to abscisic acid (ABA) is growth inhibition. In view of its unique role as a hormone that functions as an inhibitor of growth, special attention has been paid to the role of ABA in the growth of embryos of dormant seeds. Several reports (Khan and Heit, 1969; Rudnicki *et al*., 1971; Durand *et al*., 1973; Le Page-Degivry, 1973b; Jarvis and Wilson, 1978) indicate that addition of ABA inhibits growth of dormant and nondormant embryos of cold-requiring seeds and that the inhibition can be reversed by GA or kinetin. Growth of embryo axes of chilled hazel seeds is less susceptible to ABA than growth of embryo axes of dormant seeds, possibly because of production of GA during chilling (Jarvis and Wilson, 1978). In line with the role of ABA as an endogenous growth inhibitor, a decrease in its concentration along with a simultaneous increase in the concentration of growth-promoting hormones during cold treatment of seeds has been suggested as a possible mechanism to explain the breakage of dormancy (Lewak and Rudnicki, 1977).

Incorporation of ABA into the medium also inhibits growth of embryos excised from quiescent seeds, such as those of bean. The inhibitory action of the hormone probably results from its effects on RNA synthesis (Walton *et al*., 1970; Walbot *et al*., 1975a). According to Sussex *et al*. (1975) simultaneous addition of benzyladenine not only reversed ABA-induced growth inhibition in bean embryo axes but also restored RNA synthetic activity to the same level as control. In contrast to embryos of dormant and quiescent seeds, in which embryogenesis is interrupted by a period of arrested growth, in viviparous plants there is no interruption of embryo growth terminating embryogeny and separating it from quiescence or dormancy. Surprisingly enough, concentrations of ABA that inhibit growth of bean embryo axes do not have any effect on the growth of embryos excised from the viviparous plant, *Rhizophora mangle* (Sussex, 1975). Because of the high water content of embryos of this species, development of water stress followed by ABA synthesis has been proposed as a possible sequence of events leading to dormancy.

C. Effects of Surgical Treatments

It was mentioned earlier that the typical seed embryo is a bipolar structure consisting of shoot and root apical meristems located at either ends of the axis and one or two cotyledons flanking the shoot meristem. The mature embryo therefore possesses the basic organization of the adult plant, containing interacting organs that differ in some important functions. In recent years, the culture of surgically operated embryos and of embryo segments has facilitated analysis of the relationship of the different parts of the embryo to its final form in culture.

1. *Role of Cotyledons*

Perhaps the fundamental question that has been raised in microsurgical experiments is the nature of the contribution made by cotyledons to the growth of embryo axis. The bulk of the available evidence favors the view that the uninhibited growth of the shoot and root systems of cultured embryos depends upon the continued presence of cotyledons. This is well illustrated in the work of Rangaswamy and Rangan (1971) on embryos of *Cassytha filiformis*. Here the embryo axis is ensheathed by two massive cotyledons which can be fully or partially severed without injury to the axis. By employing various decotylating patterns and subsequent culture of mutilated embryos, these authors showed that the larger the portion of the cotyledon removed, the more inhibited was growth of the plumule. Complete plumular growth was achieved when a minimum amount of cotyledonary tissue, consisting of the radicular halves of both cotyledons or the radicular half of one cotyledon together with more than one half of the other cotyledon, were left on the embryo. Monnier (1978) in work with embryos of *Phaseolus vulgaris* has reported that increase in weight of the embryo axis during culture is directly proportional to the amount of cotyledonary tissue left on it.

Cotyledons might promote growth of the embryo axis either by providing a large surface area for uptake of nutrients or by supplying critical metabolites for growth. In embryo culture investigations, the embryo is generally oriented flat on the solid medium and nutrients are allowed to diffuse through the entire surface of the explanted organ. Initiation of growth is followed by the positive geotropic orientation of the root into the medium and the negative geotropism of the shoot into air above the medium. However, some workers (Bulard, 1952; Brown and Gifford, 1958; Ball, 1959; Berlyn, 1962; Engvild, 1964) have shown that for optimum growth of gymnosperm embryos, carbohydrates and other nutrients have to be supplied through cotyledons. For this purpose, the embryo is planted with cotyledons in contact with the medium and the culture tube inverted so that the morphologically basal end of the embryo grows downward. In a study of the role of cotyledons during growth of excised pine (*Pinus lambertiana*) embryos in culture, Berlyn and Miksche (1965) found that decotylated embryos, after an initial period of slow growth, elongated nearly to the same extent as those planted with cotyledons intact. This has led to the view that cotyledons seemingly promote embryonic growth by virtue of their large surface area, facilitating the uptake of nutrients, and not by supplying hormones or critical metabolites to the embryonic axis. Thévenot and Côme (1971, 1973; Côme *et al.*, 1968) have shown that the growth of partially after-ripened apple (*Pyrus malus*) embryos is appreciably accelerated when they are planted with cotyledons in contact with the medium. In contrast, orientation of embryos in various other positions, such as with only one end in contact with the medium, slows down growth. Although a role for cotyledons in transmitting nutrients to the elongating embryo axis is implied by these results, the situation is complicated by the asymmetrical growth of cotyledons and the presence of hydrolyzable inhibitors of germination in them.

2. *Morphogenesis of Embryo Segments*

Another rationale for microsurgical experiments is the need to investigate the capacity of constituent parts of embryo to regenerate whole plants, as it has been felt that this can be best achieved with individual parts removed from the complex integration of the embryo and cultured separately. From a survey of published reports on the culture of shoot, root, and hypocotyl segments of embryos (Lee, 1955; Furuya and Soma, 1957; Bajaj, 1966), it appears that only embryo segments consisting of the shoot apical meristem are able to regenerate whole plants in culture, whereas the root and hypocotyl segments are so completely dominated by their respective habits that they never form shoots.

In some instances, the regenerative ability of embryo segments in culture, combined with the developmental potential of bud initials on them, has offered the possibility of producing multiple seedlings from a single embryo. In *Cajanus cajan,* bud initials that did not grow out in the intact embryo sprang into growth when embryo segments without plumule or without plumule and radicle were cultured. It was possible to obtain up to three seedlings from a single embryo by culturing separately the plumule and the two axillary bud primordia of the cotyledonary internode (Kanta and Padmanabhan, 1964). In a similar way, up to 20 shoot buds were formed when a decapitated plumule of *Azadirachta indica* embryo was cultured (Rangaswamy and Promila, 1972). Another report on the embryos of *A. indica* (Padmanabhan and Muneeswaran, 1975) indicated that culture of plumular segment led to the formation of an incipient, intercotyledonary internode, resulting in an altered phyllotaxy of cotyledons, from opposite to alternate.

In summary, it can be said that segmentation of the embryo releases the full potentialities of its individual parts to express themselves in culture. What is now clear is that in such segments the regional differences become amplified and are translated into regionally specific patterns of morphogenesis.

D. Culture of Embryos and Seeds of Parasitic Plants

If culture of seed embryos of any one group of plants has been exclusively attempted, that distinction belongs to parasitic plants. The usual situation in parasitic plants is that at least during a part or probably whole of their life cycle, they depend upon the release of substances from a growing donor plant. Depending upon the part of the donor plant that is preyed upon, obligate phanerogamic parasites may be broadly classified into root and stem parasites. Obligate root parasites are included mainly in Scrophulariaceae, Orobanchaceae, Santalaceae, Loranthaceae, Balanophoraceae, and Cynomoriaceae, although only genera belonging to the first four families, such as *Alectra, Buchnera, Castilleja, Striga, Sopubia* (Scrophulariaceae); *Cistanche, Orobanche* (Orobanchaceae); *Exocarpus, Santalum* (Santalaceae); and *Nuytsia* (Loranthaceae), have been studied in any great detail. At least in *Orobanche* (Rangaswamy, 1967), *Cistanche* (Rangan, 1965), and *Alectra* (Okonkwo, 1975), at the time of shedding, mature seeds harbor an undifferentiated embryo that is litte more than a globular mass of cells. Although seeds of obligate root parasites have defied attempts at germination by conventional methods in the absence of contact with the host plant, aseptic culture of seeds has convincingly shown that the requirement for donor stimulus can be overcome by a variety of supplements to the medium. These include kinetin and related purines (Worsham *et al.*, 1959; Williams, 1961; Rangaswamy and Rangan, 1966); GA (Nash and Wilhelm, 1960; Williams, 1961); strigol, a compound related to GA (Cook *et al.*, 1966); coumarin derivatives (Worsham *et al.*, 1962); yeast extract (Nash and Wilhelm, 1960; Kumar, 1977b); casein hydrolyzate along wth IAA (Bhojwani, 1969); casein hydrolyzate, coconut milk, or both (Rangaswamy, 1967; Rangaswamy and Rao, 1963; Rangan, 1965; Rangan and Rangaswamy, 1968); and ethylene and 2-chloroethylphosphonic acid (Egley and Dale, 1970). In some species, seed germination is achieved by incorporation of an exudate of the host root into the medium (Okonkwo, 1966) or by exposing cultures to light (Okonkwo and Nwoke, 1974; Shivanna and Rangaswamy, 1976).

Among the most interesting modifications of embryo morphogenesis documented in cultured seeds of obligate root parasites are those that result from the addition of growth hormones or complex additives, such as coconut milk and casein hydrolyzate, to the medium. For example, Rangaswamy and Rangan (1966) were able to demonstrate that when seeds of *Striga euphrasioides* were germinated in a medium fortified with kinetin, the radicle elongated normally, but the cotyledons formed a highly friable callus. However, the massive callus formed on seeds of *Cistanche tubulosa* germinated in a medium containing coconut milk had its origin on the radicular end of the globular embryo. Differentiation of shoot buds occurred on the callus when it was transferred to a medium supplemented with IAA, GA, and kinetin and kept in the dark (Rangan, 1965; Rangan and Rangaswamy, 1968). Regeneration of shoot and root initials from a callus of radicular origin has also been described in seeds of *Exocarpus cupressiformis* cultured in a medium containing casein hydrolyzate and IAA (Bhojwani, 1969).

Studies by Kumar (1977a,b) on the germination of seeds of *Orobanche aegyptiaca* have implicated both hormones and complex additives in the morphogenesis of the enclosed embryo. When seeds were germinated in a medium containing coconut milk or yeast extract, a typical monopolar germination occurred, leading to the formation of the seedling by the division of the radicular end of the embryo. Here, following germination, the radicular pole of the embryo formed a callus-like mass of cells (tubercle) from which both root and shoot buds appeared, the plumular pole remaining dormant. In a medium containing IAA, GA, kinetin, or strigol, in contrast, seeds germinated forming a conventional root system from the radicular end and a shoot system from the plumular end. A typical bipolar type of germination accompanied by the formation of a primary root has also been reported in seeds of *Santalum album* grown in a medium containing both casein hydrolyzate and coconut milk (Rangaswamy and Rao, 1963; Rao and Rangaswamy, 1971).

In contrast to the behavior of embryos encased in seeds, bipolar germination was invariably the rule in excised embryos of *Nuytsia floribunda* planted in a vertical orientation in a medium containing adenine, indolebutyric acid, and casein hydrolyzate; however, in occasional embryos in which cotyledons came in contact with the medium, callus differentiation and regeneration of embryoids were noted (Nag and Johri, 1969, 1976).

Obligate stem parasites belong mainly to Loranthaceae, Cuscutaceae, and Lauraceae. There are several features of morphogenetic interest in embryos of stem parasites, such as their massive size and the presence of a haustrium at the radicular end, which forms a graft with the stem of the host plant. Culture of embryos of stem parasites has been mainly concerned with comparisons between *in vivo* and *in vitro* morphogenesis and the effects of growth hormones and other additives on the latter. In cultured embryos, greatest variation has been observed in the growth of the radicular end. In the embryos of *Cuscuta gronovii* (Truscott, 1966) and *Cassytha filiformis* (Rangan and Rangaswamy, 1969) the radicle failed to organize a primary root and the functional root system of the germinated embryo was constituted of a series of adventitious roots. Embryos of *Cuscuta reflexa* (Maheshwari and Baldev, 1961) and *Cuscuta lupuliformis* (Guzowska and Zenkteler, 1969) also failed to form primary roots *in vitro;* neither did they develop any adventitious roots. Embryos of loranthaceous parasites, such as *Arceuthobium pusillum* (Bonga and Chakraborty, 1968), *Dendrophthoe falcata, Amyema pendula, Amyema miquelii, Amylotheca dictyophleba, Scurrula philippensis* (Johri and Bajaj, 1964; Bajaj, 1967), and *Scurrula pulverulenta* (Johri and Bhojwani, 1970), showed a pattern of morphogenesis similar to that occurring *in vivo* by forming haustoria from their radicular end. In cultured embryos of *Phoradendron tomentosum,* however, haustoria were conspicuously absent (Bajaj, 1970).

In much of the published work on the culture of embryos of stem parasites, a medium component found essential for all species was casein hydrolyzate to which hormones were occasionally added. For example, for normal growth and morphogenesis of embryos of *D. falcata* (Bajaj, 1968; Johri and Bajaj, 1962), *A. dictyophleba, A. pendula,* and *A. miquelii* (Bajaj, 1967, 1970; Johri and Bajaj, 1964), IAA must also be present in the medium, whereas for embryos of *A. pusillum,* auxin need appears less certain (Bonga, 1969). For growth of embryos of *P. tomentosum,* addition of kinetin to a medium containing casein hydrolyzate and IAA was necessary (Bajaj, 1970). High concentrations of IAA, indolebutyric acid, 2,4-dichlorophenoxyacetic acid, naphthaleneacetic acid, and 2,4,5-trichlorophenoxyacetic acid are reported to cause copious callus formation in cultured embryos of *D. falcata, Taxillus vestitus* and *Taxillus cuneatus* (Nag and Johri, 1976).

Although embryo growth of obligate root and stem parasites in the absence of host tissue has been obtained through seed and embryo culture techniques, expect in isolated cases, serious attempts to integrate morphogenesis of embryos with specific nutritional or hormonal factors in the medium do not appear to have been made. However, knowledge of the mode of growth of embryos in culture and their responses to different supplements to the medium is likely to set the stage for future investigations on the chemical control measures of the parasites.

E. PRECOCIOUS GERMINATION

The early growth of an embryo *in ovulo* is characterized by a period of rapid cell divisions, unaccompanied by cell elongation or cell enlargement. As a result, cell size becomes progressively reduced at each division until a mass of small cells is formed. As the embryo develops through a period embracing tissue differentiation, cell divisions become restricted to certain areas of the cellular mass predictable by their position in the cell lineage. In a mature embryo, cell divisions are even further limited and occur mainly in the meristems and primoridia. An important principle governing germination of the seed is the gradual increase in size of the embryo and in the proportionality of growth of its different parts to produce a seedling. This developmental expression is due to the activity of the meristems and primordia, producing a population of small, nonvacuolate cells that undergo limited elongation, and is sustained by stored food material present in the surrounding tissues.

Although the essentially expected steps in the growth of embryos in culture are completion of embryogenesis along the paths so faithfully followed by embryos *in ovulo,* accompanied by initiation and differentiation of tissues and organs and evolution of a transplantable seedling, cultured embryos exhibit what is known as precocious germination. During precocious germination, the embryo tends to skip the normal stages of embryogenesis and acquires the characteristics of a malformed seedling. At the cellular level, precocious germination is not fundamentally different from normal germination except that mitoses are limited to meristems and that cells formed undergo vacuolation and elongation resulting in unusually long stems, leaves, and roots.

Although precocious germination has been repeatedly observed in embryo culture studies, it is only in recent years that a clearer understanding of its causes and the extent to which it can be modified by nutritional, hormonal, and environmental factors has been obtained. One aspect of the interesting series of studies on the causes of precocious germination is recognition of the fact that it is intimately related to the endogenous hormonal control mechanisms operating during seed dormancy. According to Andrews and Simpson (1969), embryos excised from dormant caryopses of wild oat (*Avena fatua*) failed to germinate precociously in a liquid medium in which embryos excised from nondormant strains germinated. The inability of dormant embryos to germinate has been interpreted as a consequence of the presence of an inhibitor that blocks the availability of endogenous GA necessary for germination. The finding that a requirement for GA to promote germination of dormant embryos can be eliminated if the embryos are allowed to "leach" on an agar medium is in agreement with this interpretation. The presence of a leachable inhibitor also apparently interferes with the germination of embryos excised from non-after-ripened seeds of *Taxus baccata*. When embryos were initially cultured for a period of 8 days in a liquid medium and then transferred to an agar medium, they germinated, signifying that those factors which prevented germination were removed from the embryos during their adaptation in the liquid phase (Le Page-Degivry and Garello, 1973). One of the likely candidates for the role of inhibitor of germination is ABA (Le Page-Degivry, 1973c).

A further insight into the causes of precocious germination was made by culturing embryos of barley under conditions that normally suppressed it, such as high osmolarity in the medium, high intensities of light, and moderately high temperatures. Addition of GA or kinetin induced precocious germination in embryos cultured under these conditions, whereas ABA suppressed it and counteracted the effects of GA (Norstog, 1972). These interesting interactions suggest a role of endogenous hormones not only in controlling embryo dormancy but also in preventing viviparous germination (precocious germination *in vivo*) of embryos enclosed in the seed.

Studies from Dure's laboratory (for review, see Walbot *et al.*, 1975b) on embryogenesis in cotton have helped to elucidate the causes of precocious germination in a molecular framework. Ihle and Dure (1969, 1970) found that the synthesis of mRNA for such enzymes as carboxypeptidase, isocitrase, and protease, required to mobilize stored food reserves, are transcribed by embryos at about 30 days before maturation when they have attained three-fifths of their final size. Translation of mRNA and synthesis of enzymes, however, are delayed until the embryo begins to germinate under normal conditions while enclosed in the seed or, precociously, when excised from the seed and cultured. Because removal of the embryo from the confines of the ovule triggered germination and enzyme synthesis, it appeared that the ovule tissue might transmit the factor that prevented translation of mRNA for the germination enzymes. This was indeed found to be the case, as shown by the discovery that an aqueous extract of the ovule as well as ABA delayed both precocious germination and enzyme appearance. Thus, in cotton seeds, supply of ABA from the ovular tissues apparently prevents vivipary by precluding the translation of mRNA for the synthesis of germination enzymes.

In attempting to control precocious germination of embryos, recourse can be made to manipulations of the culture medium and to changes in the physical conditions of culture, such as light intensity, temperature, and oxygen tension. The first systematic attempt to prevent precocious germination was made by Kent and Brink (1947), who found that supplementation of a regular agar medium containing mineral salts and sucrose with casein hydrolyzate, tomato juice, and other natural plant extracts forestalled germination and prolonged embryonic growth of immature barley embryos. The role of casein hydrolyzate in inhibiting germination suggested that utilization of this compound may arise from one or more of its major components, such as inorganic phosphate, sodium chloride, or the amino acid complement. Subsequently, Ziebur *et al.* (1950) found that inhibition of precocious germination resulted from the high osmotic pressure of casein hydrolyzate, which was mainly produced by the sodium chloride and amino acid fractions. Therefore, the major effect of casein hydrolyzate in inducing normal growth of barley embryos was not nutritional in the sense of providing cumbustible materials for metabolism; this was also suggested by the observation that germination could be delayed in the absence of casein hydrolyzate by supplementing the medium with high levels of mannitol or sucrose. It must be emphasized, however, that high osmolarity in the medium is more effective in inducing growth in relatively immature embryos than in preventing precocious germination of seed embryos (see Section III).

The experiments of Norstog and Klein (1972) with barley embryos appear to provide convincing evidence for the control of precocious germination by extracellular factors. These workers found that in addition to the presence of an osmoticum in the medium, reduced O_2 tension, elevated temperature, and high light intensities were among the more useful means of suppressing precocious germination. From these results it can be argued that a complex interplay of extracellular and intercellular control mechanisms may operate to prolong embryonic growth *in vitro* and reduce the gap between what transpires *in ovulo* and what occurs in culture. The effect of light in suppressing precocious germination in cultured embryos presents a paradox, because growth of the embryo normally in the ovule without direct contact with light serves the same end.

III. Culture of Proembryos

The term "proembryo" is used to refer to those developmental stages of the embryo that precede cotyledon initiation. Early division phase embryos, such as globular and heart-shaped stages, are appropriately classified as proembryos, although in some cases the dividing line between the end of proembryo phase and the beginning of maturation phase is somewhat arbitrary. *In vivo* observations suggest that during their growth within the confines of the embryo sac, proembryos are dependent not only upon the metabolites present in their cells but also upon those diffusing from the surrounding endosperm. This, in turn, indicates that in contrast to the relatively autotrophic nature of seed embryos, proembryos are heterotrophic in nature.

Some comments are in order regarding the constituents of the endosperm that provide the nurture and nutriment to the developing proembryos. Present knowledge on the chemical composition of coconut endosperm (coconut milk) has been summarized by Raghavan (1976). This unique endosperm, which exists as a liquid within the vesiculate embryo sac during the early stages of embryogenesis, contains an array of vitamins, sugars, amino acids, and growth hormones. From the point of view of growth induction of the proembryo by specially balanced nutrients and stimuli, the main components of coconut milk are probably inositol, auxins, and cytokinins. Although endosperms of other plants have yielded their chemically defined components only in part, available data show that they mainly contain a complement of amino acids and hormonal substances. Quite possibly, a delicately controlled interaction between the hormones and between the hormones and one of the other compounds is involved in sustaining proembryo growth *in vivo*.

The heterotrophic dependence of proembryos upon the complex nutrient substances of the endosperm suggests that nutritional requirements for their growth in culture are bound to be more exacting than those for seed embryos. This has been borne out by experiments to be described later in this section. Based on these experiments, it is possible to trace three main themes on the nutrition and growth of proembryos: (1) studies on the effects of natural plant extracts; (2) studies on the requirement for high osmoticum, and (3) studies on the role of hormones and other stimuli.

A. EFFECTS OF NATURAL PLANT EXTRACTS

A breakthrough in the culture of proembryos was obtained with the use of coconut milk by van Overbeek *et al.* (1942). These workers found that whereas torpedo-shaped and heart-shaped embryos of *Datura stramonium* grew well in a medium enriched with an organic addendum containing glycine, nicotinic acid, thiamine, pyridoxine, ascorbic acid, adenine, pantothenic acid, and succinic acid, still smaller embryos failed to grow or grew feebly before becoming miniature calluses. A dramatic increase in growth of proembryos was obtained, however, when the above medium was enriched by the addition of nonautoclaved coconut milk. A phenonmenon exposed in these studies was the hormonal nature of the factor from coconut milk, collectively designated as "embryo factor," that triggered embryo growth. This factor, which was obtained in a sufficiently pure form, promoted growth of proembryos of *D. stramonium* when added to the basal medium in a dilution of 1:19,000 (van Overbeek, 1942; van Overbeek *et al.*, 1942). Other natural extracts that have been used are water extracts of date, bananas, wheat gluten hydrolyzate, and tomato juice (Kent and Brink, 1947); diffusates from young seeds of *Lupinus luteus* and *Sechium edule* (Matsubara, 1962; Matsubara and Ogawa, 1963); and extracts of pumpkin (*Cucurbita maxima* and *Cucurbita moschata*) (Nakajima, 1962).

In some cases, successful growth of small embryos has been achieved by keeping them in contact with the crushed endosperm serving as a nurse tissue (Ziebur and Brink, 1951; Kruse, 1974). Rarely, as shown by Thomas (1972), undifferentiated callus may also function as nurse tissue for growth of proembryos. In this work, it was found that when cleavage stage embryos of pine (*Pinus mugo, Pinus sylvestris* and *Pinus nigra*) were cultured in close contact with a parenchymatous callus of embryonic origin, division of the embryonal cells occurred.

The search for specific chemical constituents in coconut milk and other plant extracts concerned with growth promotion of proembryos has yielded some interesting results. DeMaggio and Wetmore (1961) found that myoinositol and sorbitol, known to be the chief constituents of the neutral fraction of coconut milk (Pollard *et al.*, 1961), could replace the need for coconut milk in the growth of early division phase embryos of the fern *Todea barbara*. According to Nakajima (1962), growth of small embryos of pumpkin induced by their endosperm extracts was duplicated by a mixture of IAA, 1,3-diphenylurea, and casein hydrolyzate. Studies by Norstog and Smith (1963) have emphasized that the requirement for coconut milk for growth of proembryos of barley can be met by a phosphate-enriched White's medium, at an optimum pH of 4.9 and fortified with glutamine and alanine as major nitrogen sources and leucine, tyrosine, phenylalanine, cysteine, and tryptophan as minor nitrogen sources. In a further refinement of this medium, it was found that the survival value of embryos was considerably enhanced by a five- to tenfold increase in the concentrations of KCl and KNO_3 and of certain organic components in the medium (Norstog, 1967, 1973). The suitability of the modified medium for culture of embryos as small as 0.25 mm has been confirmed using five different varieties of *Hordeum distichum* (Cameron-Mills and Duffus, 1977). The results of these studies, which are restricted in the number of species used, nevertheless represent important steps toward the formulation of completely synthetic media for the culture of proembryos.

B. Requirement for High Osmolarity of the Medium

One of the attributes of the amorphous liquid endosperm bathing proembryos is its high osmolarity (Ryczkowski, 1960; Mauney, 1961; Smith, 1973). Of significance in this context is the fact that cells of proembryos, being osmotically very active, will take up water and expand in a medium of low osmotic potential; consequently, growth processes may become hindered. On the assumption that a nutrient medium isotonic with the natural environment of the ovule would be a promising possibility for improving growth of small embryos, several investigators have designed such media for embryo culture. As might be anticipated, increased growth of proembryos was obtained in media of high osmolarity (Rietsema *et al.*, 1953a; Matsubara, 1964; Rijven, 1952; Veen, 1963; Norstog, 1961). In this context, studies by Rietsema *et al.* (1953a) are of interest as one of the first demonstrations of a relation between the osmotic value of the medium and growth of embryos of different ages. These workers found that whereas mature embryos of *Datura stramonium* grew even in the absence of sucrose in the medium, pre-heart-shaped embryos required a medium containing 8–12% sucrose for growth. However, the latter grew equally well in a medium containing mannitol adjusted to be isotonic with 8–12% sucrose, thus reinforcing the thought that sucrose was functioning to stabilize the osmotic pressure of the medium rather than as a carbon energy source. In physiological terms, the success of a high osmoticum in the culture of proembryos can be interpreted as a consequence of the small embryos being able to effectively control the flow of metabolites and inorganic ions into their cells.

C. Effects of Hormones and Other Stimuli

As mentioned earlier, growth of proembryos in complex media containing coconut milk and similar undefined plant extracts suggests that some hormones are required to initiate division in their cells. Evidence for the involvement of such hormones in the growth of proembryos of *Capsella bursa-pastoris* was provided by Raghavan and Torrey (1963). Although previous workers (Rijven, 1952; Veen, 1963) successfully cultured heart-shaped and smaller embryos of *C. bursa-pastoris* in liquid media of high osmolarity secured by the addition of 12–18% sucrose, Raghavan and Torrey were able to grow satisfactorily heart-shaped embryos in a relatively simple medium containing inorganic ions, vitamins, and 2% sucrose, solidified with agar. To grow still smaller globular embryos, it was necessary to supplement a liquid medium of this composition with IAA, kinetin, and adenine sulfate. Although there was no apparent requirement for a high osmotic pressure of the medium in the presence of a balanced mixture of growth hormones, the effects of the latter were also reproduced by a high sucrose concentration or by a 10-fold concentration of major salts. The involvement of high salt concentration in growth induction in proembryos of *C. bursa-pastoris* has been brought into a sharper focus by the more recent studies of Monnier (1970). From a comparative analysis of different inorganic media, this worker found that the high-salt medium of Murashige and Skoog (1962) was best for proembryos of *C. bursa-pastoris,* although the use of this medium was frustrating because of the small number of embryos that survived the culture period.

Nearly all workers who have attempted culture of proembryos have encountered toxicity of the medium resulting in inexplicably low rates of survival. This occurred despite the fact that embryos were excised and cultured with extreme care. Admittedly, the trauma resulting from excision of the embryo from its natural environment of the ovule and its transfer to an artificial milieu might account for a certain number of failures. Clearly, it would be highly advantageous to study not only growth and morphogenesis but also nutrition and metabolism of proembryos to formulate a nontoxic medium that allows their growth and survival. Reference was made in the previous paragraph to the work of Monnier (1970), who found that Murashige–Skoog medium, which promoted growth of proembryos of *Capsella bursa-pastoris,* was nonetheless toxic. In further work (Monnier, 1976b), by determining the most effective level of each major and minor salt of Murashige–Skoog medium, it was shown that a modification of this medium, involving an increase in the concentration of K and Ca and a decrease in the concentration of NH_4NO_3, led to excellent growth and high rate of survival of proembryos of *C. bursa-pastoris.* Enhanced growth and survival of proembryos were also obtained by supporting them on a bed of polyacrylamide instead of agar (Monnier, 1975) or by increasing the partial pressure of oxygen in the medium (Monnier, 1976a). This recognition of the role of both chemical and physical conditions of culture for the continued growth of proembryos may hint at lines of future work that may be pursued in the culture of still smaller embryos and, ultimately, of the single-celled zygote. In embryo culture studies, this is an area that begs immediate attention.

D. PROEMBRYO MORPHOGENESIS IN CULTURED OVULES

In recent years some information that bears upon the growth and morphogenesis of proembryos has been obtained by culture of excised ovules. Because the embryo is an integral component of the ovule, ovule culture has proved particulary rewarding in studying the behavior of proembryos of certain plants that hitherto have defied attempts at excision and culture. However, compared to the extensive research on the culture of isolated embryos, culture of isolated ovules has received little attention. Much of the early work, which was initiated at the University of Delhi, has been reviewed (Johri, 1962; Maheshwari and Rangaswamy, 1965; Raghavan, 1976).

The prospect of using ovule culture as an alternate method to study the nutritional requirements of young embryos was indicated by the work of Maheshwari (1958; Maheshwari and Lal, 1961). Culture of isolated ovules of *Papaver somniferum* containing zygotes or two-celled proembryos in a mineral salt medium containing 5% sucrose first met with only limited success, and it was not until the medium was fortified with casein hydrolyzate, yeast extract, or kinetin that rapid growth of the enclosed embryo was obtained. Addition of GA induced a marked inhibition of growth, and the addition of IAA stimulated embryo growth only slightly, if at all. In another study, Kapoor (1959) reported that viable seeds were obtained from the ovules of *Zephyranthes* with included zygotes cultured in a medium containing coconut milk or "casamino acids." This report also showed that the favorable effect of casamino acids was completely replaced by arginine, histidine, leucine, or to some extent by valine.

As a preparatory step toward formulating a suitable medium for *in vitro* culture of proembryos of cotton, some investigators have attempted culture of ovules with included embryos. Joshi and Johri (1972) showed that when 6-day postanthesis ovules enclosing 12-celled proembryos were cultured on White's medium supplemented with kinetin, the latter grew slowly and attained but one-third the size of *in situ* embryos. Somewhat similar results were obtained with 5-day postanthesis ovules harboring 2- to 10-celled proembryos were cultured on Murashige–Skoog medium (Eid *et al.*, 1973). This work, however, indicated that the response of cotton proembryos encased in cultured ovules depended upon the availability of moderate amounts of NO_3^- and NH_4^+ in the medium. Based on this, Stewart and Hsu (1977) subsequently demonstrated that cotton ovules containing even the undivided zygote could be cultured with reasonable success leading to the production of a large number of mature embryos by supplementing a high-salt, GA-enriched medium with 15 mM NH_4^+ supplied as either NH_4NO_3, NH_4Cl, or $(NH_4)_2SO_4$. Detailed analyses of the responses of the cultured ovule in the different media showed that the poor development of the ovule in the basal medium, resulting in space restrictions for the growth of the embryo, probably led to a failure of the embryo to develop to normal size. Addition of GA enabled the ovule to grow to its usual *in vivo* size, and NH_4^+ proved critical for continued growth and division of the zygote.

From these results we cannot generalize on the requirements for growth of proembryos of cultured ovules other than to say that no unique hormonal substances appear to be involved in the process. The question to be answered is whether some hormones are transmitted to the growing embryo from the extraembryonal tissues of the ovule. Although the basic requirements in all species studied are somewhat similar and include hormones and a liberal supply of reduced nitrogen, there are also sufficient differences between the species so that we cannot safely extrapolate from one species to another.

IV. Concluding Comments

Attempts to gain more understanding of the control of growth of the fertilized egg in the microenvironment of the ovule into an embryo have resulted in the development of embryo culture techniques. Clearly, in the last few years, as a result of improvements in culture techniques and in defined media, there has been notable success in growing progressively smaller embryos of angiosperms in culture. Progress has also been made in establishing working ranges of concentrations of various ingredients of culture media empirically effective in inducing normal growth in mature and differentiated seed embryos of a number of plants. Additionally, studies on the culture of seed embryos have yielded valuable information on their metabolism and morphogenetic control mechanisms. Although culture of proembryos has reached a seeming impasse, at which breakthrough must depend upon further refinements of the medium and manipulative skills, in the next few years cultured seed embryos may be anticipated to be used increasingly to unravel the biochemistry of dormancy and germination, as it is fairly clear that biochemical problems in embryo dormancy and germination confronting the investigator can be more accurately studied in isolated embryos than in embryos encased in the seed.

Although the practical applications of embryo culture have hardly been touched on in this chapter, it is obvious that by culturing embryos it is possible to obtain hybrids from inviable crosses, to overcome seed dormancy, and to reduce the breeding cycle of plants where delayed seed germination is a major handicap (Raghavan, 1977a). In this respect it is satisfying to think that embryo culture as a tool has come a long way in a short span of time. It is hoped that because of its relative simplicity, the embryo culture technique can become an important complement to the standard techniques available to the plant breeder.

REFERENCES

Abraham, A., and Thomas, K. J. (1962). *Indian Coconut J.* **15**, 84–88.

Amemiya, A., Akemine, H., and Toriyama, K. (1956a). *Nogyo Gijutsu Kenkyusho Hokoku D* **6**, 1–40.

Amemiya, A., Akemine, H., and Toriyama, K. (1956b). *Nogyo Gijutsu Kenkyusho Hokoku D* **6**, 41–60.

Anagnostakis, S. L. (1977). *HortScience* **12**, 44.

Andrews, C. J., and Simpson, G. M. (1969). *Can. J. Bot.* **47**, 1841–1849.

Andronescu, D. I. (1919). *Am. J. Bot.* **6**, 443–452.

Augusten, H. (1956). *Planta* **48**, 24–46.

Badizadegan, M., and Carlson, R. F. (1967). *Proc. Am. Soc. Hortic. Sci.* **91**, 1–8.

Bajaj, Y. P. S. (1966). *Can. J. Bot.* **44**, 1127–1131.

Bajaj, Y. P. S. (1967). *N. Z. J. Bot.* **5**, 49–56.

Bajaj, Y. P. S. (1968). *Can. J. Bot.* **46**, 429–433.

Bajaj, Y. P. S. (1970). *Z. Pflanzenphysiol.* **63**, 408–415.

Ball, E. (1959). *Am. J. Bot.* **46**, 130–139.

Bartels, H. (1957). *Naturwissenschaften* **44**, 595–596.

Barton, L. V. (1956). *Contrib. Boyce Thompson Inst.* **18**, 311–317.

Baskin, J. M., and Baskin, C. C. (1970). *Planta* **94**, 250–252.

Baskin, J. M., and Baskin, C. C. (1971). *Planta* **100**, 365–369.

Berlyn, G. P. (1962). *Am. J. Bot.* **49**, 327–333.

Berlyn, G. P., and Miksche, J. P. (1965). *Am. J. Bot.* **52**, 730–736.

Bewley, J. D., and Fountain, D. W. (1972). *Planta* **102**, 368–371.

Bhojwani, S. S. (1969). *Curr. Sci.* **38**, 6–8.

Bianco, J., and Bulard, C. (1977). *Trav. Sci. Parc Natl. Vanoise* **8**, 147–155.

Bonga, J. M. (1968). *In* "Biochemistry and Physiology of Plant Growth Substances" (F. Wightman and G. Setterfield, eds.), pp. 867–874. Runge Press, Ottawa.

Bonga, J. M., and Chakraborty, C. (1968). *Can. J. Bot.* **46**, 161–164.

Bonner, J. (1938). *Plant Physiol.* **13**, 865–868.

Bonner, J., and Axtman, G. (1937). *Proc. Natl. Acad. Sci. U.S.A.* **23**, 453–457.

Bonner, J., and Bonner, D. (1938). *Proc. Natl. Acad. Sci. U.S.A.* **24**, 70–75.

Bouvinet, J., and Rabéchault, H. (1965a). *Oleagineux* **20**, 79–87.

Bouvinet, J., and Rabéchault, H. (1965b). *C. R. Acad. Sci. Ser. D* **260**, 5336–5338.

Bradbeer, J. W., and Pinfield, N. J. (1967). *New Phytol.* **66**, 515–523.

Brink, R. A., and Cooper, D. C. (1947). *Bot. Rev.* **13**, 423–541.

Brown, C. L., and Gifford, E. M., Jr. (1958). *Plant Physiol.* **33**, 57–64.

Brown, N. A. C., and van Staden, J. (1973). *Physiol. Plant.* **28**, 388–392.

Buckner, G. D., and Kastle, J. H. (1917). *J. Biol. Chem.* **29**, 209–213.

Buffard-Morel, J. (1968). *C. R. Acad. Sci. Ser. D* **267**, 185–188.

Bulard, C. (1952). *C. R. Acad. Sci. Ser. D* **235**, 739–741.

Bulard, C. (1967a). *Soc. Bot. Fr., Colloq. Morphol. Exp.* 119–130.

Bulard, C. (1967b). *C. R. Acad. Sci. Ser. D* **265**, 1301–1304.

Bulard, C., and Le Page-Degivry, M.-T. (1968). *C. R. Acad. Sci., Ser. D* **266**, 356–359.

Button, J., Bornman, C. H., and Carter, M. (1971). *J. Exp. Bot.* **22**, 922–924.

Cameron-Mills, V., and Duffus, C. M. (1977). *Ann. Bot. (London)* **41**, 1117–1127.

Carew, D. P., and Schwarting, A. E. (1958). *Bot. Gaz. (Chicago)* **119**, 237–239.

Chen, S. S. C. (1970). *Planta* **95**, 336–340.

Colonna, J.-P., Cas, G., and Rabéchault, H. (1971). *C. R. Acad. Sci. Ser. D* **272**, 60–63.

Côme, D., and Durand, M. (1971). *C. R. Acad. Sci. Ser. D* **273**, 1937–1940.

Côme, D., Thévenot, C., and Tissaoui, T. (1968). *Rev. Gen. Bot.* **75**, 611–626.

Cook, C. E., Whichard, L. P., Turner, B., and Wall, M. E. (1966). *Science* **154**, 1189–1190.
Cummings, D. P., Green, C. E., and Stuthman, D. D. (1976). *Crop Sci.* **16**, 465–470.
Cutter, V. M., Jr., and Wilson, K. M. (1954). *Bot. Gaz. (Chicago)* **115**, 234–240.
Degivry, M. T. (1966). *Bull. Sci. Bourgogne* **24**, 57–87.
De Guzman, E. V. (1969). *Philipp. Agric.* **53**, 65–78.
De Guzman, E. V., Del Rosario, A. G., and Eusebio. E. C. (1971). *Philipp. Agric.* **53**, 566–579.
Del Tredici, P., and Torrey, J. G. (1976). *Bot. Gaz. (Chicago)* **137**, 262–268.
DeMaggio, A. E., and Wetmore, R. H. (1961). *Am. J. Bot.* **48**, 551–565.
Dieterich, K. (1924). *Flora* **117**, 379–417.
Durand, M., Thévenot, C., and Côme, D. (1973). *C. R. Acad. Sci., Ser. D* **277**, 53–55.
Dure, L. S., and Jensen, W. A. (1957). *Bot. Gaz. (Chicago)* **118**, 254–261.
Egley, G. H., and Dale, J. E. (1970). *Weed Sci.* **18**, 586–589.
Eid, A. A. H., De Langhe, E., and Waterkeyn, L. (1973). *Cellule* **69**, 359–371.
Engvild, K. C. (1964). *Physiol. Plant.* **17**, 866–874.
Esenbeck, E., and Suessenguth, K. (1925). *Arch. Exp. Zellforsch. Besonders Gewebezuecht.* **1**, 547–586.
Fisher, J. B., and Tsai, J. H. (1978). *In Vitro* **14**, 307–311.
Furuya, M., and Soma, K. (1957). *J. Fac. Sci. Univ. Tokyo Sect. III. Bot.* **7**, 163–198.
Gorter, C. J. (1955). *Proc. K. Ned. Akad. Wet. Ser. C* **58**, 377–385.
Green, C. E., and Phillips, R. L. (1975). *Crop Sci.* **15**, 417–421.
Guttenberg, H. V., and Wiedow, H.-L. (1952). *Planta* **41**, 145–166.
Guzowska, I., and Zenkteler, M. (1969). *Bull. Soc. Sci. Lett. Poznan, Ser. D* **9**, 37–48.
Hannig, E. (1904). *Bot. Ztg.* **62**, 45–80.
Harris, G. P. (1956). *New Phytol.* **55**, 253–268.
Helmkamp, G., and Bonner, J. (1953). *Plant Physiol.* **28**, 428–436.
Honma, S. (1955). *Proc. Am. Soc. Hortic. Sci.* **65**, 405–408.
Ihle, J. N., and Dure, L., III (1969). *Biochem. Biophys. Res. Commun.* **36**, 705–710.
Ihle, J. N., and Dure, L., III (1970). *Biochem. Biophys. Res. Commun.* **38**, 995–1001.
Jarvis, B. C., and Wilson, D. (1977). *New Phytol.* **78**, 397–401.
Jarvis, B. C., and Wilson, D. A. (1978). *Planta* **138**, 189–191.
Jarvis, B. C., Wilson, D. A., and Fowler, M. W. (1978). *New Phytol.* **80**, 117–123.
Johri, B. M. (1962). *In* "Proceedings of the Summer School of Botany" (P. Maheshwari, B. M. Johri, and I. K. Vasil, eds.), pp. 94–105. Minist. Sci. Res. Cult. Aff., New Delhi.
Johri, B. M., and Bajaj, Y. P. S. (1962). *Nature (London)* **193**, 194–195.
Johri, B. M., and Bajaj, Y. P. S. (1964). *Nature (London)* **204**, 1220–1221.
Johri, B. M., and Bhojwani, S. S. (1970). *Ann. Bot. (London)* **34**, 685–690.
Joshi, P. C., and Johri, B. M. (1972). *Phytomorphology* **22**, 195–209.
Kamiński, W., and Pieniazek, J. (1968). *Bull. Acad. Pol. Sci., Ser. Sci. Biol.* **16**, 719–723.
Kanta, K., and Padmanabhan, D. (1964). *Curr. Sci.* **33**, 704–706.
Kapoor, M. (1959). *Phytomorphology* **9**, 313–315.
Kent, N. F., and Brink, R. A. (1947). *Science* **106**, 547–548.
Kepczynski, J., Rudnicki, R. M., and Khan, A. A. (1977). *Physiol. Plant.* **40**, 292–295.
Ketring, D. L. (1977). *In* "The Physiology and Biochemistry of Seed Dormancy and Germination" (A. A. Khan, ed.), pp. 157–178. North-Holland Publ., Amsterdam.
Khan, A. A., and Heit, C. E. (1969). *Biochem. J.* **118**, 707–712.
Kögl, F., and Haagen-Smit, A. J. (1936). *Hoppe-Seyler's Z. Physiol. Chem.* **243**, 209–226.
Kruse, A. (1974). *Hereditas* **77**, 219–224.
Kruyt, W. (1954). *Acta Bot. Neerl.* **3**, 1–82.
Kumar, U. (1977a). *Biol. Plant.* **19**, 353–359.
Kumar, U. (1977b). *Can. J. Bot.* **55**, 2613–2621.

Lammerts, W. E. (1942). *Am. J. Bot.* **29**. 166–171.

Lee, A. E. (1952). *Bull. Torrey Bot. Club* **79**, 59–62.

Lee, A. E. (1955). *Bot. Gaz. (Chicago)* **116**, 359–364.

Le Page-Degivry, M.-T. (1973a). *Biol. Plant.* **15**, 264–269.

Le Page-Degivry, M.-T. (1973b). *Z. Pflanzenphysiol.* **70**, 406–413.

Le Page-Degivry, M.-T. (1973c). *C. R. Acad. Sci., Ser. D* **277**, 177–180.

Le Page-Degivry, M.-T., and Garello, G. (1973). *Physiol. Plant.* **29**, 204–207.

Lewak, S., and Bryzek, B. (1974). *Biol. Plant.* **16**, 334–340.

Lewak, S., and Rudnicki, R. M. (1977). *In* "The Physiology and Biochemistry of Seed Dormancy and Germination" (A. A. Khan, ed.), pp. 193–217. North-Holland Publ., Amsterdam.

Maheshwari, N. (1958). *Science* **127**, 342.

Maheshwari, N., and Lal, M. (1961). *Phytomorphology* **11**, 307–314.

Maheshwari, P., and Baldev, B. (1961). *Nature (London)* **191**, 197–198.

Maheshwari, P., and Rangaswamy, N. S. (1965). *Adv. Bot. Res.* **2**, 219–321.

Matsubara, S. (1962). *Bot. Mag.* **75**, 10–18.

Matsubara, S. (1964). *Bot. Mag.* **77**, 253–259.

Matsubara, S., and Ogawa, J. (1963). *Bot. Mag.* **76**, 440–445.

Mauney, J. R. (1961). *Bot. Gaz. (Chicago)* **122**, 205–209.

Mehra, A., and Mehra, P. N. (1974). *Bot. Gaz. (Chicago)* **135**, 61–73.

Miflin, B. J. (1969). *J. Exp. Bot.* **20**, 810–819.

Mitra, G. C., and Kaul, K. N. (1964). *Indian J. Exp. Biol.* **2**, 49–51.

Mitra, J., and Datta, C. (1951). *Sci. Cult.* **16**, 428–430.

Monin, J. (1964). *C. R. Soc. Biol.* **158**, 2283–2286.

Monnier, M. (1970). *Rev. Gen. Bot.* **77**, 73–83.

Monnier, M. (1975). *C. R. Acad. Sci., Ser. D* **280**, 705–708.

Monnier, M. (1976a). *C. R. Acad. Sci., Ser. D* **282**, 1009–1012.

Monnier, M. (1976b). *Rev. Cytol. Biol. Veg.* **39**, 1–120.

Monnier, M. (1978). *C. R. Acad. Sci., Ser. D* **286**, 461–464.

Murashige, T., and Skoog, F. (1962). *Physiol. Plant.* **15**, 473–497.

Nag, K. K., and Johri, B. M. (1969). *Phytomorphology* **19**, 405–408.

Nag, K. K., and Johri, B. M. (1976). *Bot. Gaz. (Chicago)* **137**, 378–390.

Nakajima, T. (1962). *Bull. Univ. Osaka Prefect., Ser. B* **13**, 13–48.

Narayanaswami, S. (1959). *Phytomorphology* **9**, 358–367.

Narayanaswami, S., and Norstog, K. (1964). *Bot. Rev.* **30**, 587–628.

Nash, S. M., and Wilhelm, S. (1960). *Phytopathology* **50**, 772–774.

Nataraja, K. (1975). *Curr. Sci.* **44**, 136–137.

Nataraja, K., Chennaveeraiah, M. S., and Girigowda, P. (1973). *Curr. Sci.* **42**, 577–584.

Norstog, K. J. (1956). *Bull. Torrey Bot. Club* **83**, 27–29.

Norstog, K. (1961). *Am. J. Bot.* **48**, 876–884.

Norstog, K. (1967). *Bull. Torrey Bot. Club* **94**, 223–229.

Norstog, K. (1969a). *Phytomorphology* **23**, 235–241.

Norstog, K. (1969b). *Trans. Ill. State Acad. Sci.* **62**, 312–315.

Norstog, K. (1970). *Dev. Biol.* **23**, 665–670.

Norstog, K. (1972). *Phytomorphology* **22**, 134–139.

Norstog, K. (1973). *In Vitro* **8**, 307–308.

Norstog, K., and Klein, R. M. (1972). *Can. J. Bot.* **50**, 1887–1894.

Norstog, K., and Smith, J. (1963). *Science* **142**, 1655–1656.

Okonkwo, S. N. C. (1966). *Am. J. Bot.* **53**, 679–687.

Okonkwo, S. N. C. (1975). *Physiol. Plant.* **34**, 378–383.

Okonkwo, S. N. C., and Nwoke, F. I. O. (1974). *Ann. Bot. (London)* **38,** 409–417.

Ozsan, M., and Cameron, J. W. (1963). *Proc. Am. Soc. Hortic. Sci.* **82,** 210–216.

Padmanabhan, D., and Muneeswaran, A. (1975). *Curr. Sci.* **44,** 402–403.

Pinfield, N. J., and Stobart, A. K. (1972). *Planta* **104,** 134–145.

Pinfield, N. J., Davies, H. V., and Stobart, A. K. (1974). *Physiol. Plant.* **32,** 268–272.

Pollard, J. K., Shantz, E. M., and Steward, F. C. (1961). *Plant Physiol.* **36,** 492–501.

Purvis, O. N. (1944). *Ann. Bot. (London)* **8,** 285–314.

Rabéchault, H., Ahée, J., and Guénin, J. (1970). *C. R. Acad. Sci. Ser. D* **270,** 3067–3070.

Radforth, W. N., and Pegoraro, L. C. (1955). *Trans. R. Soc. Can., Sect. 5* **49,** 69–82.

Raghavan, V. (1966). *Biol. Rev. Cambridge Philos. Soc.* **41,** 1–58.

Raghavan, V. (1967). *In* "Methods in Developmental Biology" (F. H. Wilt and N. K. Wessells, eds.), pp. 413–424. Crowell, New York.

Raghavan, V. (1976). "Experimental Embryogenesis in Vascular Plants." Academic Press, New York.

Raghavan, V. (1977a). *In* "Applied and Fundamental Aspects of Plant Cell, Tissue, and Organ Culture" (J. Reinert and Y. P. S. Bajaj, eds.), pp. 375–397. Springer-Verlag, Berlin and New York.

Raghavan, V. (1977b). *In* "CRC Handbook Series in Nutrition and Food" (M. Rechcigl, Jr., ed.), Vol. IV, Sect. G, pp. 361–413. CRC Press, Cleveland, Ohio.

Raghavan, V., and Torrey, J. G. (1963). *Am. J. Bot.* **50,** 540–551.

Raghavan, V., and Torrey, J. G. (1964). *Plant Physiol.* **39,** 691–699.

Randolph, L. F., and Khan, R. (1960). *Phytomorphology* **10,** 43–49.

Rangan, T. S. (1965). *Phytomorphology* **15,** 180–182.

Rangan, T. S., and Rangaswamy, N. S. (1968). *Can. J. Bot.* **46,** 263–266.

Rangan, T. S., and Rangaswamy, N. S. (1969). *Phytomorphology* **19,** 292–300.

Rangaswamy, N. S. (1967). *Phytomorphology* **17,** 477–487.

Rangaswamy, N. S., and Promila. (1972). *Z. Pflanzenphysiol.* **67,** 377–379.

Rangaswamy, N. S., and Rangan, T. S. (1966). *Nature (London)* **210,** 440–441.

Rangaswamy, N. S., and Rangan, T. S. (1971). *Bot. Gaz. (Chicago)* **132,** 113–119.

Rangaswamy, N. S., and Rao, P. S. (1963). *Phytomorphology* **13,** 450–454.

Rao, P. S., and Rangaswamy, N. S. (1971). *Biol. Plant.* **13,** 200–206.

Rappaport, J. (1954). *Bot. Rev.* **20,** 210–225.

Rietsema, J., Satina, S., and Blakeslee, A. F. (1953a). *Am. J. Bot.* **40,** 538–545.

Rietsema, J., Satina, S., and Blakeslee, A. F. (1953b). *Proc. Natl. Acad. Sci. U.S.A.* **39,** 924–933.

Rijven, A. H. G. C. (1952). *Acta. Bot. Neerl.* **1,** 157–200.

Rijven, A. H. G. C. (1955). *Proc. K. Ned. Akad. Wet., Ser. C* **58,** 368–376.

Rijven, A. H. G. C. (1956). *Aust. J. Biol. Sci.* **9,** 511–527.

Ross, J. D., and Bradbeer, J. W. (1968). *Nature (London)* **220,** 85.

Rudnicki, R., Kamiński, W., and Pieniazek, J. (1971). *Biol. Plant.* **13,** 122–127.

Ryczkowski, M. (1960). *Planta* **55,** 343–356.

Sanders, M. E., and Burkholder, P. R. (1948). *Proc. Natl. Acad. Sci. U.S.A.* **34,** 516–526.

Sanders, M. E., and Ziebur, N. K. (1963). *In* "Recent Advances in the Embryology of Angiosperms" (P. Maheshwari, ed.), pp. 297–325. Int. Soc. Plant Morphol., Delhi.

Sankhla, N., Sankhla, D., and Chatterji, U. N. (1967). *Z. Pflanzenphysiol.* **57,** 198–200.

Shivanna, K. R., and Rangaswamy, N. S. (1976). *Z. Pflanzenphysiol.* **80,** 112–119.

Sircar, S. M., and Lahiri, A. N. (1956). *Proc. Natl. Inst. Sci. India, Part B* **22,** 212–225.

Skene, K. G. M. (1969). *Planta* **87,** 188–192.

Smith, J. G. (1973). *Plant Physiol.* **51,** 454–458.

Stewart, J. McD., and Hsu, C. L. (1977). *Planta* **137,** 113–117.

Stingl, G. (1907). *Flora* **97**, 308–331.
Stokes, P. (1953). *J. Exp. Bot.* **4**, 222–234.
Stoltz, L. P. (1971). *Proc. Am. Soc. Hortic. Sci.* **96**, 681–684.
Sussex, I. (1975). *Am. J. Bot.* **62**, 948–953.
Sussex, I., Clutter, M., and Walbot, V. (1975). *Plant Physiol.* **56**, 575–578.
Thévnot, C., and Côme, D. (1971). *C. R. Acad. Sci., Ser. D* **273**, 2515–2517.
Thévenot, C., and Côme, D. (1973). *C. R. Acad. Sci., Ser. D* **277**, 401–404.
Thomas, M.-J. (1972). *C. R. Acad. Sci., Ser. D* **274**, 2655–2658.
Torrey, J. G. (1973). *In* "Tissue Culture: Methods and Applications" (P. F. Kruse, Jr. and M. K. Patterson, Jr., eds.), pp. 166–170. Academic Press, New York.
Truscott, F. H. (1966). *Am. J. Bot.* **53**, 739–750.
Tukey, H. B. (1938). *Bot. Gaz. (Chicago)* **99**, 630–665.
Tzou, D.-S., Galson, E. C., and Sondheimer, E. (1973). *Plant Physiol.* **51**, 894–897.
van Overbeek, J. (1942). *Cold Spring Harbor Symp. Quant. Biol.* **10**, 126–133.
van Overbeek, J., Conklin, M., and Blakeslee, A. F. (1942). *Am. J. Bot.* **29**, 472–477.
van Overbeek, J., Siu, R., and Haagen-Smit, A. J. (1944). *Am. J. Bot.* **31**, 219–224.
van Staden, J., Brown, N. A. C., and Button, J. (1972a). *J. S. Afr. Bot.* **38**, 211–214.
van Staden, J., Webb, D. P., and Wareing, P. F. (1972b). *Planta* **104**, 110–114.
Veen, H. (1963). *Acta Bot. Neerl.* **12**, 129–171.
Walbot, V., Clutter, M., and Sussex, I. (1975a). *Plant Physiol.* **56**, 570–574.
Walbot, V., Harris, B., and Dure, L. S., III (1975b). *In* "The Developmental Biology of Reproduction" (C. L. Markert, ed.), pp. 165–187. Academic Press, New York.
Walton, D. C., Soofi, G., and Sondheimer, E. (1970). *Plant Physiol.* **45**, 37–40.
Wang, P.-J., and Huang, L.-C. (1976). *In Vitro* **12**, 260–262.
Wardlaw, C. W. (1965). *In* "Encyclopedia of Plant Physiology" (W. Ruhland, ed.), Vol. XV/1, pp. 424–442. Springer-Verlag, Berlin and New York.
Wardlaw, C. W. (1968). "Morphogenesis in Plants." Methuen, London.
Wareing, P. F., and Saunders, P. F. (1971). *Annu. Rev. Plant Physiol.* **22**, 261–288.
Webb, D. P., van Staden, J., and Wareing, P. F. (1973). *J. Exp. Bot.* **24**, 105–116.
Williams, C. N. (1961). *Nature (London)* **189**, 378–381.
Worsham, A. D., Moreland, D. E., and Klingman, G. C. (1959). *Science* **130**, 165.
Worsham, A. D., Klingman, G. C., and Moreland, D. E. (1962). *Nature (London)* **195**, 199–201.
Wright, J. E., and Srb, A. M. (1950). *Bot. Gaz. (Chicago)* **112**, 52–57.
Ziebur, N. K., and Brink, R. A. (1951). *Am. J. Bot.* **38**, 253–256.
Ziebur, N. K., Brink, R. A., Graf, L. H., and Stahmann, M. A. (1950). *Am. J. Bot.* **37**, 144–148.

Chapter 19

The Future

GEORG MELCHERS

Max-Planck-Institut für Biologie, Tübingen, Federal Republic of Germany

I. Introduction

Why has the oldest person among the authors of these volumes been chosen to write the concluding chapter? Perhaps because prophets are always presumed to be old men? Or was it intended that he should write about the future in a less responsible if more amusing manner, because he stands less of a chance of witnessing the failure of his false prophecies than would a younger man? Or is it thought that such an old man is no longer an active researcher, and that, therefore, more complacent and mild judgments may be expected of him than from a recklessly forward-striving, if sometimes less self-critical, youth? The reader of this chapter will be disappointed, for I can fulfill none of these contradictory expectations. I will not set forth any particularly daring prospects of an El Dorado attainable through unconventional methods of cell and tissue culture of higher plants—there are already enough nearly unfillable promises of this sort; neither will I present a summary, tempered by the wisdom of maturity, of that which today particularly excites the authors of the other 18 chapters. Rather often, and readily, we have the opinion—but sometimes wrongly—that the future is predestined mostly by the facts we consider to be the most important ones in this infinitely small borderline between past and future, the present.

Regarding genuinely basic research, we cannot with any certainty speak of future results, for as C.F.v. Weizsäcker, the physicist and philosopher by profession, often says, the past consists of facts, but the future is comprised of possibilities. It is hoped that our 19 chapters provide a comprehensive and yet critical account of the facts pertaining to our discipline; for not everything that is published represents reproducible facts. Secondary literature such as this volume is only justifiable when its authors sort out the primary literature according to such criteria as (1) genuinely reproducible facts, (2) findings that are single but uncertain because of unsound methodology, and (3) unconfirmed assertions. These may stimulate new experiments. Given these criteria, we may more or less intelligently speculate about future possibilities, for if we do not proceed as if we are speaking of facts—which in any case can only make the past accessible—we are not yet damned. Particularly abundant in this context are speculations that, even if not always clearly identifiable as such, are held to justify research programs and are so presented to the organizations that finance research.

Genuinely basic research is rarer than some would like to believe. Because of the prestige that, rightly or wrongly, accrues to basic research as opposed to applied research (I shall not debate this point), not a few individuals who propose a project profess this or that significance for basic research. Genuinely basic research is often identifiable by the fact that it leads to a discovery that has not been its immediate goal. The discovery by Guha and Maheswari (1966) seems to me to be an example of this. They showed that at certain stages and under certain conditions, microspores, instead of developing normally into pollen grains, grow into haploid cell clusters (calluses) or even into plants. This discovery awakened many, but fulfilled few, hopes for applied research in plant breeding. It was almost as unintentional as Columbus' discovery of America. ''One may make a discovery without wanting to, but not without being aware of it,'' Hans Spemann once said, after somebody maintained that the discovery of the ''obere Urmundlippe'' as the organizor of amphibian development was not Spemann's discovery but the discovery of someone else, who had not, according to Spemann, been aware of it. Even such accidental discoveries only come to pass when the observations leading to them are made by researchers with enough experience to be capable of combining observation and experience in a meaningful way.

Favorable prerequisites for basic research may be organized. The experience of experimental biology during the last decades has demonstrated that the opening of contacts between hitherto isolated disciplines can lead to success. A good example of this is the development of molecular biology, which arose out of close cooperation between mathematicians, physicists, chemists, and geneticists. Basic research cannot be constrained by ''planning'' in the true sense of the word. To be efficient, however, applied research not only may be, but must be thoroughly planned.

There are certain impulses inherent in human nature and in scientific societies, impulses that work against the optimal conditions for basic research and that can be set up from without. The more narrowly the boundaries of a specialized discipline are circumscribed, and the smaller is made the "ecological niche" for researchers working in the field, the greater the probability that mediocre or insignificant talents will be able to assert themselves. For us to brave the wind that rises when we collaborate with mathematicians, physicists, and biochemists therefore goes against our nature. It also suits the sometimes barely recognized need for security felt by the majority of biologists, which remains, for that moment, unsatisfied. A defense reaction against the new is, in that moment, being unleashed. I speak of the facts of the past. However, the establishment of molecular biology is also a fact of the recent past. In many ways it merely indicates that now, in plant physiology, for example, models provided by molecular biology are also being drawn upon as explanations.

II. Some Problems of Basic Research at the Boundary of the Future

At this time I will not speculate about the future of "advances in plant tissue culture"—an expression, which, although indeed adopted by us, is barely appropriate—because, aside from one tissue, namely the tissue that in some respects resembles scar tissue, we are completely unable to cultivate truly differentiated tissues, such as epidermis, mesophyll, cortex, phloem, or storage tissue (Melchers, 1965). If only we could, we would understand more about differentiation, ontogeny, and determination than we actually do at present. I therefore make no prediction as to whether we sooner or later will be able to cultivate true tissue. I will say only that we would understand more about the important questions of developmental physiology if we would be able to observe the development, preservation, and functioning of distinct and well-described tissues in culture. Nevertheless, Kohlenbach (1977; Kohlenbach and Schmidt, 1975) has made preliminary attempts to achieve growth of specialized cells in culture *en masse*.

I do not exclude other possibilities for a deeper understanding of differentiation and dedifferentiation, even if an analysis of morphologically describable tissue formation is lacking. This means that, as fine and promising as it may be for the attainment of a better understanding of differentiation, ontogeny, and determination, if we can cultivate epidermal, mesophyll, cortical, phloem, or storage tissue as such, then there is perhaps a possibility (in the future!) that even without these successes we may come to a deeper, and this means a truly molecular biologic, understanding of these processes. If one isolates and examines a particular alternative feature of differentiation, e.g., hormonal autotrophy or heterotrophy, one frequently finds that the site of hormonal autotrophy in plants is at the shoot tips or also at the leaf meristem. One trait of differentiation is the development of hormone heterotrophy in the "adult" parts of plants. In explant cultures, the addition of hormones to the medium is usually necessary. In this special character the established callus cultures are also "differentiated," for they remain hormone heterotrophic, as they were before explantation. We still know too little about what, since Gautheret (1948), has been called "accoutumance a l'heteroauxine" (habituation). This is the state in which callus cultures are hormone autotrophic, as are their predecessors in the meristems of shoot tips.

For the moment I have the impression that habituation (hormone autotrophy) and its loss in the plants regenerated from these callus cultures, which once again leads to hormone-autotrophic meristems and hormone-heterotrophic adult parts of plants, is easier to achieve experimentally than is the cultivation of really differentiated tissues. However, this is not certain and therefore there are at least two possibilities (in the future!) for progress toward an understanding of differentiation and dedifferentiation.

Hormone autotrophy can develop not only "spontaneously" through epigenetic habituation, but also through transformation, i.e., the transport of genetic information by the Ti plasmid of *Agrobacterium tumefaciens*. It may be understandable, for some people self-evident, that habituation (the epigenetic change from the hormone-heterotrophic to the hormone-autotrophic state) is reversible. Changes of gene activities could be the molecular biologic model. Although in the period before molecular biology we could give only barely understandable explanations for many phenomena in genetics and developmental physiology, today plausible models are almost always at hand. Indeed, today we are almost always confronted by several formally plausible models, and the question is which model is most in agreement with facts. Can we perhaps, through the most intimate comparison between epigenetic habituation and transformation by plasmids, i.e., through the addition of genetic information, come closer to the problem of differentiation, which is normally an epigenetic phenomenon, or will this comparison lead us astray? With respect to hormone autotrophy, transformation is also completely reversible (Sacristán and Melchers, 1977). Other plasmid-induced traits (e.g., octopin production) can be retained in plants regenerated from callus culture, in any case up until the development of the reproductive organs (for review, see Schell, 1979).

I believe that this scant clue to a complex problem, which, through new experimental findings, may very quickly and in significant ways take on a new aspect, makes clear the senselessness of trying to foretell the results that may appear within only a few years. It would be clearest if we were to imagine for a moment that we were in the year 1969 rather than the year 1979. At that time, almost nothing was known about TIP (the tumor inducing principle) of *Agrobacterium tumefaciens*. Of course there were several hypotheses, and, to some extent, alleged but nonreproducible experimental findings. The disposition of the problems that we faced, as it appeared to me at that time, is discussed by Sacristán and Melchers (1969). In the interim it has become clear that the callus cultures, Ta, which Gautheret thought were habituated, are in reality also transformed through the Ti plasmid, as is Tc. Therefore, several points of discussion are eliminated. The problem of habituation still remains, however, and so the problems arising from the comparison of habituated and transformed tissues remain open and relevant to developmental physiology. We now know with certainty what TIP is. Important findings follow one after the other in rapid succession. Through cooperation with a few willing cell biologists, bacterial geneticists and biochemists (molecular biologists) have so far succeeded in obtaining quick and sure (reproducible!) results. At least one highly safe prediction can be made regarding the future: This advance is not yet at an end. As I have said above: The possibilities for development can be discussed. The majority of those who would like to make progress in this field will develop working hypotheses. The best hypotheses are those that can be definitely and finally disproved by experiments. An example that we should bear in mind when related problems arise in the near future is the experience with the habituation of a callus culture of *Crepis capillaris* (Sacristán and Wendt, 1971). These cultures were clearly marked karyologically, for they did not have the usual chromosomal arrangement with two satellite chromosomes in the diploid state. This culture was designated "Unisat" because on one chromosome the satellite was missing. The obvious hypothesis, that the lack of this satellite is the cause of hormone autotrophy, can be easily refuted. In plants regenerated out of callus culture, one satellite is also missing, and this rules out the selection of a "normal" cell type that is perhaps present in chimeric calluses as an explanation for the "reversion" to normal differentiated growth. Secondary callus cultures from the regenerated plants are hormone heterotrophic. Therefore, this means that this karyological abnormality, a chromosomal mutation that is morphogenetically relevant to other traits (sterility of flowers, leaf form), is not the cause of hormone autotrophy but that we are dealing with an epigenetic habituation that probably only coincidentally occurs at the same time.

Of course, when genuinely basic research is being conducted, we cannot with certainty reckon with only one alternative to a hypothesis. Not only habituation but also transformation is reversible. In the case of habituation, we might say in the jargon of contemporary molecular biology that gene activity is derepressed and repressed. Does this also hold true for reversible transformation? In this case, the development of hormone autotrophy is certainly connected with the insertion of *Agrobacterium* plasmid DNA into plant cells, for this can be demonstrated in plant cells (Chilton *et al.*, 1977, 1978). However, what takes place in the reversion to a hormone heterotrophic state? Is the transforming segment of plasmid DNA possibly lost? In the instances that we have examined (Sacristán and Melchers, 1977) we have found no indication that plants regenerated out of transformed cells retain any type of "memory." The situation is different, however, in the reverted teratomas of Braun and Wood (1976). These morphologically not completely transformed teratomas also develop through the takeover of plasmid DNA of another variety of *A. tumefaciens*. Although in instances where we have a reversion from hormone autotrophy to a hormone-heterotrophic state the loss of transforming plasmid DNA seems to be the simplest explanation, to date this has not been proved in a reproducible way. If octopine, for example, can still be detected in regenerated plants, then the hypthesis that the information for hormone autotrophy is lost is also not conclusively disproved, for we may be dealing with the loss of a segment of DNA that codes only for hormone autotrophy and not for octopine production. For the time being, another possible hypothesis is that the segment of transforming DNA coding for hormone autotrophy, which is also present, is repressed during regeneration. Because in normal development and in cases of habituation and of reversion derepressions and repressions of genes are not unreasonable assumptions, the hypothesis can only be excluded as an explanation for the recovery of tumor hormone autotrophy if the loss of the responsible DNA segment is demonstrated. In connection with this train of thought it is useful to bear that Unisat phenomenon in mind; however, it should not be overlooked that the loss of one part of the chromosome has been suspected to be the genetic cause of habituation, whereas the genetic efficiency of the plasmid DNA that is taken up by plant cells is proved. At present, it is no longer impossible to demonstrate acquisition and loss of even small segments of Ti plasmid. Today we can already say that the transforming segment of plasmid DNA in strains which induce only tumor growth but no synthesis of "opines" is shorter than segments which also induce opine synthesis in plant cells.

I will not say here that there are not also other areas of our discipline in which basic research has recently progressed. However, I know of no area where this progress has been so extensive and far reaching as in the field of plant cell transformation through *Agrobacterium tumefaciens*. Here cell biologists have also been involved in some small measure. The decisive factor for the tempo of progress, however, has been that some of them have not remained isolated in their "ecological niche" but have entered into a close working association with bacterial geneticists and molecular biologically oriented biochemists.

The first somatic plant hybrids, which were developed through protoplast fusion, also could be made in the conventional sexual way. At that time we often heard the criticism: Why be so complicated when it can also be done simply? The new method would become interesting only when sexually unproducible hybrid plants could be made in this way. Today there are some hybrid plants that could hitherto be made only somatically: *Datura* species, *Nicotiana sylvestris* × *N. knightiana*, *Lycopersicon esculentum* × *Solanum tuberosum*, and *Arabidopsis thaliana* × *Brassica campestris* (see Chapter 11, this volume). We may certainly expect more to be produced in the future. Naive with respect to systematics and taxonomy, and as if their categories had not been created by man but had been given by nature, the farther apart the partners in a hybrid were from one another, the greater the pride with which the authors described their successes. Our "pomatoes" or "topatoes" (Melchers *et al.*, 1978; Melchers, 1980) are generic hybrids according to the present system (*Lycopersicon esculentum* + *Solanum tuberosum*) but are only species hybrids according to a system that was accepted until recently. Then the tomato was still called *Solanum lycopersicum*. The fact that *Arabidopsis* + *Brassica* cell hybrids can be regenerated into plants even though they belong to different "tribes" within the family Cruciferae is perhaps an indication that the systematic categories within Cruciferae may be more "artificial" than hitherto assumed. Cell fusion, and above all the general behavior of the fusion products, will possibly become further criteria in taxonomic research. Fusion itself may not be a suitable criterion for this, for even animal and plant cells can be fused (Lima-de-Faria *et al.*, 1977; Dudits *et al.*, 1976).

Is nothing, then, to be expected from hybrid plants out of fusions that in principle can also be made sexually? When one considers that in many sexual plant hybrids the chromosomes in fact come half and half from each parent—and that only when dioecious plants with sex chromosomes are not concerned—and that the plastids and possibly also mitochondria are maternally inherited, one immediately sees the possibilities arising from the fusion of somatic cells. The fertilized egg cell begins its development with a hybrid nucleus having a diploid chromosome number, and very often with proplastids and mitochondria only from the mother. In the simplest case, when there are not fusins between more than two cells and/or partners with chromosome numbers diverging from the diploid, whereby the situation at the outset becomes greatly complicated, somatic cell fusions begin their development with a tetraploid hybrid nucleus and a mixture of plastids and mitochondria from both parents. In sexual hybridization of higher plants, cases where proplastids are also transmitted by the pollen tubes are quite rare. However, some contemporary authors seem to assume that this either never occurs or has occurred in only one or two cases. In the *Handbuch der Vererbungswissenschaft 1937* (Correns, 1937) the transmission of plastids through pollen is indicated for *Epilobium hirsutum, Erodium cicutarium, Hypericum acutum, Hypericum perforatum, Phaseolus vulgaris, Pelargonium zonale, Pisum sativum, and Viola arvenis;* for the species hybrids *Campanula carpatica* × *Campanula pelviformis, Geranium bohemicum* × *Geranium deprehensum* and additional *Geranium* hybrid species, *Hypericum acutum* × *Hypericum montanum, Hypericum hirsutum* × *Hypericum pulchrum, Hypericum acutum* × *Hypericum quadrangulum;* for hybrid species of *Oenothera* and *Pelargonium;* and for *Silene otites* × *Silene pseudotites.* With the exception of *Oenothera,* little use has hitherto been made of these facts in the analysis of the relationship between genome, plastome, and the genetic structure of mitochondria. Now Gleba (1979) and Belliard *et al.* (1978) have presented analyses of a combination of genome–plastome and mitochondrial genetics that, without protoplast fusion would have been difficult, if not impossible. To expect more analyses of this sort and their deepening through molecular biologic methods is not a particularly daring prophecy.

III. Applications to Industrial Use and Plant Breeding

Not only because of the limited space available but because the problems discussed in some of the preceding chapters are not close enough to this author, it is impossible for me to discuss the other chapters on basic research as we have discussed the problems mentioned in the first part of this chapter. Purely biochemical questions, such as intermediary metabolism, will certainly play a role in the future, but so will the biosynthesis of end products, such as cell walls and all types of secondary plant material. In this type of work it seems reasonable to me to avoid competition with faster growing and more readily manageable microorganisms (algae, fungi, bacteria). Only what can be worked out using exclusively the cells of higher plants should be worked out with them. Moreover, in the course of this work more attention must be paid in the future than has been to date to determining that the cells can be regenerated into plants, above all when it is supposed that we are dealing with mutant cell strains. Without the assistance of conventional genetics, a precise evaluation of the nature of a variant developed and isolated from a cell culture is difficult, if not impossible (see Chapter 9, Part A). If an *in vitro* culture grown under the same conditions yields identical products and intermediate products, a purely biochemical approach to the question may often ignore the particular possibilities of plant cell and callus cultures, namely the possibility of regenerating plants. That it would be particularly exciting to determine what biochemical data we might rediscover in intact plants regenerated out of a tissue culture requires no particular comment.

The ability to regenerate intact plants from callus cultures is, of course, impossible to ignore when the alterations elicited in cell cultures are induced only to get mutated plants more quickly and using less space than is possible using conventional F_2 selection. The experiments on submersed callus cultures of haploid *Antirrhinum* that we began as early as the 1950s (Melchers and Bergmann, 1959) could not succeed because at that time we were unable to regenerate plants out of callus. In the interim, several models through which these plans may be realized have been developed (Maliga *et al.*, 1973, 1975). However, to my knowledge, with the exception of up to two *Helminthosporium*-resistant cultivars of sugar cane, there is still no cultivar that has been bred and successfully grown using these methods. In my opinion this is because there have not been enough carefully and thoroughly planned experiments. As I said at the beginning of this chapter, although basic research leaves little room for real planning, planning is all the more necessary when fundamental results are at hand and problems regarding applications are being considered. If we wish to apply mutation and selection in cell cultures to the breeding of, for example, a resistant cultivar, we must first determine whether this breeding objective justifies the proposed expenditure. Obviously, a connected question is whether this goal can be more effectively, i.e., more cheaply, if more slowly, attained in a conventional manner. If future planning is as unenergetic in this respect as it has been up to now, financial backers will certainly not support such work for an unlimited period. If it is maintained that this or that experiment is also important for basic research, then it must be better explained than has usually been the case in the past. Experiments with potatoes in the callus or cell culture stage, for example, to select for resistance to *Phytophthora infestans*, seem to me to be promising. This goal seems to be particularly worthwhile when we bear in mind the mutability of these fungi into new biotypes. The first results demonstrate that calluses that are resistant to culture filtrates of the fungi also produce regenerated shoots that are resistant to the filtrate (Behnke, 1980). However, the small extent of these experiments is hardly justifiable. The same is true for experiments aimed at finding resistance against parasitic fungi (*Plasmidiophora, Phoma*) in *Brassica* shoot cultures (Sacristán and Hoffmann, 1979). Financial support has not been sufficient to produce, within a reasonable time, findings that allow us to judge whether these methods are quicker and surer than conventional methods of selection through the F_2 generation and back-crossing.

The use of somatic hybridization through fusion of protoplasts is in some respects now sufficiently developed to be used in plant breeding. Among somatic hybrids between *Datura* species are some that are difficult, if not impossible, to produce through conventional sexual methods. Schieder (personal communication) has found that some somatic hybrids have a greater yield of scopolamine than plants cultured up to now for obtaining this pharmaceutically interesting alkaloid or in plants collected in the wild. Because the cost of producing hybrids is not high, such efforts are justified even if the world demand for scopalamine is not particularly great. Pharmaceutical companies should certainly consider the fact that it is not enough to produce such a hybrid only once. Plant breeding is a never-ending task. This will not be any different with the application of modern, unconventional methods. The interested industrial firms do not seem to have a particularly great understanding of this state of affairs, and private plant breeders obviously do not see any exciting market in the improvement of medicinal plants.

The employment of mass cultures of submersed calluses and cells, toward which great efforts have been directed in the laboratories of several industrialized countries, has not yet resulted in any commercially profitable product. "The industrial use of plant cell culture is a possibility," concludes M. H. Zenk (1978), in his article "The Impact of Plant Cell Culture in Industry," and so is, completely in accord with our argument, a task of the future. Zenk's article seems to me to be particularly important for the future, for it also examines the financial problems pertaining to the application of these methods and does not fail to remind us of how in the 1950s and 1960s great efforts were made in vain by C. Pfizer and Co., and that, therefore, confidence in this type of work became very poor in the industrial world.

Similar setbacks can also be expected for unconventional methods in plant breeding if in the near future reasonable goals in competition and in cooperation with conventional methods are not sought and attained. For the moment, the potato seems to me to be the most suitable object. Although smaller than it would be for corn, rice, and wheat, the potential world market should be greater than that for medicinal plants and is worth every effort. Through anther culture, but above all through crosses with wild species, especially *Solanum phureja,* the tetraploid, highly heterozygous cultivars may be developed into dihpaloid and now also into mono-haploid lines. This means that now an essentially efficient combination breeding has become more possible than in preceding centuries, when potato breeding on a tetraploid basis resembled a lottery rather than an systematic endeavor. In the near future, if experiments are carried out with sufficient planning and funding, the potato is sure to serve as a test of how far resistance breeding through the cultivation of cell and callus cultures by means of mutation and selection may be pursued. The same holds true for qualitative compounding of amino acid pools through application of analog resistance. The proportion of free amino acids in the total nitrogen content of these plants opens up completely different possibilities than a modification of the amino acid content of reserve proteins (e.g., in cereals). The potato is also a cultivated plant on which the applications of "genetic engineering" can be tested immediately: (1) One can easily infect potatoes with *A. tumefaciens*. (2) Transformed traits can also be used, even when they are lost in sexual reproduction (vegetative reproduction through tubers is normal in potato).

The possibility of utilizing protoplast fusion in potato breeding has been indicated repeatedly (Straub, 1976; Melchers, 1978; see also Chapter 11, this volume). There is no other way in which two dihaploid strains that are heterotic because of heterozygosity can be brought together so quickly and surely, and without destroying the homogeneity of the material, as through fusion of protoplasts. How we can combine analytic methods of haploidization with artificial polyploidization and fusion in order to combine qualitative traits of existing culture types has just been shown by Wenzel *et al.* (1979; see also Chapter 11, this volume). This work of the future, whereby one can systematically cultivate such well-defined traits as disease resistance, heterosis of dihaploid lines, and other qualitative traits for the creation of new varieties has already been started in the Max-Planck-Institut für Züchtungsforschung at Köln-Vogelsang, and this shows the necessity for the combination of conventional methods of selection and crossing with the new methods of haploidization and fusion of protoplasts. Here my otherwise necessary reticence in predicting future successes can be set aside. Now there remains the question of finding reasonable breeding objectives and organizing our work so as to follow the path outlined here and achieve good results. It is also probable that these breeding methods will lead us to our objective more quickly than extensive conventional methods of breeding new potato cultivars. This method is thoroughly described in Schieder and Vasil's contribution to this volume (Chapter 11).

Already extant successes demonstrate the applicabilty of the so called anther or pollen culture to haploid breeding. Not only do several newly authorized tobacco cultivars exist in China and Japan, but in China there exist new cultivars of rice and wheat (Anonymous, 1978). What is still missing from these endeavors, however, is as close as possible a comparison between the new methods and conventional breeding aimed at the same breeding goals. Such investigations have been started at two institutions in China. With such plants as rice and wheat, where extensively developed methods of conventional breeding have already been employed and where there exists a very great variety of productive strains suited to specific conditions, it is still questionable, in light of their still small yields of "pollen plants" and considering the expenditure necessitated by their production through "anther cultures," whether the new methods offer clear-cut advantages. In the case of objects with high yields of pollen plants, such as *Datura* or tobacco, the foreseeable advantages of breeding by anther culture, above all with regard to the time necessary for obtaining a certain breeding goal, are in my opinion unambiguous, and the failure to utilize them commercially is unjustifiable. All efforts to systematically increase the yield of pollen plants through anther culture deserve generous support. It has been shown again and again that even in the case of a plant whose yield in pollen plants seemed at first to be hopelessly low, yields can be increased through systematic and well-organized efforts. However, a considerable outlay for good laboratory equipment and technical personnel is unavoidable.

To the extent that such results as we may expect from basic research in the future are unforeseeable, we must resign ourselves to genuine surprises, positive and negative. In applied research utilizing combinations of microbiologic methods for breeding research we not only have achieved practical results but may expect more, and this means, as we have seen, results that will be attained through clear planning and a sufficiently large expenditure.

I reiterate here what I have said before: there is no better way to waste money than to invest too little money over a long period of time (Melchers, 1977).

In this concluding chapter I have not considered vegetative reproduction through meristem, organ, or callus cultures (see Chapter 6, Part A). In the past, basic and applied research have led to great successes in this area and today such methods already play a major role in industrial plant production and provide real competition for conventional methods.

Use of the unconventional methods discussed here requires immediate, close integration with conventional methods, and preferably, as we have seen, in the hands of the same people, or at least in closely adjacent laboratories and experimental greenhouses and in jointly cultivated experimental plots. Not only in basic research, but in applied research as well, we should make even greater efforts in the future to eliminate the development of "ecological niches" for "tissue culturists."

Am I then indeed a prophet? If I cannot provide prophecies I can at least provide an exhortation: μετανοεῖτε! Transform thy spirit! Prophets such as this are not amusing, for they make themselves unpopular. These are experiences and facts extending back for thousands of years; but are the exhortations of old people easier to bear than those of youth?

REFERENCES

Anonymous (1978). *Proc. Symp. Plant Tissue Cult., Peking.*

Behnke, M. (1980). *Theor. Appl. Genet.* **80,** 151–152.

Belliard, G., Pelletier, G., Vedel, F., and Quetier, F. (1978). *Mol. Gen. Genet.* **165,** 231–238.

Braun, A. C., and Wood, H. N. (1976). *Proc. Natl. Acad. Sci. U.S.A.* **73,** 496.

Chilton, M. D., Drummond, H. J., Merlo, E. J., Sciaky, D., Montoya, A. L., Gordon, M. P., and Nester, E. W. (1977). *Cell* **11,** 263.

Chilton, M. D., Drummond, H. J., Merlo, E. J., and Sciaky, D. (1978). *Nature (London)* **275,** 147.

Correns, C. E., (1937). *In* "Handbuch der Vererbungswissenschaft," Vol. II H. Borntraeger, Berlin.

Dudits, D., Rasko, G., Hadlacky, G., and Lima-de-Faria, A. (1976). *Hereditas* **82,** 121–124.

Gautheret, R. (1948). *C. R. Soc. Biol.* **142,** 774–775.

Gleba, Y. Y. (1979). "Plant Cell and Tissue Culture. Principles and Applications." Ohio State Univ. Press, Columbus.

Guha, S., and Maheswari, S. C. (1966). *Nature (London)* **212,** 97–98.

Kohlenbach, H. W. (1977). *In* "Plant Tissue Culture and its Bio-technologica Application" (W. Barz, E. Reinhard, and M. H. Zenk, eds.), pp. 355–366. Springer-Verl., Berlin and New York.

Kohlenbach, H. W., and Schmidt, W. (1975). *Z. Pflanzenphysiol.* **75,** 369–374.

Lima-de-Faria, A., Eriksson, T., and Kjellen, L. (1977). *Hereditas* **87,** 57–66.

Maliga, P., Breznovits, A., and Marton, L. (1973). *Nature (London), New Biol.* **244,** 401–403.

Maliga, P., Breznovits, A., Marton, L., and Joo, F. (1975). *Nature (London)* **255,** 401–403.

Melchers, G. (1965). *Ber. Dtsch. Bot. Ges.* **78,** 21–29.

Melchers, G. (1977). *Naturwissenschaften* **64,** 184–194.

Melchers, G. (1978). *In* "Production of Natural Compounds by Cell Culture Methods" (A. W. Alfermann and E. Reinhard, eds.), pp. 306–311.

Melchers, G. (1980). *In* "Advances in Protoplast Research" (L. Ferenczÿ and G. L. Farklar, eds.), pp. 283–286.

Melchers, G., and Bergmann, L. (1959). *Ber. Dtsch. Bot. Ges.* **71,** 459–473.

Melchers, G., Sacristán, M. D., and Holder, A. (1978). *Carlsberg Res. Commun.* **43,** 203–218.

Sacristán, M. D., and Hoffmann, F. (1979). *Theor. Appl. Genet.* **54,** 129–132.

Sacristán, M. D., and Melchers, G. (1969). *Mol. Gen. Genet.* **105,** 317–333.

Sacristán, M. D., and Melchers, G. (1977). *Mol. Gen. Genet.* **152,** 11–117.

Sacristán, M. D., and Wendt, M. F. (1971). *Mol. Gen. Genet.* **110,** 355–360.

Schell, J. (1979). *In* "Nucleic Acids in Plants" (T. Hall and J. Davis, eds.). CRC Press, Cleveland, Ohio.

Straub, J. (1976). *Ber. Landwirtsch., Z. Agrarpolitik Landwirtsch,* **54,** 377–387.

Wenzel, G., Schieder, O., Przewozny, T., Sopory, S. K., and Melchers, G. (1979). *Theor. Appl. Genet.* **55,** 49–55.

Zenk, M. H. (1978). *In* "Frontiers of Plant Tissue Culture, 1978" (T. A. Thorpe, ed.), pp. 1–13. Univ. of Calgary Press, Calgary, Alberta.

Subject Index